Aid, Ownership and Development

One of the key principles for effective aid programmes is that recipient agencies exert high degrees of ownership over the agendas, resources, systems and outcomes of aid activities. Sovereign recipient states should lead the process of development. Yet despite this well-recognised principle, the realities of aid delivery mean that ownership is often compromised in practice.

Aid, Ownership and Development examines this 'inverse sovereignty' hypothesis with regard to the states and territories of the Pacific Island region. It provides an initial overview of different aid 'regimes' over time, maps aid flows in the region, and analyses the concept of sovereignty. Drawing on a rich range of primary research by the authors and contributors, it focuses on the agencies and individuals within the Pacific Islands who administer and apply aid projects and programmes. There is indeed evidence for the inverse sovereignty effect; particularly when island states and their small and stretched bureaucracies have to deal with complex and burdensome donor reporting requirements, management systems, consultative meetings and differing strategic priorities. This book outlines important ways in which Pacific agencies have proved adept not only at meeting these requirements, but also asserting their own priorities and ways of operating. It concludes that global agreements, such as the Paris Declaration on Aid Effectiveness in 2005 and the recently launched Sustainable Development Goals, can be effective means for Pacific agencies to both hold donors to account and also to recognise and exercise their own sovereignty.

John Overton is a geographer who has worked on development issues and the Pacific region for over thirty years. His Pacific education began with a position at the University of the South Pacific in the mid-1980s and has continued through working with many students and colleagues from the Pacific Island region. He is currently Professor of Development Studies at Victoria University of Wellington, New Zealand.

Warwick E. Murray gained a PhD in geography at the University of Birmingham, UK and served as a lecturer in geography at the University of the South Pacific (1997–2000). Taking up a university role in Wellington in 2001 he has focused his research on development geographies of the Pacific Island region and Rim (including Latin America). An Editor of *Asia Pacific Viewpoint* since 2002, he is currently Professor of Human Geography at Victoria University of Wellington.

Gerard Prinsen teaches Development Studies at Massey University New Zealand, after a professional career in development practice. Most of his research revolves around local health and education services as spaces where small, rural, or remote communities negotiate their relationships with big metropolitan powers.

Avataeao Junior Ulu is an international development practitioner and has worked in the wider Pacific region since 2002. His master's degree focused on aid sovereignty in Samoa, and excerpts from his thesis can be found in this book. Junior is currently completing his PhD on migration, education and development with reference to Samoa. He also does development consultancy work in his spare time.

Nicola (Nicki) Wrighton spent many years working in the Pacific Island region for aid agencies, as a consultant and for Pacific organisations, including the Government of Tuvalu. Her master's degree in 2010 on aid and Tuvalu helped shape the Marsden Fund grant that funded the research for this book. She was working on her PhD on aid and sovereignty in the Pacific at Victoria University of Wellington when illness struck and she died tragically in 2014.

Aid, Ownership and Development

The Inverse Sovereignty Effect in the Pacific Islands

John Overton, Warwick E. Murray, Gerard Prinsen, Avataeao Junior Ulu and Nicola (Nicki) Wrighton

LONDON AND NEW YORK

First published 2019
by Routledge
2 Park Square, Milton Park, Abingdon, Oxon OX14 4RN

and by Routledge
605 Third Avenue, New York, NY 10017

First issued in paperback 2020

Routledge is an imprint of the Taylor & Francis Group, an informa business

British Library Cataloguing-in-Publication Data
A catalogue record for this book is available from the British Library

Library of Congress Cataloging-in-Publication Data
A catalog record for this book has been requested

ISBN 13: 978-0-367-73379-7 (pbk)
ISBN 13: 978-0-367-00052-3 (hbk)

Typeset in Sabon
by Apex Covantage, LLC

Visit the eResources: www.routledge.com/9780367000523

IN MEMORY OF NICKI WRIGHTON

Contents

Figures

Tables

Research insight boxes

Authors and contributors

Authors

John Overton is a geographer who has worked on development issues and the Pacific region for over thirty years. His Pacific education began with a position at the University of the South Pacific in the mid-1980s and has continued through working with many students and colleagues from the Pacific Island region. He is currently Professor of Development Studies at Victoria University of Wellington, New Zealand.

Warwick E. Murray gained a PhD in geography University of Birmingham, UK and served as a lecturer in geography at the University of the South Pacific (1997–2000). Taking up a university role in Wellington in 2001 he has focused his research on development geographies of the Pacific Island region and Rim (including Latin America). An Editor of *Asia Pacific Viewpoint* since 2002, he is currently Professor of Human Geography at Victoria University of Wellington

Gerard Prinsen teaches Development Studies at Massey University New Zealand, after a professional career in development practice. Most of his research revolves around local health and education services as spaces where small, rural, or remote communities negotiate their relationships with big metropolitan powers.

Avataeao Junior Ulu is an international development practitioner and has worked in the wider Pacific region since 2002. His master's degree focused on aid sovereignty in Samoa, and excerpts from his thesis can be found in this book. Junior is currently completing his PhD on migration, education and development with reference to Samoa. He also does development consultancy work in his spare time.

Nicola (Nicki) Wrighton spent many years working in the Pacific Island region for aid agencies, as a consultant and for Pacific organisations, including the Government of Tuvalu. Her master's degree was in 2010 on aid and Tuvalu helped shape the Marsden Fund grant that funded the research for this book. She was working on her PhD on aid and sovereignty in the Pacific at Victoria University of Wellington when illness struck, and she died tragically in 2014.

Contributors

Potoae Roberts Aiafi is currently Samoa Country Representative on the Pacific Leadership Programme and she teaches public policy at the National University of Samoa. She completed her PhD at Victoria University of Wellington in 2016.

Séverine Blaise specialises in international and development economics. She gained her PhD in economics in 2004 at Aix-Marseille University. She worked at the Australian National University before becoming senior lecturer at the University of New Caledonia in 2007.

Adele Broadbent completed her Master of Development Studies at Victoria University of Wellington in 2012 looking at media development in Solomon Islands. Adele worked for the Council for International Development in Wellington.

Karly Christ completed her Master of Development Studies at Victoria University of Wellington in 2012 which investigated the sovereignty movement in Rapa Nui with a particular emphasis on the role of women.

Peter C.L. Eafeare graduated with Bachelor of Education degree from the University of Papua New Guinea, a postgraduate Diploma in Library Studies, Victoria University of Wellington, and a diploma in Foreign Service, PNG Department of Foreign Affairs. He worked for twenty-nine years with PNG Foreign Affairs and is now a private consultant.

Finbar Kiddle graduated with a master's in Development Studies from Victoria University of Wellington in 2014 which focused on RAMSI in Solomon Islands.

Hannah Mackintosh completed her master's degree in Development Studies at Victoria University of Wellington in 2011. Her thesis considered how the universal concept of human rights is being engaged with and interpreted by Māori communities in Aotearoa New Zealand.

Alexander Mawyer is a cultural and linguistic anthropologist whose work has focused on the Mangarevan community in French Polynesia's Gambier Islands for over two decades. Associate Professor of Pacific Studies at the University of Hawai'i at Mānoa, he is currently editor of *The Contemporary Pacific: A Journal of Island Affairs*.

Julien Migozzi is a PhD candidate in urban and economic geography at the University of Grenoble Alpes, and currently a Research and Teaching

Assistant at the *École Normale Supérieure* (Paris). With fieldwork based at Victoria University of Wellington, his master's focused on aid and urban development in New Caledonia (awarded by the *École Normale Supérieure* de Lyon).

Helen Mountfort completed a Master of Development Studies at Victoria University of Wellington in 2012 and her thesis focused on aid sovereignty in Tonga. She is currently a Programme Manager in the Risk and Resilience team at the Overseas Development Institute, UK.

Mattie Geary Nichol completed her master's degree in Development Studies at Victoria University of Wellington in 2014. Her thesis looked at monitoring and evaluation in government departments and non-governmental organisations in Vanuatu. Mattie is currently the international programmes coordinator for Family Planning New Zealand.

Pedram Pirnia gained a PhD in Development Studies from Victoria University of Wellington in 2016. He is the Officer responsible for promoting the SDSGs for the UN Association of New Zealand. He has travelled extensively in the Pacific region and currently lectures at Auckland University of Technology.

Sisikula Sisifa completed her PhD in Management at the University of Auckland Business School in 2016 and studied the project management practices used in development projects in Tonga. She is currently working as a Post-Doctoral research fellow at the University of Auckland Business School.

Felicia Pihigia Talagi completed her master's degree in Development Studies at Victoria University of Wellington in 2017 and studied aid relationships in Niue. She is currently working for the Government of Niue.

Faka'iloatonga Taumoefolau is an Australia National University graduate with a master's degree in International Affairs and with experience of working for government agencies in Tonga. He is currently pursuing his PhD in Development Studies at Victoria University of Wellington.

Klaus Thoma completed his master's degree in Development Studies at Victoria University of Wellington in 2014. He is currently establishing an NGO with the main interest of making electro-mobility accessible to small island nations which formed the topic of his action-research based thesis based on Samoa.

Abbreviations

AAA	Accra Agenda for Action
ABC	Australian Broadcasting Corporation
ACP	African, Caribbean and Pacific countries
ADB	Asian Development Bank
AMD	Aid Management Division (Tonga)
ASEAN	Association of South East Asian Nations
AusAID	Australian Agency for International Development
AVID	Australian Volunteers for International Development
CDC	Cabinet Development Committee (Samoa)
CIDA	Canadian International Development Agency (disestablished)
CNMI	The Commonwealth of the Northern Mariana Islands
COFA	Compacts of Free Association (between the USA and Palau, the RMI and the FSM)
COP	Conference of the Parties (to the UN)
CSO	Civil Society Organisation
CSR	Colonial Sugar Refining Company
DAC	Development Assistance Committee (of the OECD)
DFAT	Department of Foreign Affairs and Trade (Australia)
DFID	Department For International Development (UK)
DWCP	Decent Work Country Programme
ECHR	European Court of Human Rights
EEC	European Economic Community
EIB	European Investment Bank
ELR	Employment Law Reform (Vanuatu)
EPAs	Economic Partnership Agreements
EU	European Union
EVs	Electric Vehicles
FAO	Food and Agriculture Organisation (of the UN)
FAP	Forward Aid Programme (Niue)
FATDC	Foreign Affairs, Trade and Development Canada
FDI	Foreign Direct Investment
FMS	Financial Management Systems
FSC	Fiji Sugar Corporation

FSM	Federated States of Micronesia
GBS	General Budget Support
GDP	Gross Domestic Product
GFC	Global Financial Crisis
GFC	Green Climate Fund
GNI	Gross National Income
HIPC	Heavily Indebted Poor Countries
IEOM	Institut d'emission de l'Outre-Mer
IFAD	International Fund for Agricultural Development
IFC	International Finance Corporation
ILO	International Labour Organisation
IMF	International Monetary Fund
IR	International Relations
ISEE	Institut de la statistique et des études économiques (New Caledonia)
JBIC	Japan Bank for International Cooperation
JCfD	Joint Committee for Development (Government of Niue/ Government of New Zealand)
JICA	Japan International Cooperation Agency
LES	Locally Engaged Staff
MDGs	Millennium Development Goals
MFAT	Ministry of Foreign Affairs and Trade (New Zealand)
MIGA	Multilateral Investment Guarantee Agency
MIRAB	Migration, Remittances, Aid and Bureaucracy
MoFNP	Ministry of Finance and National Planning Tonga
MSG	Melanesian Spearhead Group
MSGTA	Melanesian Spearhead Group Trade Agreement
NAFTA	North American Free Trade Agreement
NDA	National Designated Authority (Green Climate Fund)
NDS	National Development Strategy (Solomon Islands)
NECC	National Energy Coordinating Committee (of Samoa)
NEMS	National Environmental Management Strategies
NGO	Non-Governmental Organisation
NZAID	New Zealand Agency for International Development (reintegrated into MFAT)
OCTs	Overseas Countries and Territories (of the EU)
ODA	Official Development Assistance
OECD	Organisation for Economic Cooperation and Development
OIA	Office of Insular Affairs (USA)
PACC	Project and Aid Coordination Committee (Tonga)
PACER	Pacific Agreement on Closer Relations
PACTAM	Pacific Technical Assistance Mechanism
PEFA	Public Expenditure and Financial Accountability
PIFS	Pacific Islands Forum Secretariat
PNG	Papua New Guinea
PRC	People's Republic of China

PROFIT	People, Resources, Overseas para-diplomacy, Finance/taxation and Transportation
PRS	Poverty Reduction Strategies
PRSP	Poverty Reduction Strategy Paper
RAMSI	Regional Assistance Mission to the Solomon Islands
RMI	Republic of the Marshall Islands
RSE	Recognised Seasonal Employer (New Zealand)
SAPs	Structural Adjustment Programmes
SBS	Sectoral Budget Support
SDGs	Sustainable Development Goals
SDL	Soqosoqo Duavata ni Lewenivanua (United Fiji Party)
SDP	Strategic Development Plan (Fiji)
SDS	Strategy for Development (Samoa)
SIDS	Small Island Developing States
SIG	Solomon Islands Government
SOLMAS	Solomon Islands Media Assistance programme
SPARTECA	South Pacific Regional Trade and Economic Cooperation Agreement
SPC	Pacific Community (formerly South Pacific Commission)
SPREP	South Pacific Regional Environmental Programme
SWAps	Sector Wide Approaches
SWP	Seasonal Worker Programme (Australia)
TA	Technical Advisors
TERM	Tonga Energy Road Map
TESP	Tonga Education Support Programme
TSDF	Tonga Strategic Development Framework
TVET	Technical and Vocational Education and Training programme (Tonga)
UN	United Nations
UNAIDS	UN programme on HIV/AIDS
UNCRD	UN Centre for Regional Development
UNDP	UN Development Programme
UNEP	UN Environmental Programme
UNESCO	UN Educational, Scientific and Cultural Organisation
UNFCC	UN Framework Convention on Climate Change
UNFPA	UN Population Fund
UNGA	UN General Assembly
UNHCR	the Office of the UN High Commissioner for Refugees (also known as the UN Refugee Agency)
UNICEF	UN International Children's Emergency Fund (known as the UN Children's Fund)
UNRISD	UN Research Institute for International Development
UNTA	UN Technical Assistance
US-GAO	United States Government Accountability Office
VSA	Volunteer Service Abroad (New Zealand)
WTO	World Trade Organisation

Acknowledgements

The authors wish to acknowledge and thank many people and institutions who contributed to the genesis and production of this book. The substantive research originated in a project in 2011–2014 on the inverse sovereignty effect supported by the Marsden Fund of the Royal Society of New Zealand. This was complemented by a grant from the Fonds Pacifique of the French Government which allowed us to build research collaborations with colleagues in New Caledonia.

We were greatly assisted also by the continuous support and encouragement of Luamanuvao Dame Winnie Laban, the Assistant Vice Chancellor (Pasikifa) of Victoria University of Wellington. Winnie helped us communicate with many stakeholders in the aid world of the Pacific, particularly diplomatic representatives based in Wellington. Professor Vijay Naidu of the University of the South Pacific was also a key resource and Associate Professor Sailau Suaalii-Sauni helped shape our early thinking notably through her co-supervision of Nicki Wrighton's PhD thesis.

Adele Broadbent, as well as a contributor of material to the book, provided sharp and accurate proof reading. We also particularly thank Helen Mountford and Alex Mawyer who provided suggestions and material for various chapters as well as their named research insights boxes. All contributors to the book's research insights boxes provided more than words on a page. Some as masters and PhD researchers were part of the research team's meetings and deliberations as our ideas evolved and contributed much of the richness of the book through their fine field research. Others we met during the course of our work and we would particularly acknowledge the work of a cohort of excellent Pacific doctoral scholars (Potoae Roberts Aiafi, Sisikula Sisifa and Faka'iloatonga Taumoefolau, as well as Junior Ulu).

Finally, this research would not have been possible without the active participation of scores of Pacific Island officials, NGO workers, politicians, diplomats and community leaders and members who shared their experiences and views on aid and sovereignty in Oceania.

Tourists and Moai statues, Rapa Nui

1 Aid in the Pacific in historic and geographic context

Aid in the Pacific Islands

There is considerable debate about the impact of aid in the Pacific. Critics of aid range from those who see it as a form of neo-colonial dominance to those who argue that it is at best ineffective and, at worst, a severe constraint on economic growth. The critics on the political left see aid as a tool of control over Pacific Island states by major donors (Stephen 2001). It is, thus, one of a number of instruments which diminish Pacific self-determination, belittle the ability of Pacific Island people to manage their own affairs, and frame the region and its residents as passive, in need, fragile and helpless. On the other hand, critics from the right of politics point to the very large volumes of aid flowing into the region and yet very poor economic performance: a situation that leads to the conclusion that 'aid has failed the Pacific' (Hughes 2003).

Yet despite these criticisms of aid in the region and the fact that many of its countries and territories have much more favourable development indices – in terms of literacy, income and life expectancy - than elsewhere in the Global South, aid levels have been maintained or increased in the past two decades. This has been the case even though the global financial crisis of 2007–2008 had severe impacts on the economies of the major donors to the region: USA, France, Australia, Japan and New Zealand. Furthermore, the aid landscape of the region has been continually transformed over recent years: China and other non-traditional donors have become more active; there have been fundamental shifts in the driving principles of established donors; and the climate change issue is becoming a more prominent feature of the narratives concerning aid and donor-recipient relationships.

The mainstream view of aid relationships, often expressed in documents by the OECD or UN, often builds on a set of stereotypes and truisms concerning the region and its 'inherent' geo-economic and geopolitical, and environmental vulnerability. This view encompasses the following assertions:

- The island states and territories[1] of the Pacific comprise the most aid-dependent region of the world.

- Most of these entities are small in size, with populations numbering from thousands to less than one million; their land areas are fragmented in small islands generally over vast swathes of sea; and they occupy a marginal position in the global economy, hampered by very high unit transport costs and limited economies of scale.
- Colonial relationships persist with some territories remaining as integrated parts of metropolitan countries (Hawai'i, Rapa Nui) and others with partial degrees of self-determination (New Caledonia, Tokelau). Such relationships have particular implications for sovereign decision-making and aid volumes.

Whilst some of these assertions are empirical realities and do have detrimental implications for mainstream development indices, it is our contention that the broader outcomes of such factors are complex and contested. Although there are significant problems associated with political-economic power asymmetries, the Pacific is not necessarily as vulnerable or passive as orthodox analyses often portray.

The superficial view of aid relationships touched on above in the region disguises considerable difference and debate with regard to the way Pacific Island people experience, negotiate, manage and benefit from aid. It also hides significant questions about what aid is, why it is given and received and how it relates to processes of development. Experiences of aid in the region vary considerably. For example, although Papua New Guinea receives much more official aid than any other Pacific Island state – $US 656 million in 2013 (see Table 1.1), its per capita aid receipts ($US 89) are by far the lowest. On the other hand, the territory of Tokelau received only about $US 24 million in 2013 but this amounted to over $US 20,000 for each one of its 1200 residents.

The relationships that shape aid patterns are being played out through a complex set of agreements, and protocols as well as both institutional and personal networks. From the donor side, although the years 2000–2008 were marked by a strong promotion of poverty alleviation as the driving principle for aid, this did not displace the underlying self-interests of donors in engaging with the region. Various donors have recognised a long-standing obligation to some islands resulting from a colonial past; most have sought in some way to gain political support (as in the voting rights of Pacific states in international forums) through aid diplomacy; yet others have sought to use aid as a means of opening and building commercial relationships with the region to steer aid revenues back to their own economies through overt or subtle forms of what is described as 'tied aid'. For Pacific Island states, aid has become a critical and seemingly permanent element of their economies, paying for development projects (roads, airports, water supplies, agricultural schemes), supporting capacity building (training, education), and financing government expenditure (health, education). For many, aid has become, like remittances from citizens working overseas, a key determinant of local living standards.

Table 1.1 Countries and territories of the Pacific Islands[2]

	Population (estim 2013)	*Land area (km²)*	*Net ODA[3] (2013 $US mill)*	*ODA per capita ($US)*
American Samoa	56,500	199	n.a.	n.a.
Cook Islands	15,200	237	15.29	1,006
Federated States of Micronesia	103,000	701	143.16	1,390
Fiji	859,200	18,333	91.24	106
French Polynesia	261,400	3,521	n.a.	n.a.
Guam	174,900	541	n.a.	n.a.
Hawai'i	1,374,810	16,634	n.a.	n.a.
Kiribati	108,800	811	64.58	594
Marshall Islands	54,200	18	93.91	1,733
Nauru	10,500	21	28.78	2,741
New Caledonia	259,000	18,576	n.a.	n.a.
New Zealand	4,439,000	268,107	–457.31	–103
Niue	1,500	259	18.30	12,200
Norfolk Island	1,895	35	n.a.	n.a.
Northern Mariana Islands	55,700	457	n.a.	n.a.
Palau	17,800	444	35.46	1,992
Papua New Guinea	7,398,500	462,840	656.54	89
Pitcairn	45	47	n.a.	n.a.
Rapa Nui	5,700	164	n.a.	n.a.
Samoa	187,400	2,934	118.18	631
Solomon Islands	610,800	28,000	288.32	472
Tokelau	1,200	12	24.06	20,050
Tonga	103,300	749	81.15	786
Tuvalu	10,900	26	26.80	2,459
Vanuatu	264,700	12,281	90.89	343
Wallis and Futuna	12,200	142	105.53	8,650

Source: South Pacific Commission, and OECD.Stat.

When examining aid in the Pacific then, it is clear that there are marked inequalities in power. Donors, in the form of governments negotiating bilateral arrangements, or international agencies representing many donors, appear to have much relative power. They can dictate the terms by which aid is received and spent and they can request or demand concessions, both economic and political. Donor power flows from the money they can offer but also from the way they can access or impose technical advice and expertise. Recipients, especially those from small countries, have small overstretched bureaucracies and the principle of being able to interact on an equal basis with donor countries is an illusion. Aid in the Pacific, despite all the rhetoric about partnerships, consultation and recipient ownership characteristic of global 'aidspeak', remains driven by donor systems, donor objectives and donor ways of operating.

The research on which this book is based began with this view above. Nicola Wrighton, our lead researcher, had spent many years working in

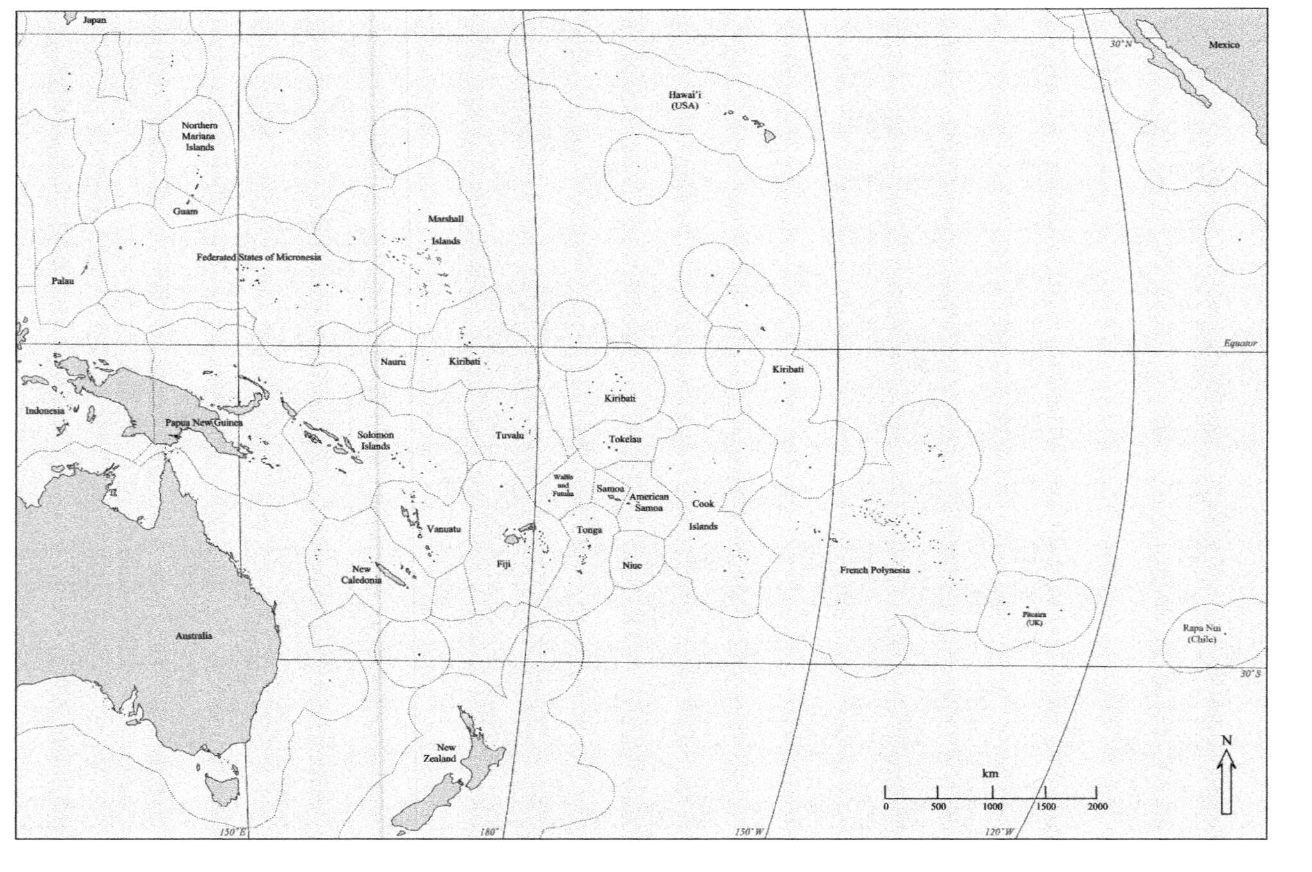

Figure 1.1 The Pacific Islands

the Pacific Islands, both for donor agencies (New Zealand) and, latterly, as an advisor for the Government of Tuvalu. In Tuvalu and Kiribati she was struck by the pressures faced by her local friends and colleagues (Wrighton 2010a, 2010b; Wrighton and Overton 2012). They had to receive an endless stream of diplomatic missions, journalists, academics and consultants, all visiting the islands to find information, consult, sign agreements, advise and 'do development'. They also had to represent their country at numerous meetings overseas, often necessitating them leaving their desks for weeks at a time. And amongst all this, they had to prepare reports (often to comply with donor requests for information) and, somehow, conduct their usual daily jobs, all within a context where they were expected to be readily accessible to local people. Box 1.1 illustrates the conclusions Wrighton drew from Tuvalu on the basis of her research there. Out of her work this project sought to examine the issue on a wider Pacific stage. Its central concern was the way officials and institutions in the Pacific could exert – or not – a degree of control over their own development given the resources and constraints that came with aid.

Box 1.1 Arrivals in Tuvalu

Nicola Wrighton

A survey of visitor arrivals took place over a three week period in 2008. Several tourists and a dozen or so unidentified visitors arrived during this period. However, at least thirty-four people entered the country as part of approximately twenty-five development-related missions or visits over the period. The list at the end of this box gives an idea of the development-related personnel who arrived in Tuvalu over the three week period.

As far as is known, apart from the Pacific Regional Infrastructure Facility mission, none of the other missions represented more than one donor or agency. This is a surprising result given that this is an environment where donors have committed to harmonisation through agreements such as the Paris Declaration on Aid Effectiveness and the Pacific Aid Effectiveness Principles.

This level of arrivals is consistent with annual visitor statistics in 2008, collected by the Tuvalu government, which show that approximately 900 people entered Tuvalu on development-related business that year. This is approximately 10% of the total Tuvalu population (9729 at 1 July 2008 according to the Secretariat of the Pacific Community). Against the total population of Aotearoa New Zealand, this would roughly equate to 400,000 people coming into Wellington annually to discuss development-related matters. The equivalent in the UK would be approximately 6.5 million foreigners visiting government departments in London and in the USA a staggering 32 million foreigners knocking on government doors in Washington DC!

I undertook further 'back of the envelope' calculations. If each of the development personnel arriving in 2008 had meetings lasting an average of six hours with two Government officials in attendance (as a conservative estimate), five full-time staff would be needed to deal with well over 200 hours of meeting time per week.

Given the small size of Tuvalu's civil service, the probable squashing of a large proportion of these meetings into the short time period between the arrival of the Tuesday plane and the departure of the Thursday plane, and the general opinion expressed by officials that missions are becoming more frequent, this level of engagement must have a significant impact on the amount of time Tuvalu Government officials have to do their own jobs.

Identified development-related personnel that arrived over the three week period:

- the researcher
- an environmental filmmaker
- an environmental researcher
- several consultants related to the same ADB mission on restructuring State-Owned Enterprises
- two consultants relating to metrology (from different agencies on different missions)
- two consultants relating to tax (from different agencies on different missions)
- four consultants on the same labour and employment mission
- at least six people from four different missions from different agencies relating to the environment
- two Tuvalu Trust Fund Advisory Committee members
- an AusAID placement under Pacific Technical Assistance Mechanism (PACTAM)
- the PACTAM programme manager
- a consultant for the Pacific Regional Infrastructure Facility (PRIF) (jointly funded by AusAID, NZAID, ADB and World Bank)
- an ADB consultant working on the Integrated Framework for Trade
- a consultant working on the government accounts system
- an AusAID official
- a group of three people from the Australia Pacific Technical College
- a consultant with the Tuvalu Public Works Department
- a consultant working with Customs.

Source: Adapted from Wrighton (2010b: 9)

Aid, ownership and sovereignty

Issues of aid are inextricably linked to questions of sovereignty. Although Pacific Island states and territories recognise the value of aid for their incomes and welfare, they are also strongly cognisant of the link between aid and dependence. Relying on donors, particularly on a single donor, for a large component of their economies means making compromises regarding power and control over their financial and other resources. Donors require a return on their 'investment' whether that be economic or political. The political or economic price of aid for Pacific Island governments is therefore a significant consideration to be kept in mind, even if it is not always made explicit by donors.

This concern about aid relationships between donors and recipients and the impact on sovereignty is not confined to the Pacific by any means. Indeed, we have identified it as a key element in determining how effective aid can be in addressing key development objectives in general (Murray and Overton 2011a). The global aid environment in the first decade of the present century was dominated by two key global agreements which underpinned the approach of most established donors, those who were members of the OECD's Development Assistance Committee (DAC). The first was the adoption of the Millennium Development Goals by the United Nations in 2000. These articulated eight key goals and associated targets with the aim of alleviating poverty on a global scale by 2015. Subsequently, the MDGs began to guide many, if not most, development assistance. Aid programmes – whether by multilateral agencies such as the World Bank or the UNDP or by major Western donor countries – adopted this 'poverty agenda' and during the following years, many committed more and more funding these MDGs. It meant that, with the alleviation of poverty as the key objective, regions such as the Pacific did receive more aid but the expanded aid programmes were now redirected to activities with an explicit poverty focus as defined by the MDGs: primary education, maternal and child health, gender equality in schools, etc.

The second key global agreement was the Paris Declaration on Aid Effectiveness of 2005. This was the second of ultimately four 'high level forums' on aid effectiveness brokered by the OECD. The Paris Declaration was the pivotal agreement and its five principles (ownership, alignment, harmonisation, managing for results, and mutual accountability) became central to development policy and practice worldwide. The principles were essentially concerned with improving the effectiveness of aid delivery yet, in retrospect, the core principles represented a remarkable degree of consensus about what made aid work. The ownership principle was the most important and, potentially, the most revolutionary. Despite thorny questions about ownership and legitimacy of different political entities, it acknowledged that the key lesson learnt from decades of aid delivery was that aid and development do not work when they are imposed on the supposed beneficiaries. Instead people, communities and states need to define their own development needs and prioritise resource allocation. They should develop their own strategies and solutions, they should be responsible for implementation and monitoring, and they need to live with, and be accountable for, the results.

Supported by the alignment and harmonisation principles, the Paris Declaration had the potential to rewrite global development aid practice. Reversing previous neoliberal approaches to limit and diminish the role of the state in development, it saw a lead role for recipient governments in developing poverty reduction strategic plans and overseeing long-term programmes through government budgets and institutions. And rather than replacing or replicating such state-led activities, donors agreed to support such state efforts, co-funding development efforts through Sector Wide Approaches (SWAps) or, ideally, General Budget Support (GBS). State institutions became critical and, in principle, donors were to step back and let local institutions take the lead.

As with the MDGs, the Paris Declaration had profound effects on the aid environment in the Pacific. Pacific governments were expected to articulate strategic three to five-year poverty reduction plans, all government departments were to implement and monitor these plans and a range of civil society organisations were to be consulted and be involved, along with government, in reducing poverty. In theory, the ownership principle would empower state institutions. Ideally, recipients would lead and donors follow: this would be a remarkable inversion of the apparent power imbalances noted above and national sovereignty, in making development policies, would be respected and encouraged.

However, the reality of aid turned out to be somewhat different. Understandably, donors were reluctant to merely pass on funding without robust systems for ensuring that funds were spent appropriately. They also wanted to ensure that governments consulted widely with their populations and were both transparent in the way their revenues were spent and accountable for any failures or misspending. Donors were keen to ensure that systems, plans and audits conformed to international standards and templates to make the global aid system consistent, workable and understandable. The result of these donor concerns was a plethora of systems, practices, policies, plans, agreements and consultations, all in the name of aid effectiveness but with the end result of a set of new conditionalities imposed upon recipients. Unlike the heavy policy conditionalities imposed during the neoliberal era of the 1980s and 1990s – most notably through much-criticised Structural Adjustment Programmes – these were 'process conditionalities'. They were to do with the way officials and governments operated – sets of practices that were required as the price of higher aid receipts.

The inverse sovereignty hypothesis

So whilst the Paris Declaration discussed above made much of 'ownership', raising the prospect of enhanced local sovereignty over development, the practice of aid threatened to challenge the independence and flexibility of Pacific governments. Supporting or restoring the recipient state's sovereignty was hailed as the overarching goal, but the actual exercise of sovereignty was curtailed and prescribed by donors. This we termed the '**inverse sovereignty effect**': the Paris Declaration aimed to reinforce policy sovereignty, but it was undermining and reversing it in practice (Overton et al. 2012; Murray and Overton 2011b).

This inverse sovereignty hypothesis guided our research project in the Pacific. Here it was to be given particular form, shaped by the complex issues of scale and political statuses in the Pacific. Nicki Wrighton's observations in Tuvalu (2010b) were a starting point: do small states such as Tuvalu struggle with these new demands and conditionalities more than larger ones? How do local officials cope with these demands in practice? How relevant are aid agendas at the global level to the realities of small island states? Furthermore, these questions could be broadened and reconstituted elsewhere in the Pacific: do those states that are less politically independent face fewer or more demands from their donors, along with higher aid volumes? Are there examples of some Pacific states or regional Pacific institutions that are able to use these new principles better than others? How do Pacific Island officials perceive and work within the aid policy environment in different ways? We hoped that if the inverse sovereignty hypothesis was confirmed, then it would be possible to inform policy at national, regional and global levels so that ownership and sovereignty could be better respected and enhanced.

In pursuing this research, the issue of sovereignty became a central concern. We were influenced by two different sets of debates in the Pacific. One involved the contentious matter of political independence. The Pacific region had been colonised by Western states during the nineteenth and early twentieth centuries and decolonisation occurred rather later and more slowly than in many other parts of the world. Independence was sought and fought for in the region. In some cases, it came relatively smoothly, and full independence was granted to countries such as Papua New Guinea and Solomon Islands (see Table 1.2 below). Elsewhere, the move to independence seemed to stall. In Kanaky/New Caledonia some strong indigenous calls for independence in the 1980s were resisted by other residents. In smaller countries, such as Cook Islands, Niue or Palau, questions about viability of an independent state were tackled by adopting a partial model, where full internal self-government was coupled with a voluntary 'free association' arrangement where the former colonial power retained some powers and responsibilities and citizens were able to retain some rights within the metropole. And in yet other places, such as Tokelau, moves to full independence from New Zealand have been debated, but not followed through. Other Pacific islands, such as Wallis and Futuna, have never really considered independence from France. Political sovereignty remains a contested and complex notion in the Pacific Islands region and the relationship between national sovereignty and development is one we return to in chapter 4.

The second approach to sovereignty that informed this research came from two directions. Firstly, given that most of the research team lived and worked in Aotearoa/New Zealand, the debates there about Maori sovereignty – *tino rangitiratanga* – provided insights into a concept that is much more intricate and complex than the macro level and Eurocentric concept of state sovereignty (see Box 1.2). Instead issues of self-determination and control over one's own development at community or *iwi* (tribal) level arise and suggest that sovereignty should be seen as many-layered, multi-scalar and inherently complex. Secondly the work of Epeli Hau'ofa, in suggesting a Pacific world that extends beyond national borders and spans the ocean in networks of kin (Hau'ofa 1993), was very influential in highlighting the critical agency of Pacific Island peoples and kinship groups in identifying and exploiting a range of opportunities. For Hau'ofa, Pacific cultures and environments are not constraints, and adjectives such as 'small', 'traditional', 'communal' or 'maritime' are not limiting descriptors but help define a range of resources that Pacific people have proven adept at using.

Box 1.2 *Tino rangatiratanga* and Maori sovereignty

Hannah Mackintosh

"Indigenous scholars have argued that sovereignty emanates from the unique identity and culture of peoples and is therefore an inherent and inalienable right of people to the qualities customarily associated with nations" (Barker 1995: 3). From this perspective sovereignty is connected to identity and is defined within the context of particular local realities which are the result of specific historical experiences. This more holistic view moves sovereignty away from absolute power towards the ability to assert a common identity based on cultural practices, language and shared historical experiences.

The term 'sovereignty' itself is not often used within the Māori sovereignty movement. For many Māori, sovereignty is a European concept that carries the weight of a colonial history (Durie 1998; Moon 2000) and does not truly convey the meaning behind Māori aspirations and demands. For this reason, the more favourable term is *tino rangatiratanga* or self-determination[4] (Melbourne 1995). Article Two of the Treaty of Waitangi of 1840 assured Māori they would retain their *tino rangatiratanga* over "lands, villages and all things precious" (Orange 2004). Sovereignty for Māori in Aotearoa, therefore, revolves around the assertion of *tino rangatiratanga.*

Māori calls for *tino rangatiratanga* in Aotearoa are often viewed as a Māori secessionist claim. Most Māori, however, view it as neither antagonistic nor threatening but rather it is a call for autonomy and control over all decision-making relating to or impacting Māori (Dixon 2006; Durie 1998; O'Sullivan 2006). It represents the freedom to be able to engage meaningfully in all aspects of society as Māori and to be respected as a partner in the Treaty of Waitangi. This is therefore not necessarily a direct threat to the sovereign power of the state. However, it does call for the rearrangement of the current social order, "It is about forging a new social contract for living together differently" (Maaka and Fleras 2005: 103).

Not all Māori agree on the meaning or the aims of *tino rangatiratanga*. In reality, it has different meanings for different individuals and collectives (Maaka and Fleras 2000). This has resulted in a movement which embraces a whole range of social, cultural, economic, spiritual and political aspirations (Chile 2006). Upon considering *tino rangatiratanga*, Mason Durie (1998: 239) sums up the overall aspirations of Māori as:

> Economic self-sufficiency, social equity, cultural affirmation and political power, stand alongside a firm Māori identity strengthened by access to *whānau*, *hapū* and *iwi* and confirmation that future generations of Māori will be able to enjoy their lands and forests, rivers and lakes, harbours and the sea and the air.

Thus an appreciation of the debates concerning *tino rangatiratanga* can provide insights into wider understandings of sovereignty. Moving away from a simple state-centred view of absolute political power, we can appreciate that sovereignty can, and should, encompass how people and communities can exert ownership and control over their own resources, cultures and identity. Such forms of self-determination, then, have to do with politics and institutions but also with how sovereignty is practiced, negotiated and felt on an everyday basis over a range of systems, interactions and relationships.

Extracted and modified from: Mackintosh (2011)

Putting these ideas together, this research adopts a view of sovereignty that seeks to move beyond a simple notion of the supposed costs and benefits of political independence at national level – though that remains important – and instead suggest an additional view of sovereignty that stresses its malleable, diverse and multi-layered nature. Within the context of international aid, sovereignty is practiced and performed on an everyday basis and is not just given and accepted as an absolute state of (political) being. We place particular emphasis on the role of key agents in the aid world: the officials and politicians from both sides, the consultants, researchers, journalists and

NGO workers who are engaged in aid negotiations, communication, implementation and evaluations.

Despite the clear asymmetries, the crucial issue is how recipients negotiate and interact in overt and less explicit ways in order to challenge the power inequities and capability problems that undermine sovereignty. As a consequence, the costs and benefits of aid to all parties involved are manifested in complex ways across the region. Three major factors affect the interactions and negotiation. Firstly, size and scale are important. Larger states receiving aid may have more to offer by way of resources and have better capacity to deal with the requirements for compliance and implementation that come along with aid inflows. Smaller states may struggle to deal with the burden of aid administration – as we shall see – yet they can also use their size to advantage, often being able to attract more from donors in per capita benefits than their larger neighbours. Secondly, political status is also a key factor. There is a clear correlation between dependence and aid: sovereign independent states generally attract much less aid than those territories that maintain higher degrees of association with a metropolitan 'patron' with whom they have a colonial history. And finally, culture is critical. This is not a simple anthropological matter of shared language, rituals, understandings and history – though these factors are of influence – but a more general issue of political and bureaucratic practice in the ways Pacific Island officials deal in different ways, through their attitudes, behaviours and policies, with donor institutions and officials.

If ownership and – by extension – sovereignty are to promote the efficacy of aid and development, then we need to examine it not only as a set of complex political and technical agreements and arrangements but also as an even more intricate set of practices, attitudes, protocols and relationships that work through dynamic personal and institutional networks that have evolved over a long history. In order to contextualise our research we turn in the subsequent section to a brief examination of colonial and postcolonial Pacific history emphasising the diverse and dynamic nature of sovereignty across space and time.

A brief history of colonisation and sovereignty in the Pacific

The interaction of the Pacific Islands with outside powers has moulded its sovereignty over a long period of time. Colonialism and negotiating the influence of external powers over the sovereignty of local polities on Pacific islands did not begin with the establishment of the formal European empires of the 1800s, although this is when it was first codified in a Western legal and political sense. The Pacific was first settled up to 40,000 years ago by humans moving into what is now Papua New Guinea. These people also spread subsequently into the islands of Melanesia (Solomon Islands, Vanuatu, New Caledonia and Fiji). Later, more extensive settlement was carried out by Austronesian people with groups crossing through the Indonesian archipelago and Papua New Guinea into what is now Melanesia around 4000BP (Before Present), including Fiji by approximately 3000BP. The settlement of the broad area of Polynesia began around 2500BP with the last areas of the triangle to be settled being Hawai'i (1500 BP), Rapa Nui (1000BP) and Aotearoa (perhaps 800–1000BP) (Thorne and Raymond 1989). Conflicts over control of resources – at the time probably expressed through tributary obligation, but in today's terms associated with what we term sovereignty – were likely to have been many and, to an extent, drove the onward migrations of the Pacific peoples.

European contact was made first in the early 16th century with Dutch, Spanish and Portuguese explorers. It was in the 1700s and 1800s that European influence was consolidated through the actions of missionaries, traders and whalers that had built over previous centuries (Campbell 1989; Howe 1984, also Spate 1979). Competition between the two great maritime powers of the time, the British and the French, partly drove this. Firth (2000) refers to this early group of colonists as the 'buccaneers' of global capitalism. The South Pacific became an important strategic location given the political rivalries that existed during this phase of European expansion as it moved from the mercantile to the industrial phases of colonialism. The European powers brought with them their own particular concepts and philosophies such as, for example, Christianity, private property, money and moral codes regarding race, class and gender. Colonists and their supporters back in Europe considered these concepts as tools and as signifiers of progress and civilisation.

While there were economic as well as political interests that drove European colonisation of the Pacific – economic resources such as copra, fruit and sugar, and later minerals, served as inputs in the European industrial complex – resource extraction did not represent the main driver of colonisation in the Pacific. In fact, early colonisation throughout the Pacific in the first half of the 19th century was a rather slow process, still largely driven by mercantile trading interests both of European and Pacific people. We would describe the early colonisation process as one of polycolonialism: a complex and geographically differentiated colonisation process of the Pacific region by different actors, for different motives, at different scales and driven by an exogenous as well as endogenous dynamics.[5] The exogenous dynamic came from early European traders and missionaries who – in order to support their business and proselytisation activities – called upon the governments of their mother countries to follow them and establish 'law and order'. In other words, the early settlers asked for political and material support from their governments to secure their often tenuous footholds in the Pacific and regulate the conflicts between themselves, as well as with Pacific peoples.

Generally speaking, the European metropoles provided this support rather reluctantly and certainly not in the volumes asked for by the early settlers. If anything, European governments of the time embraced policies emanating from Adam Smith's classic liberalism – free trade and minimal state intervention – and sought the expansion of global trade by private trading houses. As such, the first dynamic of European colonisation in the Pacific was propelled by the establishment of trading posts of these trading houses, for which their shareholders alone bore the costs. One form of support European governments could give these trading houses was to strengthen their chief traders' positions by designating them their local diplomatic representatives. As representatives of their countries, these 'consuls' could then call upon their nations' battleships to settle business disputes with local adversaries or competing trading houses. For example, the trader John Brown Williams became the US representative in Fiji in 1846 and in 1849 he called in US warships to assert and expand his trade network against local chiefs. Similarly, the network of German trading house Godeffroy – having Chancellor Bismarck as a shareholder – represented Germany throughout the Pacific, with trading posts-cum-consulates in Samoa, Tuvalu and Tahiti.[6]

The endogenous dynamic that facilitated the formalisation of European colonisation resulted from the diverse set of formal treaties, business deals, and informal arrangements between local chiefs and political elites throughout the Pacific and – on the other hand – the European missionaries and traders-cum-consuls. These local Pacific elites used these relationships with European powers to increase their trade revenue and military capabilities and subsequently dominate or conquer neighbouring polities, or to engage in dynastic struggles for local power within Pacific polities. New Zealand's northern Māori polities, for example, were the first to acquire firearms through European traders and they began a violent and dramatic expansion of their influence throughout the entire North Island and as far as the Chatham Islands. These 'Musket Wars' took most of the first half of the 19th century and left tens of thousands of Maori dead or enslaved, overhauling the political map of Aotearoa/New Zealand. An example of a Pacific dynastic struggle that was exacerbated was the First Samoan Civil War between 1886 and 1894, characterised by increasingly violent warfare among rival local factions, each faction having successfully obtained support from equally rivalrous trading houses from Germany, the USA and Britain. This support started with the supply of guns and credit, but later expanded to military training and, after that, combat troops and even warships.

While the encounter between Pacific and European peoples and powers initially resulted in what we termed a polycolonial process, driven by endogenous and exogenous actors alike and each for their own motives, it generally ended by the late 19th century with a clear hegemony or full formal control over Pacific islands by European powers. Control or dominance over Pacific islands was ultimately determined in formal arrangements between European and American colonial powers. This process began notably with the Treaty of Waitangi between the majority of Māori chiefs and the British crown in New Zealand (1840) and, for France, with its unilateral declaration of New Caledonia as French territory in 1853. The process ended around 1900. A clear example of European and American control at the expense of Pacific political elites was the Tripartite Convention on Samoa of 1899 in which Germany, Britain and the USA partitioned the islands of Samoa – without any involvement of local parties. The three parties agreed to respect the territorial claims on Samoa by the USA and Germany and in exchange for withdrawing its claim on Samoan territories, Britain was offered the 'rights' to Tonga, from which Germany duly withdrew its claims and trading posts. Table 1.2 summarizes European colonialism and independence in the Pacific Islands. Although generally Europeans did not settle in vast numbers in the region, their influence was profound and far reaching, leading to the establishment of British, French and German colonies in the late 1800s and early 1900s. The German territories included Samoa (1900–1914) and German New Guinea (1884–1914). The French added French Polynesia and Wallis & Futuna in the 1880s and the British presence was centred on Fiji (1874), Papua and Solomon Islands. Thus, until the conclusion of the First World War, the region was controlled by three European powers though their effective power in many territories, especially in Melanesia, was relatively thin. In the late 19th century the USA added a number of territories including American Samoa and Guam, and came to play an especially important role in Micronesia mainly gaining control where the Spanish were defeated or had sold their colonial possessions. European populations settled in the region, but only in New Zealand and Hawai'i were they able to usurp local polities and establish a degree of independence from the colonial power. French settlement in Polynesia and New Caledonia was relatively extensive but the territories remained part of metropolitan France in a formal sense.

In sum, in our view, colonialism in the Pacific throughout most of the 19th century did not involve a straightforward and successively wider military conquest of Pacific territory by European and American powers. Instead, in the Pacific the contacts with newly arrived non-Pacific actors created a process of polycolonialism; a complex dynamic of trade, conflict and competition among European and North American powers, within and between neighbouring Pacific polities, and also between Pacific peoples and the increasingly present outsiders. However, at the start of the 20th century, the overall balance of power and control was concentrated in the hands of the US and European governments as imperial control over Pacific islands was formalised in international diplomatic treaties. This launched the Pacific into mercantilist production circuits at the periphery of the evolving global economy.

The impacts are summarised by Firth writing on colonialism in 2000:

> Globalisation when combined with colonial rule meant incorporation into the global economy on terms that suited the interests of the colonial powers . . . the place of the tropical world in the first globalization was to be subordinate to the temperate and developed world.
>
> (p. 184)

Colonialism was unevenly felt across the region. In some places, such as Hawai'i and Rapa Nui, the indigenous authority was overthrown, land

Table 1.2 Colonisation of the Pacific Islands

Country	*Comments on colonial history*	*Present political status*
American Samoa	After rivalry between Great Britain and Germany, the USA was granted the territory through the 1899 Tripartite Agreement. Named American Samoa in 1911	'Unincorporated' US territory (1967)
Cook Islands	European influence and Christianity spread in 1800s. British protectorate in 1888. Annexed by UK in 1900, and incorporated into the colony of New Zealand 1901	Self-governing in free association with NZ (1965)
Federated States of Micronesia	Annexed by Spain in 1885, sold to Germany in 1899 and captured by Japan in WW1. US administration established 1947 as part of the Trust Territory of the Pacific Islands. Self-governing as FSM 1979.	Independent with compact of free association with USA (signed 1986, formalised 1990)
Fiji	European settlers in the 1800s. Ratu Cakabau self-declared Tui Viti (king of Fiji) offered to cede to the UK in order to offset US and French power. Accepted by UK 1874	Independent (1970)
French Polynesia	First European settlers 1790s. 1842–1888: Island groups unilaterally declared French protectorate	'Overseas collectivity' of France (2003)
Guam	Annexed by Spain 1668, ceded to USA 1898. Occupied by Japan 1941–1944, re-occupied by USA	'Unincorporated' US territory (1950)
Hawai'i	Kingdom of Hawai'i overthrown in 1893. Republic led by US interests declared. Annexed to USA 1898, and fully incorporated following referendum in 1959	Integrated within USA (state – 1959)
Kiribati	British Protectorate of Gilbert and Ellice Islands became UK colony in 1916. Occupied by Japanese during WW2 until 1943. Ellice Islands became Tuvalu in 1978, and Gilbert Islands and other groups became Republic of Kiribati in 1979	Independent (1979)
Marshall Islands	Annexed by Spain 1874, sold to Germany 1884. Occupied by Japan in WW1 and administered by Japan as a trust territory 1919–1944. US administration 1947 as part of the Trust Territory of the Pacific Islands. Self-governing 1979	Independent with compact of free association with USA (1986)
Nauru	Annexed by Germany 1888, captured by Australia 1914, administered as a trust territory by Australia, NZ & UK 1919–1941. Occupied by Japan 1942–1945. Australia, NZ & UK trust territory 1947	Independent (1968).
New Caledonia	First European settlers, traders, missionaries and blackbirders in the 1840s. Unilaterally declared a French possession in 1853	'Special Collectivity' of France (1999)

Country	*Comments on colonial history*	*Present political status*
New Zealand	Became British territory 1840 following the Treaty of Waitangi. Self-governing 1857, dominion status 1907, 'equal status' with entities of the British Empire 1931, full separation of the UK and NZ crown in 1947 when this was ratified. 1986 Constitution Act adopted. 1987 revoked all residual UK legislative power	Independent (though some technical debate remains concerning the date of and exact nature). 1947 full sovereignty and 1986 Act led to free-standing constitutional monarchy in 1987
Niue	Acquired by UK as a protectorate in 1900, passed to NZ 1901 as part of Cook Islands. Separate status as NZ colony 1903	Self-governing in free association with NZ (1974)
Norfolk Island	Part of New South Wales colony – UK (1788). Territory of Australia 1913	External territory of Australia (1979)
Northern Mariana Islands	Acquired by Spain c.1668, sold to Germany 1899. Japanese UN mandate 1919, US administration 1947 as part of the Trust Territory of the Pacific Islands	US territory with 'commonwealth' status (1978)
Palau	Acquired by Spain 1574, sold to Germany 1899. Japanese UN mandate 1919, US administration 1947 as part of the Trust Territory of the Pacific Islands	Independent with compact of free association with USA (1994)
Papua New Guinea	Acquired by Germany (New Guinea 1884) and UK (Papua 1884 then to Australia 1906). Australia full control of Papua New Guinea 1921. Japanese occupation of most of the country 1942–1945.	Independent (1975)
Pitcairn	Early Polynesians became extinct prior to 15th century. Mutineers from the Bounty settled in 1790. UK colony (1838). 1856, entire population emigrated to Norfolk, but a small number returned to Pitcairn.	British overseas territory
Rapa Nui	European contact in 1722; Peruvian slave raids and disease greatly reduced population to less than 120 in early 1870s; annexed by Chile (1888)	Integrated within Chile. Chilean citizenship granted (1966)
Samoa	British missionaries from 1830. Two civil wars (1886–1894; 1898) reflecting Western interests (Germany, UK and USA). German Samoa annexed in 1900 under Tripartite Agreement (1899), occupied by New Zealand 1914–1962 (Western Samoa Trust Territory). Pacifist Mau resistance movement through 1900s. Independent in 1962. Renamed Samoa 1997.	Independent (1962)

(*Continued*)

Table 1.2 (Continued)

Country	*Comments on colonial history*	*Present political status*
Solomon Islands	German protectorate over North Solomons 1886 and British protectorate over southern islands 1893. German portion transferred to UK 1899. Occupied by Japan 1942. UK re-established 1945. Self-government 1976.	Independent (1978)
Tokelau	Missionaries from 1845, Peru slave raiders 1863, British protectorate after 1977, annexed by UK and included as part of Ellice Islands 1916. Placed under New Zealand administration 1925. Became incorporated into New Zealand 1949.	Part of New Zealand with self-governing legislature (1996)
Tonga	Missionaries from 1797. Unified under Tāufa'āhau in 1845 as a constitutional monarchy (constitution granted in 1875). Self-governing as a British protected state from 1900	End of protected status – fully independent (1970)
Tuvalu	Ellice Islands declared a British Protectorate in 1892 as part of British Western Pacific Territories, then as Gilbert and Ellice Islands from 1916 to 1974. Voted for independence as Tuvalu, separate from Gilbert Islands (Kiribati), in 1974.	Independent (1978)
Vanuatu	Labour 'blackbirding' and missionaries from 1860s. France and UK 'joint naval commission' 1887 to protect expatriates but no jurisdiction over indigenous inhabitants. France and UK declared a 'condominium' 1906. Confused administration with dual colonial systems.	Independent (1980)
Wallis and Futuna	First European settlers 1837, local treaties established French protectorate in Uvea (Wallis) 1887 and 1888 (Futuna and Alofi). Annexed as colony 1917 and under authority of New Caledonia. 1961–2003 French overseas territory.	French overseas collectivity (2003)

Source: Expanded and adapted from Murray and Overton (2015: 355)

was seized, and people displaced. In others, there was profound change in Pacific cultures due to the introduction of Christianity, wage labour, diseases and new material goods. In Fiji, Aotearoa/New Zealand, New Caledonia and French Polynesia, new populations from Europe and Asia were settled alongside Pacific societies, with major implications for national identity. In these places, colonialism was dominant, visible and disruptive. Yet, in many rural regions of some territories such as Papua New Guinea, Solomon Islands and parts of Vanuatu, daily life carried on largely as if colonial rule did not exist. Indeed, only with the Second World War and the ousting or threatening of the colonial powers by Japan, was there any major visible presence and disturbance of daily life for a significant proportion of the region's population.

The era of formal colonialism in the Pacific lasted until the 1960s by which time colonialism had become consolidated in general terms. There were a number of major events during this period that were of influence and which rendered the colonial map ever-more differentiated. The First World War saw the end of formal German colonies in the region, with German Western Samoa being occupied by New Zealand on behalf of the British Empire in 1914. The remaining German protectorates including New Guinea and parts of Melanesia were formally handed over with the Treaty of Versailles in 1919. Many of the Japanese controlled territories of the early 20th century were transferred to the USA following defeat in the Second World War. Given the declining relative influence of Great Britain following this latter war, and to a lesser extent France, the USA rose to assume both political and economic power motivated in large part by Cold War politics of the post-1945 period. New Zealand and Australia inherited some of the influence over various territories and countries that Great Britain left behind during second half of the 20th century.

To an extent the colonial powers prepared some parts of the Pacific Islands for independence in the post-Second World War period, by recognising, protecting and incorporating elements of indigenous authority into the state's ruling institutions. Thus, in Fiji, selected chiefs became agents of the colonial state, left largely to control indigenous Fijian lands and villages, and in Tonga the royal family and associated nobles received British protection. These were examples, common throughout many parts of the Pacific, of a policy of 'colonial conservation': the maintenance, even strengthening, of a local political and social elite left largely in control of local populations while the European colonial powers managed the commercial and urban sectors more directly. Arguably this was a form of local sovereignty that was protected and nurtured (arguably even created) by the colonial states, albeit a form that ossified social hierarchies and limited the opportunities for social and economic mobility.

When independence did begin to appear in the Pacific Islands, it eventuated later than other waves of decolonisation across the world. This can be explained by debates concerning national viability (for smaller territories), lack of cohesion and potential conflict amongst diverse populations (Fiji, Vanuatu, Papua New Guinea) and the important role that some colonies and territories played during the Cold War (French Polynesia and the American Micronesian territories especially). Colonies that moved to full independence included Samoa (1962), Nauru (1968), Fiji (1970), PNG (1975), Solomon Islands (1978), Kiribati (1979), and Vanuatu (1980) (see Table 1.2). Tonga, which had been a protectorate of Britain, withdrew from this arrangement to assume full independence in 1970.

Given the strategic role of the region and viability issues, models of semi-autonomy and other sovereignty arrangements were developed in the Pacific that at the time were unique to the world. In fact, these models were very similar to paths chosen by many other colonised islands in the Caribbean and the Atlantic and these models of unique sovereignty arrangement remain in place or continue to evolve until today – something we will explore further in Chapter 4. For example, the Cook Islands and Niue obtained 'self-governing in free association' status with New Zealand in 1965 and 1974 respectively. This sees the latter responsible for defence and much foreign and monetary policy while inhabitants of these countries have citizenship in New Zealand. More recently, in what we later discuss as 'retroliberalism' – harking back to the classic liberalism of the early 19th century – the New Zealand government has begun to refer to these territories together with fully dependent Tokelau, and other smaller islands and atolls as the New Zealand 'realm' (see Box 2.1). In the case of the USA, various 'compact in free association' agreements were reached with the Marshall Islands (1986), FSM (1986) and Palau (1994) – territories that were administered as United Nations trusts following the Second World War and during the height of the Cold War (see Box 3.4). Dependent status as 'unorganised territories' was developed in American Samoa and Guam which continue to play important military roles in the Pacific region for the USA. France – and many inhabitants of its territories – have been reticent to let go of its colonial status, despite pressure to do so both from the UN Special Committee on Decolonization and indigenous peoples of the territories. This has resulted in significant conflicts and independence movements in French Polynesia and New Caledonia particularly. The former played a very important role during the Cold War when nuclear tests were undertaken in the small atolls of the territory. The latter has one of the world's largest nickel deposits which have provided an important mineral wealth resource for export. Each of the French territories has had several different official statuses.

There are also territories that have remained fully incorporated into the administrative structures of the colonial power. Hawai'i, for example, was originally a Polynesian monarchy which was overthrown by American capitalist elites in order to maximise the gains of fruit and other exports. It became an official US colony from 1898 and incorporated as a state in 1959 becoming the 50th and newest of all the United States. Rapa Nui has also been fully incorporated to one extent or another over time following its annexation by Chile in 1888 after Europeans, including Scottish sheep farmers, had abandoned their economic interests. The current status of Rapa Nui as a *provincia* (the second tier in a four-tier spatial government system) in Region V of Chile is significant and came following the end of the Pinochet dictatorship in 1990 during which it was at the lower rank of *comuna*. Compared to other provinces, the population of the territory is very small and thus it wields relatively large influence within Chile. Despite a visible independence movement there is little prospect of change in the future, especially given recent expansion of what the central government has termed 'concessions' under Bachelet (2014–2018).

However, as we discuss in Chapter 4, the story of decolonisation in the Pacific Islands is not simply a matter of an inevitable trajectory from colonial rule to full independence. Many countries have followed this path yet others appear to remain stalled (Robie 1989), paused (Connell 1988, 1993), or under continuous renegotiation (Baldacchino 2010). To some, incomplete decolonisation appears to be the result of colonial powers being reluctant to leave – such has been the narrative of many for New Caledonia/Kanaky for example. However, there are also examples of Pacific peoples themselves being reluctant to shed strong ties with their metropolitan 'patrons'. Tokelau has rejected a degree of independence in two referenda, the result of New Zealand being pressured by the UN Special Committee on Decolonization (see Box 4.2). In various other territories (Rapa Nui, Wallis and Futuna, Guam, American Samoa for example) there does not seem to be a strong push for full independence despite some groups advocating this. Indeed, there are some examples of resistance to further decolonisation; in an interesting example of domestic inter-island rivalry, groups in the Marquesas Islands of French Polynesia have stated that, if the Tahitian independence movement was to get its way, it would seek to split from Tahiti and retain a status as a French territory.

There are good reasons for many Pacific islands to seek to maintain ties with the metropole. Beginning with independence and the postcolonial period, there has been a relatively large-scale migration of Pacific Island peoples to the Pacific Rim and beyond – especially to New Zealand, Australia and, to an extent, the USA and France. The remittances sent by migrants back to their homelands combined with high per capita levels of aid and associated government employment and bureaucracy led to the characterisation of the microstates of the postcolonial Pacific as comprised to lesser or greater degree by MIRAB economies.[7] Remittances have continued through the generations, and although some migrant flows have been stemmed recently, third and fourth generations continue to send high levels of per capita remittances by global standards. To varying degrees, these MIRAB economies are predicated on maintaining constitutional arrangements that link them to their former or present colonial metropoles. For many fully independent Pacific states, the options for migration and remittances – and for high aid levels – are much more constrained. As we discuss at length further in this book, although aid has ebbed and flowed it has remained very high in per capita global terms. As a consequence, the power of donors has remained significant, calling into question the ostensible sovereignty of a number former colonies and semi-autonomous territories.

Political sovereignty, then, is varied and relates to different economic strategies being pursued by a very diverse range of Pacific states. However, sovereignty seems to have an increasingly economic dimension. The arrival of a period driven by neoliberal policies in the 1980s (see Chapter 2) has caused some commentators to reflect on the economic and political sovereignty or otherwise of the Pacific Island countries and territories. As we explore in the next two chapters, strict conditionalities applied to aid and loans in the 1980s and 1990s attempted to impose neoliberal solutions for the Pacific Island economies, both independent and otherwise. Apart from the evolution of some export-oriented labour – or environmentally-intensive industries such as garments and agro-food exports in some of the larger economies, it has proven very difficult to obtain sustainable outward-oriented economic development. In fact, such developments have often been reliant on trade preferences which have been gradually withdrawn under the past two decades of WTO-inspired and enforced free market trade reform. Only tourism has provided sustainable economic opportunities, although the control and flow of profits is not always retained within the island economies themselves and the broader costs to sustainability can often be substantial. According to Firth, this period of globalisation, beginning in the late 1980s and early 1990s, represented a consolidation of weakened national sovereignty in the Pacific despite political independence:

> Just as the place of the Pacific Islands in the first globalisation was to be subordinate to the temperate, developed world, so their place in the second globalisation is also to be subordinate, this time to a set of international institutions that have set the rules of the global economy. The new globalization *now combined with independent sovereign rule*, means incorporation into the global economy on terms that suit the interests of the financial markets, the aid donors, and those relatively few Pacific Islanders who are in a position to benefit from this situation.
> (Firth 2000: 191)

This complex colonial and postcolonial political and economic history has, as already noted, given rise to a wide range of formal arrangements and actual outcomes in terms of the sovereignty of the region. The process of decolonisation has been partial and diverse. There have been conflicts in the French territories, and to an extent the conflicts of Melanesia (such as Bougainville and the Solomon Islands) come as consequences of the long-term unfolding of colonialism in the region and in particular the creation of independent nation states that retain diverse ethnicities in ways that perpetuate precolonial struggles. It is perhaps not surprising that such postcolonial internal and regional conflicts have occurred in countries that had experienced only a thin veneer of colonial rule and where attempts to establish a single national identity and polity were largely unsuccessful. In contrast, as we have seen in the case of some territories such as Tokelau, there have been continued calls from the indigenous population to retain constitutional ties to the colonial metropole. In this sense formal measures of independence do not tell us the full story of sovereignty – it is fluid, negotiated and more difficult to measure than legal statuses and recognised histories might suggest.

Indeed, a central argument in this book is that sovereignty is as much about the day-to-day performance of political freedoms, independence and self-determination as it about such formal arrangements. Sovereignty is about agency as much as it is about structure.

The relationships between former colonies and the colonial powers remains deeply intertwined, variable, dynamic and complex. Therefore, the postcolonial landscape is one where a variety of institutions and individuals interact in multifaceted ways. Relationships occur at many levels, not just between the nation-states. This, and perhaps every, colonial and postcolonial moment is not monolithic – a concept we referred with reference to historic patterns as polycolonialism, but which may also be extended to the present array of arrangements regarding independence and sovereignty (see Chapter 7). Global structures, signals and flows are interpreted, ignored, adjusted and reformed by agency on the ground. Relationships are numerous, complex, multifaceted and two-way: 'colonialism' (defined broadly as the deliberate interpenetration of one system by another in search of material or other advantage) thus occurs in many forms. This involves much more holistic and flexible performances of sovereignty that go beyond formal constitutional arrangements and that ultimately, in the day-to-day affairs of the region, are of at least as much influence.

Researching aid in the Pacific

This book explores the complexities of sovereignty in the Pacific through the workings of international development aid. It analyses the inverse sovereignty hypothesis posited earlier in this chapter. Does the aid regime of the first decade of the 21st-century result in diminished sovereignty of small and overstretched Pacific island states, despite its rhetoric of country ownership and donor alignment and harmonisation? We do this by developing three 'vertical' themes and questions through subsequent chapters: (1) how do size and scale affect the ability of Pacific states to engage effectively with aid donors and secure benefits from aid? (2) to what extent does political status relate to the ability to establish policy sovereignty? and (3) what 'cultures' of aid – in terms of official and unofficial policies and practices – have developed and how can these help secure policy sovereignty – agency by Pacific islands – in the region?

With support from New Zealand's Marsden Fund, a three-year research project took place largely from 2011–2014, with ongoing work undertaken since that date. Thesis work by Nicki Wrighton (PhD) and Avataeo Junior Ulu (Masters) and several others combined with the research work of the academic staff (Overton, Murray and Prinsen). Several country case studies were conducted, particularly focusing on the policies and practices of local aid offices and officials, together with an overview of aid volumes and policies in the region. For case studies in the Francophone territories, we received funding from France's *Fonds Pacifique*. Subsequent funded research work on migration and education in the Pacific has allowed some continued indirect engagement with a number of the locations and themes addressed in this book.

Although the project began with a plan to choose certain country case studies to cover a range of countries in terms of size and political status, a key principle from the outset that the work had to involve the active participation – and decision to participate – of Pacific individuals and offices. This guided the central research of Wrighton in particular who went to considerable lengths to explain her work then ask individuals and departments to join or not if they wished. It was an 'opt-in' principle and there was a strong wish not to contribute to the pressures on officials that Wrighton had noted earlier. Fortunately, officials in several countries were keen to participate and asked to join and were active and forthcoming in expressing their views and telling their stories. Through Wrighton's work and that of others on the broader team (the supervision of Suaalii-Sauni and the master's and doctoral research of Mountfort, Kiddle, Broadbent, Geary Nichol, Talagi, Thoma, Christ and Pirnia) we were able to compile a set of country-based material (Tuvalu, Cook Islands, Vanuatu, Samoa, Tonga, New Caledonia, Wallis and Futuna, Niue, Solomon Islands). We draw on this work both throughout the text of this book and also through a series of 'research insights' boxes. The latter also include contributions from colleagues from throughout the region (Aiafi, Sisifa, Blaise, Mawyer and Taumeofolau) who we met and who challenged and extended our understandings of aid and sovereignty in the Pacific.

Despite the enthusiastic involvement of many Pacific Island participants, the research team remained a group of largely *palagi* (European) researchers based in New Zealand. Many of the team had much experience in the region, some could converse in Samoan, *bislama*, French or Spanish and others spent time working in local institutions. But the view is still largely that of outsiders. This has some advantages, in terms of the ability to retain a more detached global perspective on the issues, yet we could not possibly gain the sorts of relationships, insights and understandings that Pacific Island researchers could.

Overall, the methodological approach of the research was varied. We adopted a quantitative approach for the analysis of aid statistics and policies. Yet, for much of the work, we recognised the importance of the way participants gave meaning and expression to their own experiences. This recognised a more constructivist epistemology, one that recognised that different ways of knowing and communicating knowledge exist in the region (Gegeo 1998; Gegeo and Watson-Gegeo 2001). Qualitative approaches, through the use particularly of in-depth semi-structured interviews and discussions, coupled with observation, often participant observation, were effective in allowing participants to talk widely and freely, and in many instances steer the direction and tone of the conversations. Whilst not explicitly adopting Pacific methodologies, such as *talanoa* (for example Vaioleti 2006; Prescott 2008; Farrelly and Nabobo-Baba 2014; Suaalii-Sauni and Aiolupotea 2014), we recognised that value of developing relationships through talking in a relaxed and appropriate manner.

Over the three years of the project, the central concern of the project – the inverse sovereignty effect – was questioned and adjusted. We soon found that this hypothesis gave us only a partial and sometimes misleading view of aid and aid relationships. Instead we identified and explored a sort of 'perverse' or 'reverse' sovereignty effect. Yes, we could see that the top-down compliance burdens of the new conditionalities were undermining policy sovereignty in many instances but there were also many instances of astute bottom-up strategies and practices being used to resist or reshape the nature of external control and influence. Pacific island officials were far from passive or powerless in aid relationships and, in fact, we saw many cases where the Paris Declaration principles were being followed and driven from within the region.

Overall, we reject some previous attempts to frame aid in the Pacific either as strongly negative (*pace* Helen Hughes) or marked by a simple dichotomy between dominant powerful donors and weak and passive recipients. It is the complexity and variability in practices and attitudes that we highlight, and it is in the niches and fissures of personalities, everyday practices and manifold interactions that we see the real exploration and exercise of agency by Pacific actors, expressing policy sovereignty. Nonetheless, we also return to the importance of formal and material aspects of the aid sector. Agreements, policies, statistics, review documents and signed memoranda of understanding are all important in giving substance to what is discussed, negotiated and contested.

The study is not concerned with grand theories or prescriptive approaches. We are not trying to define what is 'good aid' or appropriate sovereignty for Pacific peoples. Rather this is a largely descriptive and analytical study focused on policy, practice and the operational aspects of aid in the Pacific islands: we emphasise local agency, capabilities and resources. We aim to gather and compile ideas and examples from the region and globally to help inform how ownership/sovereignty can be given substance in the region and bring the improvements to their lives that Pacific peoples seek. Whilst focusing on the inverse sovereignty hypothesis, we are also mindful, following Hau'ofa, Helu Thaman and other Pacific scholars, of the need to be critical of academic frames of development which portray Pacific peoples simply as vulnerable, constrained, passive and powerless (Hau'ofa 1983; Thaman 1993). Thus, as well as being concerned with the way external agencies are imposing conditions and burdens on Pacific institutions and officials through the aid regime, we are keen to uncover ways in which Pacific people exercise appropriate agency in the aid world.

The book begins in the next chapter with a review of the global aid environment. Although our focus is on the period approximately between the years 2000 and 2010, marked by the overriding influence of the MDGs and the Paris Declaration, we place this period in the context of previous aid 'regimes', such as the neoliberalism of the 1980s and 1990s, and the more recent opening up of a 'retroliberal' era. Chapter 3 then paints a picture of aid in the Pacific, mapping volumes and flows of official aid over time and identifying key donors and recipients. It also attempts to place these aid flows against other forms of development assistance, such as trade preferences and migration concessions. We then seek in Chapter 4 to analyse in greater detail what is meant by sovereignty in relation to development aid. This begins with a review of conventional definitions of state sovereignty before exploring the way sovereignty is played out in Pacific island states in particular ways. This sees both a strongly island-centric and terrestrial view of sovereignty – bounded, competitive, territorial, yet contestable and tradable – and a more oceanic view, one that stresses fluidity, mobility, interaction and multiple identities.

In Chapter 5 we move explicitly to the inverse sovereignty hypothesis and assemble evidence of the way donor conditionalities combine with limited local capacity to produce often serious pressures that undermine the ability of local agencies to manage their own development effectively. However, Chapter 6 then examines the many ways in which Pacific Island agents and institutions are developing strategies and tactics to combat these pressures. This is a reverse effect, evidence of the assertion of policy sovereignty through a wide variety of policies, practices, relationships and negotiations. This demonstrates the ways in which many Pacific Island leaders and bureaucrats have proved very adept at engaging with the global aid regime and using it, through locally-adapted systems and social norms, to leverage important local gains and resist unpalatable external demands. Finally, in the concluding chapter, we attempt to pull the thematic strands of the book together, returning to our original questions and hypothesis. We also attempt to present some practical suggestions – drawn from exemplars throughout the region – for ways in which development sovereignty can be strengthened in the Pacific and help achieve the development aspirations of Pacific people.

Notes

1 Throughout this book 'island states' refers to independent island state nations (such as Fiji), and 'territories' refers to metropolitan overseas territories of which there is a range of types as we discuss elsewhere.

2 Tables 1.1 and 1.2 list island states and territories that we consider to be part of the Pacific Islands. We are aware that this is a subjective selection. We include Hawai'i and New Zealand, for example, because of their high proportional Pacific Island or Pacific Island-descended populations and their deep historical connections to the region.

3 Official Development Assistance as measured by the OECD. Negative values represent a net donor.

4 "*Tino rangatiratanga*" translates literally as "absolute chieftanship or full chiefly authority." However, in contemporary Aotearoa society it is more broadly defined as "the power to be self-determining" (Fleras and Spoonley 1999: 27).

5 This term is intended to counteract a monolithic view of colonialism that unfolds in a uniform way across the region. Similarly it could be argued that the decolonisation process, which is also highly complex and ongoing, has led to complex geographies of sovereignty, representing an extension of this earlier polycolonial process.
6 Processes such as those outlined here resonate with the later concept of the aid regime of 'retroliberalism' explored in chapter 2 which has state sponsored mercantilism at its core.
7 The acronym MIRAB refers to economies shaped by Migration, Remittances, Aid and Bureaucracy (Bertram and Watters 1985 – see Box 2.2).

Chinese aid: New government buildings under construction in Nuku'alofa, Tonga

2 Global aid regimes and the Pacific

Introduction

In this chapter we seek to connect shifting aid regimes at the global scale with changes in the aid sector in the Pacific Island region. The purpose of this is to contextualise the more recent evolution of aid relationships and to interpret them in a broader frame. Here we undertake a chronological review of shifting aid regimes in the wider world and the relationship to shifts in the region, with some commentary on the regional and local geographic impacts. First, we consider the continuous evolution of global aid regimes and discuss how these unfold differently across time and space. We then seek to apply this global framework to the particular case of the Pacific drawing similarities and differences with change at broader scales. In doing so, we begin with the colonial period and follow this with a discussion of modernisation, neoliberalism and neostructuralism. Finally, we turn to the most recent aid regime – retroliberalism – and discuss its implications, noting in particular the competition between the 'traditional' Western donors and the rise of China as a new donor. In each case we trace out the unfolding of the regimes at the global scale and then seek to draw implications for the Pacific.

Aid regimes: Pacific currents, global tides

As aid policies continue to shift and their analysis deepens it has become increasingly common to talk of 'aid regimes'. In a broad sense an aid regime conceptualises and delivers official development assistance and is characterised by a general discourse manifested in a set of guiding principles aimed towards broad goals, combined with regulatory mechanisms which deliver certain objectives. As we argue elsewhere (Mawdsley et al. 2018) wider 'regimes of accumulation' influence the accompanying aid regime significantly. While an aid regime will vary across time, the exact nature of its constitution will differ according to particular local relationships. However, arguably within a given regime there will be more factors in common than not when comparing the different roles and policies of donor and recipient countries. Each regime will be comprised of a combination of delivery mechanisms, referred as aid 'modalities'. Such modalities – whether discrete projects or integrated programmes – may well cut across specific aid relationships temporally and spatially, but it is their combination that allows us to discern between different broader aid regimes.

In Table 2.1 we trace four global aid regimes across time. This matrix assesses the unfolding of global geopolitical and geo-economic events, and political change in the West, and relates these to the principles, goals and modalities of the corresponding aid regime. Even though these are presented chronologically, once again it is important to state that reality is more complex than this – elements of each regime can persist at any given time and vary between places. These are rather like Kuhnian scientific paradigms – change involves a discernible movement between distinct intellectual and political constellations but in ways that are complex, co-deterministic, and imprecise.

These regimes shift as global circumstances evolve and as donors respond to them. The changes are partly orchestrated and partly institutionalised through particular global institutions (for example, the World Bank and the OECD-DAC) and then reproduced through national donor agencies and their policies. In this chapter, we elaborate on the framework above and we seek to draw implications for the Pacific region in particular.

As we shift through each aid regime, we see the set of overarching goals shift to an extent in concert. However, this can hide important variations. As noted, we must be wary of top-down frameworks such as this. In the unfolding of these regimes, one thing remains constant according to Mawdsley et al. (2015: 5): aid has "always been accompanied by claims to the altruistic pursuit of improving the lives of others *and* self-interest. While the precise formulations of these interests may change (e.g. during and after the Cold War) or be constructed differently by different donors (e.g. Norway and the USA) they are invariably presented as positively aligned." This, argue these authors, is not problematic in and of itself; rather it often conceals very particular geopolitical and geo-economic interests that are unique to the relationship concerned. In this sense there has never been a set of universal goals for aid – each discourse and associated regime emerges according to historico-geographical contingencies – economic, political and cultural.

This is especially important in the context of the Pacific Islands – a region that is highly heterogeneous in each of these aspects. However, as we categorise the Pacific – sorting it by cultural roots (as in Melanesia, Polynesia, Micronesia), or by its colonial history (as in French, British and US influence etc.) – these divisions often hide more than they reveal. This is surely the case in the world of aid. The relationship between any given donor and recipient is highly contingent and dynamic. It will turn on an infinitely complex array of history, geography, environment and culture. It is within this postcolonial environment that countries and territories seek to assert, negotiate and perform their agency, their policy sovereignty. As we will see the consequent

Table 2.1 Aid Regimes 1950 to the present

	Modernisation	*Neoliberalism*	*Neostructuralism*	*Retroliberalism*
	1950–1980	*1980–2000*	*2000–2010*	*c. 2010 to present*
Global Events	• Allied War victory • Truman's four-point programme • Cold War rivalries • Decolonisation • Transformations in China	• Economic crises (oil shocks and debt crises 1970s/80s, share market falls (1987 & 1997) • Fall of the USSR (1991)	• 9/11 and invasions of Iraq and Afghanistan • Period of economic growth	• Global financial crisis • The rise of China and other 'emerging powers'
Domestic Political Context in the West	• Cold War politics • Kennedy's Alliance for Progress	• Thatcherism • Reaganomics	• Rise of Tony Blair's New Labour (UK) Clinton's Democrats	• Swing back to the right: Cameron (UK), Abbot (Aus), Key (NZ), Republican control of Senate in US
Principles	• Modernist and traditional structuralist ideas concerning role of industrialisation and backwardness of rural development. • Keynesian economics • Geopolitical imperative of preventing domino effect across the Third World, based on alternative socialist modernities and ideas of dependency theorists	• Neoliberal theories and monetarist economics • The state crowds out the private sector and leads to inefficiency and corruption • The market will arrive at Pareto optimality • Benefits of export growth will trickle-down to poor through employment • Comparative advantage and trade liberalisation	• 'Third Way' - state tackles social justice but in the context of a globalised economy that remains open • Poverty and inequality are seen as consequences of the market but are responsibilities of the state • Deliver the benefits of globalisation and ensure its trickle-down	• The state exists to facilitate economic growth • The private sector should not be crowded out by the state • Government sponsors and facilitates the private sector • Ricardian comparative advantage coupled with neo-Keynesian economics to stimulate the private sector during recession

(*Continued*)

Table 2.1 (continued)

	Modernisation	*Neoliberalism*	*Neostructuralism*	*Retroliberalism*
	1950–1980	*1980–2000*	*2000–2010*	*c. 2010 to present*
Development goals	• Grow industrial sector • Promote regional alliances • Promote urbanisation and reduce rural inefficiencies	• Reduce government size • Raise productivity • Stimulate exports • Develop the private sector	• Poverty alleviation • Equality promotion • Aid effectiveness through market mechanisms	• Economic growth • Infrastructural development • Stimulate trade and investment through financing
Aid policies and modalities	• Import substitution Industrialisation • Land reform • Support for state budgets and building state capacity • Colombo Plan	• SAPs: privatisation, hollowing out of the state, reduction in social expenditure • Export-orientation • 'Good governance' • Market-based projects • Use of civil society	• MDGs • National interest and development agenda (formally) separate • Poverty Reduction Strategy Papers and poverty reduction-based projects • Sector Wide Approaches (SWAPs) and GBS • Reconstruction of the state for security	• Infrastructure development • Semi-tied aid projects • New (returnable) forms of development financing • Development for diplomacy and the rolling together of national interest and developmentalism • Partial return to project modalities

Source: Adapted from Murray and Overton (2015: 352–361); Mawdsley et al. (2018); and Murray and Overton (2016a).

strategies vary from place to place and are inherently tied to the relationship with aid donors and the history therein. Notwithstanding this, there are clearly identifiable waves of aid that have washed across the Pacific; distinguishing between and drawing such lines around them allows us to throw significant light on this geographic differentiation and historical dynamism.

Colonial transfers and the foundations of 'aid' in the Pacific

Having outlined the chronological history of colonialism in the Pacific in the first chapter, this section revisits that with the aim of understanding the roots and foundations of international development assistance in the region. As noted, colonialism in the Pacific followed a somewhat different chronology to other areas of the Global South, to the extent that Pacific-wide patterns can be identified (see Table 1.2 in the previous chapter). We need to understand in particular the later part of the colonial period as a precursor to subsequent aid regimes where relationships were entrenched, and discourses seeded. Three important points warrant reiteration in that regard.

Firstly, the movements for political sovereignty and independence were generally less violent than they were elsewhere in the world's decolonisation process. There were indeed moments of conflict over sovereignty, in Samoa for example. And there still are occasionally violent calls for independence in New Caledonia, and vocal independence movements in Vanuatu, French Polynesia and Rapa Nui. In addition, there have been bloody conflicts in settings (as in Bougainville) concerning the nature and shape of postcolonial states. Yet, in most cases, independence in the Pacific has not had to be fought for as forcefully as elsewhere in the former colonies. Secondly, the movements and consequent independence arrived later in the Pacific than elsewhere. While the first colonised territory to obtain independence – Western Samoa in 1962 – gained that status in the midst of the global decolonisation processes of the 1950s and 1960s, most other Pacific territories that became independent did so only in the 1970s. The relatively late arrival of decolonisation reflected the important strategic role of the islands in the context of the Cold War as well as the relatively muted calls for independence among some of the territories themselves compared to other parts of the world. Although different in each case, this was, at least in part, due to concerns both within and outside the region regarding the viability of small Pacific Island territories in the face of global economic and political forces. Could a small island state such as Tuvalu, with a population of around 10,000 people ever successfully follow the decolonisation trajectory of states such as India or Nigeria? This latter question leads to a third point: colonialism has not fully ended in the Pacific. Whereas other former colonised regions also have seen colonial enclaves remain, none (with the possible exception of the Caribbean) is subject to such significant continued colonial presence as the Pacific Islands region. As outlined in Chapter 1, the nature of these constitutional associations varies between full integration (as in the case of Rapa

Nui) to territories in free association (as in Niue). However, and as noted in Chapter 1, we see a region that is highly complex in terms of the range of relationships that exist between metropole and associated territories both now and in the past.

Following the Second World War, the peoples of most colonial territories began clamouring for independence and by the late 1950s, most colonial powers acknowledged the legitimacy of the claims and the inevitability of decolonisation. The year 1960 marked a turning point in that it saw the UK Prime Minister Harold Macmillan's speech on the 'winds of change' and the UN's Declaration on the Granting of Independence to Colonial Countries and Peoples. Most colonial powers – particularly Britain and France – began to prepare for independence of their colonies in various ways. This eventually had an effect in the Pacific with undertakings from Australia with regard to Papua New Guinea. This country, because of its relative size, became a central concern. Australia had directly paid for much of the cost of the colonial administration and maintained the budget. Such payments were continued and increased to provide infrastructure and build the ability for local leaders and institutions to govern themselves. Before independence, these transfers to promote development and welfare were not counted as ODA (Official Development Assistance) as such and represented an extension of the colonial transfers of previous periods, but they were a precursor to later aid regimes. When independence was granted to PNG in 1975, these colonial support funds eventually became 'aid': transfers of funds into the government coffers to support the cost of government and undertake development projects. This involved direct budget support (later reinvented under the aid rubric as General Budget Support, though with rather more stringent conditions). During the transition to independence across the region, great care was taken to ensure that Western oversight and, crucially, capitalism and integration into the global economy would remain central to the development of Pacific states in the context of the Cold War.

As we noted above, decolonisation has not been completed in the Pacific and several formal colonial relationships persist that have implications for the nature of 'development' aid transfers. In some cases, funds from the metropolitan power are counted as internal flows supporting the cost of government. In French Polynesia and New Caledonia, such transfers were captured by ODA statistics until 1999 and, as we see in Chapter 3, these amounted to amongst the highest volumes of aid in the Pacific, along with Australia's postcolonial aid to PNG. Then, however, a political decision was made to cease to enumerate these transfers as 'aid' – a symbol that the territories were part of metropolitan France rather than states in their own right. Similar arrangements to re-label statistics on development aid are in place for American Samoa, Guam and the Northern Mariana Islands, where financial transfers from the USA to support the costs of government, welfare and development are not counted as ODA, even though a direct colonial relationship may be said to exist. Yet in other territories, for example Wallis and Futuna where the degree of incorporation into the metropole is very high, transfers continue to be counted as development aid. Here, and in semi-autonomous states such as Tokelau, Niue and Cook Islands (New Zealand's 'realm' states – see Box 2.1), aid transfers are transparent and counted as ODA. However, the reality is that these are financial transfers that reflect continuing quasi-colonial relationships and responsibilities – attempts to provide public services and governance similar to the metropole rather than predicated on the usual ODA-inspired quest to promote development programmes that have local ownership.

Box 2.1 New Zealand and the 'Realm' states

John Overton

In 2010 a New Zealand Parliamentary Committee undertook review of the country's relationships with Pacific countries (NZ Parliament 2010). The committee, chaired by MP John Hayes, produced a report which was subsequently largely overlooked in terms of direct policy formulation (though it probably helped raise aid levels to the countries concerned) yet it reveals an interesting perception by New Zealand politicians of one group of countries. It used the term 'realm states':

> *The people of three countries – Tokelau, the Cook Islands, and Niue – hold New Zealand citizenship, share the Queen as Head of State, and are part of the Realm of New Zealand.*
>
> (p. 21)

The term actually refers to a group of countries that share the 'Queen of New Zealand' as Head of State. The term has to do primarily with the name and domain of the monarch; it does not refer to the relationship between the elected Government of New Zealand and the other countries. However, the Parliamentary report sketched a reworking of that relationship and there was a dismissal of the 'self-government in free association' model:

> *There needs to be an acknowledgement that the way the decolonisation experiment has been applied to the Cook Islands and Niue (it is still on offer to Tokelau) has largely fallen short of the expectations of all parties. (p. 21) . . . The self-government model has resulted in administration that cannot be sustained by local communities and which, especially in Niue, is choking any local entrepreneurial spirit. It is likely to do so in Tokelau.*
>
> (p. 27)

Instead, what was recommended was a change in aid strategies:

> *We propose a fundamental re-think of New Zealand's assistance strategy, aimed at improving standards and delivery of basic services – such as education, health, policing and justice – for communities in Tokelau, the Cook Islands, and Niue, so as to bring them into line with New Zealand standards over time. (p. 9) . . . It is important that all communities of New Zealand citizens receive services – especially in education, health, infrastructure and security – on a par with those enjoyed by similarly sized population centres in New Zealand.*
>
> (p. 14)

What was not spelled out was what this meant in terms of sovereignty. By using terms such as 'communities of New Zealand citizens' the report seemed to hint at closer integration and direct subsidy (and perhaps even direct administration) from Wellington. Interestingly, this would cast Cook Islands, Niue and Tokelau closer to French or American constitutional arrangements in places such as Guam or Wallis and Futuna and mark a return to quasi-colonial systems of sovereignty.

Perhaps such a change could be favoured by some residents of these communities if they were to receive higher levels of services and social welfare equivalent to New Zealand, even if it meant a loss of a degree of political independence. That the recommendation of this model came from National Party members of the committee in New Zealand and not from any explicit call from the three territories themselves is instructive.

Modernisation and the development project

At the global scale, an aid regime based on 'modernisation' was constructed and ran from the period of decolonisation following the Second World War until the 1980s. During this period the concept of 'development' as a policy objective was invented (Escobar 1995; Rist 1997; Kothari 2005). Philip McMichael has termed this era of development the 'development project' (McMichael 2017), a time when there was much emphasis on newly independent states being supported to undertake welfare and development projects. It was state-centred, it had a strong emphasis on infrastructure and industrialisation, and it sought to address welfare issues through modern education and health provision. In post-war speeches, the American President Truman posited the need for international development to prevent conflict and to raise living standards. This was backed by economics and sociological work from the likes of Walt Rostow and Talcott Parson who hypothesised stages of growth and social modernisation. Implicit in this, and explicit in the subtitle of Rostow's treatise, was the need to resist and prevent the rise of socialism and communism across the periphery. Yet it was an approach that also owed an intellectual debt to Keynes and, arguably to structuralist theorists such as Prebisch, for it actively promoted a leadership and interventionist role for the state. The market was to be fostered and encouraged, yet the state maintained both a regulatory oversight and the ability to intervene directly, as in infrastructure, when the private sector was seen as incapable. It was a development discourse that sought to transform – even eliminate – old tribal and traditional ways of life and build modern developmental states, globally integrated market economies, and urban societies of consumers living in nuclear families.

In the Pacific, however, the chronology of this modernisation mission was altered. Cold War politics were important in the Pacific, as elsewhere in periphery at the time, and the region became a theatre of the Cold War though generally without open conflict. Limited attempts by the Soviet Union or Communist China to establish a diplomatic presence in the region were treated with suspicion and, if they appeared, they were usually responded to with thinly-disguised Western aid generosity to the Pacific nation involved. The strategic importance of the region in the context of Cold War geopolitics combined with the relatively small-scale of the islands led to large aid transfers. Pacific Island governments were able to bargain for high levels of aid partly because of their geopolitical comparative advantage. From this period onwards, and indeed into the present, the Pacific Islands came to rank among the highest per capita aid recipient territories in the world. Again, as noted above, this had little to do with the generosity of donors or a positive postcolonial spirit of reparation. Rather, important strategic advantages could be maintained through such linkages and continued control of recipients' sovereignty – even in newly independent countries – could be exerted. Furthermore, with recent independence, donors felt compelled to sustain and strengthen governments of Pacific countries so they would not disintegrate or fall into the hands of Cold War rivals.

As part of the push towards modernisation, colonial states emphasised an aid regime that promoted Western models of urbanisation and industrialisation, often overlooking the suitability of such models for island economies and societies. In a broader cultural sense, the conscious modernisation attempt to shift away from putatively more 'primitive' forms of (under) development led to maintaining and expanding the large-scale bureaucracies that were inherited from the colonial governments. This saw the nation-state and its bureaucracy at the centre of social, economic and cultural transformation. Local elites dominated the political arena, and many were able to secure a livelihood in the expanding sphere of government employment. Given the typically small size of local economies on Pacific islands, the state sector provided the main source of employment, and much of this was paid for by aid revenues. Furthermore, this large and expanding public sector was also an accepted function of the state in the modernisation era. States were expected to lead development planning – akin to the developmentalist approach elsewhere across the periphery. Economic planning, manifested across the region in the 1950s-1970s in a series of national five-year plans, attempted to promote economic development and spatial and social integration. It aimed to put in place the necessary foundations for this, often in the forms of roads, ports and airstrips, energy projects, and regional development schemes, as well as plans to ensure that people had access to reasonable education and health facilities. Again, such plans had a direct link to aid, for the five-year plans provided a convenient list of projects and programmes that could be justified to, and funded by, donors keen to see in place prosperous and stable governments and societies. This state-centred modernisation model in the Pacific has been referred to by Overton (1999) as *vaka matanitū* utilising a Fijian term adapted and used to signify 'government'.[1]

Whilst this developmentalist *vaka matanitū* approach suited some of the larger states of the region that had become independent – PNG, Fiji, Vanuatu and Solomon Islands – it was rather less appropriate for smaller countries and territories that lacked a large physical resource base, and which had only tiny (though relative to their small populations, very large) bureaucracies. Here the push to full independence was not as strongly pursued and important ties to the metropole were maintained, through such arrangements as the free association model of Cook Islands and Niue or the similar and later compact models between Micronesian states and the USA. In these instances, government bureaucracies also provided a critically important and sometimes only form of wage employment and they too lay at the centre of modernisation strategies for development. Even more so than in the larger independent states, these public sectors were maintained through aid in the form of often direct budget support. Allied to this, and a consequence of continuing constitutional ties with the metropole, livelihoods and development were also predicated on more direct engagement with the outside world.

The migration of many hundreds of thousands of Pacific Island peoples to the countries of the Pacific Rim was facilitated by political agreements but also by increasingly cheap air travel and a vastly denser network of international air services from the 1970s onwards. This migration and the consequent remittances from Pacific people working in New Zealand, Australia, the USA and France became a major component in the GNP figures for many countries and territories (as has remained so until today). This arguably successful combination of aid, bureaucracy, migration and remittances, underpinned by the maintenance of open relationships with forms of connection on past and continuing colonial powers, was observed in the 1980s by Bertram and Watters (1985, 1986; Bertram 1986) as a particular 'MIRAB' development model relevant to small island states (see Box 2.2).

Box 2.2 The MIRAB model

Warwick E. Murray

The acronym MIRAB is used to signify countries and territories where migration, aid, remittances and bureaucracy form a central political-economic and social role. Bertram and Watters (1985, 1986) coined the term, analysing enduring links in the region to current and former colonial metropoles, historical and contemporary development strategies, and resultant social, political and economic structures. They suggested that, measured principally in terms of consumption, Pacific Island populations lived better material lives than GDP per capita might predict (Bertram 1993). This was made possible by the postcolonial 'MIRAB' nature of the region. Long-lasting 'transnational corporations of kin' were established based on extensive migration (MI) from the region to current and former metropoles on the Pacific Rim. Through these networks broad family groups remit money and goods across borders (R). Together with these remittances, aid (A) formed a central proportion of the national income. At the time, this was underpinned by the geopolitical strategies of the Cold War, often couched as postcolonial 'obligations' of former powers. Having inherited relatively large colonial bureaucracies, the state sector became the major employment sector on most islands (B).

In 1986 Bertram suggested that MIRAB-based development was rational and, crucially, sustainable especially in the light of growing calls for neoliberal-inspired cuts. At the time the hypothesis was applicable to varying degrees across the region, and critics argued that it simplified and overgeneralised a complex region. A number of regional critics (such as Hau'ofa 1993) suggested that MIRAB belittled Pacific societies, normalising and legitimising dependency and stifling local agency and inventiveness to the advantage of quasi-colonial powers. The sustainability of remittances has been questioned as second and third generations of migrants lose cultural and corporeal ties with 'homelands', though there has been remarkable buoyancy in these flows.

Given the rise of neoliberalism in the late 1980s declining comparative advantage in geopolitics, there were widespread calls among donors to reduce aid, couched in terms of efficiency seeking reforms to stimulate export-led development. This involved a drastic downsizing in the state sector, and aid in the Pacific was made contingent on such reforms echoing the shifts in the aid regime at the global scale. In the 2000s, the neostructural turn in aid fostered something of a revival in the MIRAB structure of the Pacific economies – through various forms of budget support which increased state employment. Concerns about regional security and political stability in the larger states of Melanesia led donor countries to increase their support for states that were deemed fragile. Elsewhere, limited-term seasonal employment schemes allowed Pacific Islanders to travel to New Zealand and Australia and work as labourers in the agricultural sector, which added a new remittance flow.

The MIRAB development model remains a highly debated topic in the region and beyond (Poirine 1998). It could be seen as a form of 'indigenous globalisation', offering an alternative to neoliberalism (Murray 2001). Economic diversification in some parts of the Pacific have led to a revision and expansion of the construct. Growth in tourism gave birth to the TOURAB concept (Bertram 2006). In some cases, we might hypothesise a partially inverse MIRAB effect – where remittances flow from the territory to the metropole (e.g. Rapa Nui). Critics following the Hau'ofa (1993) line remain critical of the hierarchical and colonial connotation of the concept and urge us to move away from it.

Neoliberalism

Although the modernisation era of aid and development in the Pacific seemed to be taking firm hold in the region between the 1960s and 1980s – and was developing local variants such as MIRAB – it was soon overtaken by a powerful new development ideology that had its origins well outside the region. Based on theoretical models from the Chicago school of economics led by Milton Friedman and following experiments in Chile in the 1970s, neoliberal policies were adopted across the world's northern hemisphere. The West first adopted these policies first in the 1980s: for example, in the USA under Reagan's Presidency from 1980; the UK under Margaret Thatcher from 1979; and, New Zealand with the Lange Labour Government after 1984. These harsh and sudden reform packages had major social and economic impacts in the Global North but were soon diffused to the Global South in the mid-1980s and 1990s through structural adjustment policies (SAPs) applied by global institutions – the World Bank and IMF in particular. By the end of the 1980s SAPs – and the associated aid, loans and debt restructuring – became the dominant development paradigm and the vast majority of countries in Latin America, Asia and Africa had adopted such policies. In this regard the debt and oil crises of the 1970s and 1980s represented a significant watershed opportunity allowing the Western global institutions of the World Bank and IMF to use the smoke-screen of modernisation-induced debt reduction to enforce policies that would facilitate globalisation of trade and finance, while reducing protection of national industries and employment standards. Indeed, neoliberalism could be seen as a deliberate strategy to attack and undo the fundamental tenets of modernisation: state-led development, welfarism and planning.

Policy makers in the West began to talk of the potential economic and even democratic benefits of 'globalisation' (often used interchangeably with neoliberalism at the time) and, more to the point, the costs of not participating. At the same time as the benefits of open markets were emphasised and pursued during this phase of neoliberalism, most of the Western countries where these policies emanated from were kept behind high protectionist walls.[2] The neoliberal goal of maximising capital mobility clearly boosted the revenue and reach of large-scale corporations, most of which had their headquarters in Europe or North America. There was also a specific geopolitical rationale in all of this – SAPs were part of the Cold War assault on non-capitalist regulation. Countries that did not participate such as Vietnam, Cuba, Mozambique and briefly Chile, were either ridiculed, blockaded, covertly undermined or even violently invaded.

It is possible to split the neoliberal phase at the global level into two periods, and these were mirrored in terms of the neoliberal aid regime (see Murray and Overton 2016a, also Harvey 2005; Ferguson 2009). From the early 1980s until the end of the Cold War, an extreme form of neoliberalism (which some have called neoliberalism 1.0 – Hendrikse and Sidaway 2010) was instituted. This was replaced by neoliberalism 2.0 which in some ways was less aggressive and less politically overt than its predecessor, given the shifts in the geopolitical landscape following the end of the Cold War in the 1990s.

The first period saw extreme policies regarding privatisation, the downsizing of the government, free trade reform, outward orientation, cuts in health and education expenditure, cuts in income tax but the institution of, or increases in, consumption taxes and overall deregulation of the economy. Although these ideas are often seen as emanating from the UK and US, in the case of both during their so-called neoliberal period neither significantly reduced government expenditure, often off-setting expenditure cuts in welfare with increases in defence spending. These reform packages were foisted on indebted countries by both multilateral financial institutions (for example the World Bank, Asian Development Bank and IMF) and bilateral Western aid agencies. These policies saw development aid and loans increasingly conditional upon the deregulation of markets, the downsizing of the public sector, and other neoliberal reforms. They were put in place suddenly, with little or no public consultation, and they tended to be extreme in their implementation – full restructuring of the economy was vigorously pursued, rather the more gradual and piecemeal strategies adopted in countries such as Australia or Japan. It was diffused very rapidly across the Global South. By the early 1990s virtually every country in Africa, Asia and Latin America had introduced SAPs. This led to significant losses in development opportunities and welfare for large parts of the population as education and health expenditure were often aggressively cut in the ostensible search for balanced budgets and economic efficiency as adjustment that in theory would eventually yield long terms economic gains.

One of the socio-political outcomes of this period was the rise in NGOs as development actors. Because donors were less wary of the activities of NGOs and unified in a suspicion that states in the Global South tended to be ineffective at best and corrupt at worst, NGOs saw a significant increase in donor funding available to them. With the state's institutions being downsized and its responsibility for public services diminished, donors turned to NGOs and civil society organisations (CSOs) such as churches and community groups to fill the vacuum that neoliberal policies had created (Agg 2006; Howell and Lind 2008). Moreover, CSOs fitted the neoliberal discourse of individual and community responsibility for welfare and NGOs – both in the Global North and the Global South – were often brought into contracting arrangements with donors to provide basic welfare services often abandoned by the state. The market also took a much more prominent role, usually as state assets and activities were privatised. However, rather than fully replacing the state in areas such as public transport and utilities, privatised companies often sought profits through asset stripping or rationalisation. However, certain new businesses certainly developed and national economic growth eventually recovered, but the neoliberalism policies had wrought havoc on the existing public institutions and usually left a legacy of increased inequalities in the population, rising poverty and increased immiseration for many.

During the second phase of neoliberalism, from the mid-1990s, the World Bank and the IMF became more sensitive to criticisms of their policies of the earlier period. SAPs continued to be promoted but were now expected to be complemented by policies that were associated with good governance, reducing corruption and increasing the efficiency of public policy. This involved a re-regulation of the economy and the reconstruction of states in order to make them more open for, and supportive of, business. Furthermore, small-scale initiatives to increase marketisation and efficiency such as micro-credit policies were emphasised. Some have termed this a shift within neoliberalism from 'rolling back the state' to 'rolling out the state' (Peck 2010): the turn from a blunt strategy of cutting the size and scope of the state to one which rebuilt the state in a limited way to facilitate capitalism more effectively. Thus, the rolling out in terms of the latter phase must be interpreted as merely the use of the state to facilitate and secure the ascendency of the actors in the globalised or globalising market.

Neoliberalism and aid in the Pacific

Neoliberalism had profound impacts across the Pacific in terms of aid policy. Aid donors, having carried out harsh neoliberal reforms at home, believed they could not continue to support supposedly top-heavy and expensive bureaucracies in the Pacific or stifle private sector development in the islands through their aid policies (Scheyvens and Overton 1995). In contrast to most other countries in the Global South, the leverage of aid donors over Pacific governments rested not so much on the weight of government debt, but rather on the Pacific's governments' heavy dependence on development aid. That could have offered room for more careful and detailed negotiations: talking about reducing annual aid transfers should be easier than talking about reducing debt whose interest increases it every day. However, this made little difference in practice. As across the globe, donors demanded that Pacific governments introduce radical neoliberal reforms, and these reforms were instituted. Across the region there were substantial cuts in government expenditure required by donors became a significant part of policy in the 1990s. In some countries such as Niue and the Cook Islands employment in the civil service was cut by over half over very short periods (see Murray and Terry 2004) with devastating impacts in terms of employment – leading in some cases (where the opportunities were available) to further outmigration (Connell 2008). Few states were immune: the French territories were as harshly hit as those under the influence of the USA, Australia or New Zealand.

Further to this, aid policies also emphasised that recipient governments needed to have policies intended to deregulate markets, opening up local economies for international businesses and developing an export-oriented framework for local businesses (Firth 2000; Murray 2001). The new neoliberal policies also saw the rise of project-based aid across the region intended to stimulate sectors as diverse as vanilla exports from the Cook Islands and the manufacture of footballs in Niue. There were also some attempts to diversify the Fijian and relatively larger economies away from traditional exports such as sugar and mining products. It was during this period also that tourism as a viable alternative was explored and some attempts to stimulate it were made through development aid (Scheyvens 2002).

Another strategy favoured by donors at this time – but strongly resisted within the region – was the call to privatise communally-held land. In most countries and territories of the Pacific, land is not held by individuals but by kin-based communal owners. Land is neither alienable nor transferable, except through mechanisms of limited-term leasing (Crocombe 1987). Donors believed that communal land tenure was an impediment to economic growth for a variety of reasons: owners could not use land as collateral to secure credit, foreign investors could not buy it to develop as they saw most profitable, and there was no incentive for individuals to invest in rural development because benefits would have to be shared rather than accumulated by individuals. Agencies such as the Asian Development Bank tried to encourage governments throughout the region to undertake surveys and titling of land that would transfer ownership of land from communities to individuals and create a land market (Anderson and Lee 2010). But this was one neoliberal reform that did not take firm hold for resistance was both overt (there were riots in PNG) and covert as a number of Pacific political leaders dragged their feet on this policy knowing it would be highly unpopular. Yet, despite opposition, there have been such 'liberalising' of Pacific communal land tenure in Vanuatu and Samoa (Daley 2010; Iati 2010).

During the early phase of neoliberalism in the Pacific, in the early and mid-1990s there was then a significant 'hollowing-out' of state administrations. And in the later 1990s, given the rise of the good governance agenda and shifting international priorities in development aid, there were moves to promote democracy and good governance as a tool for further economic development (Kaufmann et al. 1999). The second stage of neoliberalism also heralded the start of the new poverty agenda, though it came later to the region than elsewhere in the Global South (Storey et al. 2005). Globally, the recognition of the need to target poverty had started as early as 1990 with the UNDP's introduction of an annual Human Development Report that defined 'development' not only in economic terms, but added people's life expectancy, education and purchasing power. The World Bank began acknowledging the need to address poverty alleviation – especially as it became the focus on the anti-globalisation movement that emerged in the mid-1990s. At the turn of the millennium, SAPs were replaced by requirements by donors for more detailed development plans that involved both a degree of public consultation and an explicit targeting of poverty. These were the new Poverty Reduction Strategy Papers (PRSPs) (Box 2.4). These PRSPs were a precursor to the new aid regime of the following decade.

In general, the impacts of the neoliberal regime of aid delivery in the Pacific were not resoundingly positive. State administrations were not supported and there were significant conflicts despite the good governance approach – possibly because such models were external impositions. In the economy, neoliberalism promoted a fundamentally new objective for the region. Instead of *vaka matanitū* as state-led development with a degree of insulation from the global economy (despite MIRAB), neoliberalism forced a turn to a particular form of globalisation predicated on trade liberalisation, deregulation, privatisation and new opportunities for private enterprise and investment. This meant that Pacific Islands turned – sometimes reluctantly – to the global economy that lay beyond the horizon. This new globalisation strategy might be termed *vaka vuravura*.[3] It was outwardly-oriented in many respects but internally, led to an increasing and broadening of inequality and a rise in poverty, the degradation of the environment in export-oriented primary production sectors and, above all, there is little evidence this neoliberal development aid regime delivered economic growth in any sustained way. Neoliberalism failed the Pacific significantly, some questioning whether the market model was ever applicable in the first place (see Banks et al. 2012; Murray and Storey 2003 for a review).

Neostructuralism

Voices from the Pacific were not alone in its dislike of the institutional and social disruption and often misery of neoliberal policies. At a global level, politicians, activists and academics were pushing for a more explicit focus on poverty and inequality and this was expressed by widespread popular movements in the West. Social movements had raised issues of debt and poverty for some time, finding expression, for example in Live Aid in 1985. Twenty years later the 'Make Poverty History' campaign carried some of these ideas further and there were some parallels with the 'anti-globalisation' movement which opposed the activities of global corporations and banks, as well as wars and environmental destruction. Although such global activism began before 2000, such as in the case of Live Aid in 1985, and continues today, it seemed to gain particular traction in the first decade of the 2000s with politicians such as UK Prime Minister Gordon Brown supporting a campaign to forgive Third World debt, represented by the Make Poverty History movement. We believe the early 2000s marked a turning point in development aid discourse: away from neoliberalism and towards what has been referred to as neostructuralism (Murray and Overton 2011a). Three particular events carried this shift.

First, in 2000, amidst fanfare, the World Bank published a research report titled "Voices of the poor: Crying out for change" (Narayan et al. 2000) capturing participatory poverty assessments in twenty-three countries, involving 23,000 poor people. The lead authors (Narayan, Chambers, Shah and Petesh) all had personal reputations as fierce critics of the World Bank's neoliberal policies. Arguably, this report signalled that even if the Bank was not willing to move away from its deregulation and free trade agenda, it was now willing to include attention for poverty and the poor into its discourse on development aid. Its World Development Report 2000–2001 was ominously titled 'Attacking Poverty'. Almost at the stroke of a pen, the World Bank and the IMF abandoned SAPs as the dominant framework for development aid and introduced a new framework to direct and assess development aid: the Poverty Reduction Strategy Paper (PRSP). Nations in the Global South that sought development aid now had to produce a PRSP outlining how their national policies were going to reduce poverty – next to further deregulation and the promotion of free trade. Once a PRSP was in place, the World Bank advocated that new ways to transfer aid would have to be used. Instead of transferring aid in the form of development projects or as earmarked funds to government departments, the new aid modalities were meant to ensure that international aid was added to the national public budgets of recipient governments enabling them to implement their poverty reduction strategies. The new aid modalities were defined as General Budget Support (GBS) or, if circumstances so warranted, Sectoral Budget Support (SBS) or Sector Wide Approaches (SWAps) (See Box 2.4).

Second, in the same year, all members of the UN signed up to the Millennium Development Goals (MDGs). In the international aid environment, the MDGs were extremely influential and built on a growing consensus to move beyond the neoliberal discourse on development aid. The role of the UN and the UNDP in particular cannot be underestimated in this regard. As such the targeting of poverty (although not the inequality that followed the neoliberal policies) became central, building on growing calls during the 1990s. The MDGs gave specific targets for development aid that were especially aligned with expansions and improvements in public services such as education and health.

Politicians responsible for development aid in countries in the Global North realised the need, or opportunity, for shifts in their aid policies with the changes at the World Bank, the broad political consensus surrounding the MDGs in the UN, and the associated rise of a civil society movements including diverse groups such as Jubilee 2000 and End Poverty. This led to the third, pivotal, event: a 2002 conference in Monterrey (Mexico) where countries in the Global North pledged to refocus and increase their development aid programmes to address poverty and support governments in the Global South to rebuild their public services in order to achieve the MDGs. Importantly, in the 'Monterrey Consensus' donor countries agreed, when conditions were appropriate, to channel their aid directly into the recipient government's overall budget (General Budget Support, or GBS) or into the budget for a particular sector such as the Ministry of Health or Education (the so-called Sector Wide Approach, or SWAp). The OECD – an organisation of Europe's wealthiest nations, as well as the USA, Canada, Japan, Australia and New Zealand – took over leadership from the UN and organised a series of four successive 'High Level Forums' between 2003 and 2011 to continue and expand support for increasing aid and make it more effective in achieving the targets outlined in the MDGs.

These marked changes in development aid discourse in the early 2000s placed poverty reduction and other aspects of the MDGs centre stage in terms of development policy. It seemed to shift the centre of gravity of the aid world from Washington (the neoliberal Washington Consensus) to Paris (and the leadership of the DAC). Ostensibly it looked like a revolution in development policy. Yet, in essence, whilst there was rhetorical recognition of poverty and inequality and a commitment of state funds to ameliorate these, globalisation and the market remained at the centre of the policies that were recommended for development and for aid.

Neostructuralism harked back to Latin American ideas concerning social inclusion and an active role for the state but combined this with the imperative of open economies and making the most of globalisation, together with a cursory nod to environmental sustainability. This period at the global scale has been termed in various ways – 'equity with growth' (as in the Lagos and Bachelet administration in Chile 2000–2006) or as 'Third Way Politics' (as in the Tony Blair term as UK Prime Minister 1997–2007). Defined as a means of maximising the room for deregulated globalisation whilst stimulating or ensuring a trickle-down of the benefits to the wider population, in reality it can be seen as a means of making neoliberalism more palatable to democratic electorates in a post-Cold War world (Leiva 2008). 'Inclusion' in this sense might be recast as a means of constructing and enabling an environment with more consumers and higher levels of consumption of goods produced by global corporations.

Box 2.3 MDGs in the Pacific

John Overton

When the MDGs were launched in 2000, there was some initial scepticism in the Pacific. The MDGs seemed to relate more to sub-Saharan Africa or Asia rather than relatively better-off small island states. In addition, there was some opposition to applying the term 'poverty' to a region where famine was almost non-existent and where most people had claim to customary land and resources and, thereby, the means of basic subsistence. To claim that the Pacific Island countries were characterised by poverty was to further belittle their ways of living and suggest they needed more external 'help'. Yet the MDGs did help focus attention of some indices that were of major concern in parts of the Pacific: maternal and child health, and literacy in particular (Kidu 2009; Naidu and Wood 2008; Naidu 2009). Others used the MDG framework to argue that, contrary to the scepticism, widespread poverty was an issue in the Pacific too. Oxfam, for example, used the MDGs to argue that almost half the region's population are living in poverty (Oxfam n.d.).

Furthermore, the MDGs, within the neostructural aid regime, helped increase and channel funding towards higher priority sectors (education and health) and countries (e.g. Papua New Guinea, Solomon Islands and Kiribati). And, indirectly, they contributed to strengthening state institutions as the prime delivery agencies (Overton 2009).

The MDGs also led to a major effort to gather data and monitor progress towards achieving the goals in the Pacific. A report by the Forum Secretariat in 2015 presented a wealth of data on the different targets and gave a basic report card for seven of the eight MDGs for the final year of the MDG period (PIFS 2015b). This coded countries by whether they were 'achieved' (Y), 'mixed' (M) or 'not achieved' (N). The results are summarised below.

Country	*MDG1 poverty*	*MDG2 education*	*MDG3 gender equality*	*MDG4 children's health*	*MDG5 maternal health*	*MDG6 HIV, TB & malaria*	*MDG7 environment water etc*
Cook Islands	Y	Y	Y	Y	Y	Y	Y
Niue	Y	Y	Y	Y	Y	Y	Y
Palau	M	Y	Y	Y	Y	Y	Y
Tonga	M	Y	M	Y	Y	Y	Y
Fiji	M	Y	M	Y	Y	M	Y
Samoa	M	Y	M	Y	M	M	Y

(*Continued*)

(continued)

Country	*MDG1 poverty*	*MDG2 educa-tion*	*MDG3 gender equality*	*MDG4 chil-dren's health*	*MDG5 mater-nal health*	*MDG6 HIV, TB & malaria*	*MDG7 environ-ment water etc*
Tuvalu	N	M	M	Y	Y	M	M
Vanuatu	M	M	M	M	M	Y	N
FSM	N	M	M	Y	N	M	Y
Marshall Is	N	M	M	Y	Y	M	M
Nauru	N	Y	M	M	M	Y	N
Solomon Is	M	M	M	N	M	M	N
Kiribati	N	N	M	M	M	M	M
PNG	N	N	N	N	N	N	N

Overall, for the fifteen Forum countries monitored (New Zealand and Australia were not monitored), there was much variation in terms of achievement by countries: the table above ranks countries from full achievement in Cook Islands and Niue to near complete lack of achievement in Papua New Guinea. There is a weak correlation here between achievement, size and sovereignty status: smaller dependent countries such as Niue and Palau have fared better than large independent ones, such as PNG. Yet independent Fiji and Tonga have 'performed' better than more dependent FSM and Marshall Islands.

Furthermore, there is a marked contrast in achievement across the different goals. Attention to education and children's health have yielded good results whilst gender equity, major diseases, the environment and poverty (in this case measured by possibly questionable indicators such as employment or growth rate of GDP) have been more sluggish.

In summary, poverty has been tackled in the Pacific through the lens of the MDGs and some results have been impressive but significant pockets and facets of poverty remain.

The search for aid effectiveness

Whilst the World Bank's PRSPs, the MDGs, the Monterrey Consensus, and public campaigns helped increase public support – and budgets – for aid and focus attention on key aspects of poverty alleviation, there were also strong pressures from donors in the Global North to ensure that the increased aid allocations were spent efficiently and effectively. These concerns led to a set of global agreements on the principles that should guide the way aid is delivered. The OECD-DAC had been wrestling with these issues for some time and held a first so-called High Level Forum in Rome in 2003 to address how aid could be made more effective. This was a donor-led initiative; the Rome meeting really only involved OECD members. Two years later, the next High Level Forum led to the Paris Declaration that articulated five principles for a more effective regime. This time more recipient representatives were invited to participate, and they signed the Declaration (some no doubt surprised that donors had suggested – in principle at least – to hand over control of aid delivery!). The Paris Declaration and its five principles became a manual for development aid and they constituted a key foundation of the neostructural aid regime. And, although they brought together ideas and practices that had been emerging in the previous decade, together they marked a remarkably novel shift in the way aid was to be conceived and delivered. They seemed to demonstrate that aid had moved a long way from the early days of neoliberalism and SAPs.

The first of the five principles agreed in Paris was that of ownership – that recipients would own their own development policies and that they would be developed along lines established through their own national developmental needs. This combined with a second principle objective – alignment – that sought to ensure that donor policy matched up with these objectives explicitly. In combination these two principles would hand control and sovereignty back to recipients and were the source of much optimism in terms of regaining the development agenda in the Global South. This would reverse the general dynamic that had been established in both the modernisation and neoliberal regimes and they constituted a potentially radical remit for aid, vesting control of development in the hands of recipients, not donors. Three other principles would support this redrawing of power relations embodied in the Paris Declaration: harmonisation of donor policy; 'management for results' with a focus on tangible outputs and outcomes; and 'mutual accountability' which would see an explicit consideration of the measurement of impacts and, supposedly, asked donors to share in the responsibility for long-term outcomes.

This new regime was confirmed in the Accra Agenda for Action (2008), following the third high level forum, and specific modalities were discussed for the implementation of this broad ranging agenda. In particular the state was placed at the centre of development policy once again – based on notions of good governance from the 1990s and in response to the hollowing out of the state during the neoliberal period which had raised both economic and security concerns in the wake of 9/11. At Accra, there was some recognition of the potential role of civil society in the aid effectiveness arena, but they were given little voice and states continued to drive the process.

The principal aid modalities of this era – general and sector budget support – handed significant power to recipients to define, implement, manage and measure their own agendas (see Box 2.4). In theory, this is not only a strong recognition of the sovereignty of aid recipient states but also an attempt to strengthen that sovereignty. The ostensible shift in power and control back to recipients is at the heart of this book – as case studies have shown that the reporting requirements and institutional costs associated with the shift in ownership and alignment and accountability have created pressures that in some cases in fact increase the reliance on overseas aid donors and can lead to the emulation of bureaucracies found in the West. In 2011 we argued:

> Therefore the new aid modalities have led to a quiet – but often revolutionary – restructuring of the state. What is being put in place amounts to a replication of the bureaucracies and management systems of donors to allow donors to have confidence in the way their aid is distributed and accounted for . . . The new aid regime has recreated the state in the image of the West, just as the neoliberal aid regime reproduced the market in poorer countries in the image of richer ones.
>
> (Murray and Overton 2011a: 316)

Whilst ostensibly ownership had been shifted, the costs of compliance and accountability often led to an undermining of the very principles upon which this regime was built. In this regard:

> the new aid regime is actually increasing dependent relations of power between donors and recipients in ways that substantially favour the former and 'lock in' the latter. In some ways this is the antithesis of the ostensible goals of neostructuralism, representing instead a recasting and extension of the dominance wrought through the top-down neoliberal aid regime.
>
> (Murray and Overton 2011a: 316)

Box 2.4 Aid modalities

Gerard Prinsen

In September 1999, the World Bank and the IMF decided to end their support for Structural Adjustment Programmes (SAPs) that had pushed nations in the Global South to downsize their public administrations' expenditure on education and health services, while increasing free trade. Instead, the Bank and the IMF now acknowledged "an emerging global consensus" that "the gap between the rich and the poor is large and growing" and that "unless current trends are reversed, the broadly supported International Development Goals will not be met." In particular, they were now primarily concerned about poverty, "not just in incomes but in education and health outcomes as well" (World Bank 2000: 1). The World Bank and the IMF had made an about-face.

In agreement with donor countries, the World Bank and IMF decided that the countries classified as Heavily Indebted Poor Countries (HIPC) would be given access to financial support on the condition they would prepare national Poverty Reduction Strategy Papers

(PRSP). These PRSPs would need to be directed by goals for which "the International Development Goals [later called the Millennium Development Goals] can serve as benchmarks" (World Bank 2000: 4). As we know, these MDGs had no targets for free trade or budget deficits but focused on making sure everyone has access to public education and public health services – especially the poor.

At this point in time, a remarkable and remarkably fast consensus emerged, bringing together the four groups of actors in development aid: donor countries providing the largest volume of bilateral aid, the Development Assistance Committee (DAC) of the OECD providing common aid policy frameworks, the UN promoting the MDGs, and World Bank and IMF as leaders in financial policies. They agreed to put more money into aid, to put more of this aid into the national budgets of countries in the Global South, and to transfer aid in new administrative formats. These new aid delivery methods were captured in the term 'aid modalities'. At the heart of the new aid modalities lay the principle of donors seeking to add their budgets for development aid to a recipient country's national budget in support of national policies (PRSPs) to reduce poverty, particularly using the public education and health systems. In short, the principle underpinning the new aid modalities was called 'budget support'. (The corollary was that donors agreed to reduce their financial support for development projects that were established according the donor's policies and managed by donor employees.)

The actual introduction of the new aid modalities contended with one critically important challenge: how could donors justify transferring aid into the bank account of the Ministry of Finance of a recipient nation, when the general accounting and auditing that country's public financial management did not meet the requirements of donor countries? Donors wanted to have a level of certainty that their aid would not be diverted into payments that were not part of the poverty reduction strategies. This was a particular concern for donors in the light of a parallel debate about aid, which concluded that persistent poverty was also caused by a lack of 'good governance' in many of the recipient nations, which included poor public financial management (Dollar and Pritchett 1998).

As a result, donor officials designed a range of new aid modalities, all striving towards the ultimate goal of what was defined as General Budget Support (GBS). However, when the assessment of a recipient nation's public financial management gave donors cause for concern, donors would opt for more limited, controlled, forms of budget support. When donors had serious doubt about a recipient nation's public financial management, they would pool their budgets for, for example, the education sector in a so-called Sectoral Basket Fund. This aid modality would be used to fund certain expenses in the Ministry of

Education's policies, but the money would not be transferred to the government but managed by donors. The Ministry would basically send invoices for approved expenses to donors. If donors had more confidence in the Ministry's financial management, they would transfer aid to the Ministry's bank account but control over the money would be shared between the Ministry and donors and targeted to specific aspects of the Ministry's budget. This was called an earmarked Sector Wide Approach (SWAp). One step further up the ladder of confidence, donors would agree to transfer aid to the bank account of the Ministry of Education and accept the Ministry's annual financial report to the Ministry of Finance as a record of accountability. This aid modality is characterised as a full SWAp.

In practice few donors used GBS and SWAps became the most common aid modality. Once a recipient government had seen its PRSP endorsed by the IMF – and aside from poverty reduction objectives, PRSPs also included measures to improve the financial management of the public administration – the sectoral ministries of education and health made detailed multi-annual plans. These plans were presented to donors. A SWAp is then the agreement between a group of donors and a particular sectoral ministry to jointly fund such a plan for successive period of three to four years. For a Ministry of Education, for example, such a plan normally included the expansion of student enrolments, building more schools, training and hiring more teachers, improving students' results, etc. Generally, donors committed for the full period to transfer annually fixed tranches (70–90% of the agreed amount) and the remaining 10–30% would be performance-based tranches, paid if the Ministry achieved certain targets, such as the enrolment of female students in line with the MDGs.

Amidst the use of SWAps, donor concerns about recipient countries public financial management remained. To manage these concerns, donors developed a standard global model to assess the quality and trustworthiness of recipient nations' public finance management systems: the Public Expenditure and Financial Accountability (PEFA) framework. To make a PEFA assessment, an international team of experts visits a recipient country and assesses its financial management on a set of thirty-one indicators. These indicators assess the country's public financial management on seven areas: reliability, predictability and control, accounting and reporting, transparency, management of assets and liabilities, policy-based budgeting, and external scrutiny and auditing. By 2016, all the world's poorest nations had undergone a PEFA assessment, including 61% of the countries in the Pacific region (Hadley and Miller 2016).

On balance, looking back, budget support has definitely increased the financial capability of governments in poorer nations to carry out poverty reducing policies in the health and education sectors. On the other hand, the merging of a nation's own domestic revenues with aid money has also started a practice in which financial 'inspectors' hired by donors have full and unfettered access to the same nations' financial management systems. In this light, the 'transparency' is a one-way mirror that illustrates the inverse sovereignty hypothesis.

Neostructuralism and aid in the Pacific

MDG targets became central goals in terms of aid in the Pacific. In Melanesia in particular development indicators were relatively poor (see Box 2.3). This arose partly as a consequence of ongoing and historic conflict, but also the particular challenges associated with the fact that most Melanesian countries acquired independence in the 1970s and thereby lost direct access to resources from their colonial metropoles – in contrast to many Polynesian and Micronesian islands. In addition, neoliberal policies of the 1980s and 1990s had shrunk the already limited education and health services further, which had negatively impacted the five of the eight MDG targets connected to these two public services. Attaining the MDGs became a central concern in this sub-region and Australian and New Zealand aid programmes were partly reoriented to those countries that suffered the highest levels of poverty, although this occurred within the context of already existing geopolitical and historic obligations which saw New Zealand for example continue to play a role in Polynesia where historic ties through colonialism (see Chapter 1) were especially salient (Banks et al. 2012).

As all across the globe, the shift in focus to a poverty agenda in the Pacific was accompanied by a shift in aid modalities. The neoliberal era move to use projects and NGOs to deliver aid-funded activities had by default a relatively short-term and localised outlook. Yet the new neostructural regime required that international donors and Pacific governments turn to much more ambitious and longer-term policies and programmes by Pacific governments that addressed major national concerns such as literacy or infant mortality on a broad scale because they cannot be achieved through piecemeal projects, nor by relying on CSOs or communities. Instead, it was recognised that institutions of the state in line ministries such as education and health would have to take a lead role, developing long-term strategies, planning nation-wide provision, integrating infrastructure and human resource development and monitoring progress. Thus, international aid shifted into alignment: after often long negotiations, aid budgets were approved that supported SWAps in the education and health sectors with the aim of achieving large-scale results over a term of several years. Furthermore, there was an explicit goal not to use 'tied' aid, where aid donation was linked to requirements to purchase inputs (materials, consultants, education, machinery) from suppliers in the donor country. Instead, the alignment principles demanded that recipient government procurement systems be used to purchase inputs from wherever they could be most efficiently and cheaply obtained.

There were widespread ramifications of this combination of global policy change and a consequent shift in the aid regime on the donor side. The OECD-DAC decision of 2001 to advocate for the untying of aid became a key element in the Paris Declaration. The tying of international development aid to domestic commercial interests became unlawful in the UK with in the new International Development Act of 2002. With support from popular movements such as Make Poverty History, departments for international aid across the Global North were separated – or made even more autonomous – from ministries of foreign affairs. In 2001, the New Zealand government took the management of the aid programmes out of the Ministry of Foreign Affairs and Trade by establishing a semi-autonomous aid department (NZAID) with the resources to hire specialist staff. Australia had started this process earlier (Box 2.5). Aid departments across the Global North were given clear mandates to focus on poverty alleviation in order to keep the two agendas separate. It was felt that if the national self-interest of donors, in terms of diplomatic or economic gains, was confused with poverty alleviation goals, then achievement of the latter would be compromised.

Box 2.5 Australia: one clear objective?

John Overton

In 1997 a review committee released its report on Australian overseas aid (the 'Simons' Report'), one of a series of periodic reviews of the aid programme commissioned by the government. It was a comprehensive report with many recommendations but one of its most central concerns was the way aid policy appeared to have multiple objectives and it noted that "the pursuit of commercial returns has become a driving force of Australian aid" (Australian Overseas Aid Program 1997: 73). It noted the importance of economic growth for poverty reduction but cautioned that "development strategies which focus only on maximising economic growth are inadequate" (p. 76). Its title 'One Clear Objective' encapsulated its recommendation the Australian aid programme should have a single guiding objective: poverty reduction.

It was a strong set of recommendations that had the potential to reorient and focus Australian aid in line with the emerging neostructural aid paradigm that would crystallise further with the launch of the MDGs three years later. The report was accepted by the government of the day, the administration led by Prime Minister John Howard and Foreign Minister Alexander Downer. Indeed, a year later, Downer seemed to reinforce the "overarching poverty reduction objective of

Australia's aid programme" in a speech in Fiji (Downer 1998). Yet, in practice, the potential to focus on poverty reduction was soon lost and this was apparent in other statements made by the government in the following months and, indeed, in the report itself.

The report's subtitle was 'poverty reduction through sustainable development' – an interesting precursor to the rhetoric of retroliberal aid approaches a decade later. The recommendations were a similar mix of contrasting rhetoric and policy. Economic growth was called for, yet "education, health, infrastructure and rural development" should be given priority (p. 16). Overall, though, the report argued that Australia should not use its aid programme to promote its own economic interests explicitly and that it should move to untie aid. Soon after, however, the government and its aid agency seemed to have moved away from the single poverty focus. Instead, an AusAid document in 1998 on its Pacific strategy articulated the view that "the goal of Australia's aid programme as a whole is to advance Australia's national interest" (Australian Overseas Aid Program 1998: 4) and it listed not one but five principal objectives for the Pacific Island countries as: 'better governance', 'stronger growth', 'greater capacity', 'better service delivery' and 'environmental integrity' (p. 6). Yet the language of poverty alleviation remained and there was rhetorical adherence to it as a central principle into the new millennium (AusAid 2001).

It retrospect the Simons' report encapsulated many of the contradictions that have characterised aid programmes over many years and in different countries, particularly the tension between national self-interest and economic growth on one hand and social justice and poverty reduction on the other. This report attempted to nudge Australia more in the direction of a poverty focus – what became the neostructural approach of the following decade. Yet the Howard government largely dismissed the report in practice: "The government of the day endorsed this approach, and then proceeded to take not the slightest notice of it. The 'one clear objective' idea had a half-life within government of about one hour" (McCawley 2010).

The geopolitical situation after the 9/11 terrorist attack of 2001 meant that the above poverty focus was recombined with an attention to state-building, because the Global North feared terrorism emanating from what were now being called 'fragile' or 'failed' states (May 2003). In view of the hollowing out of the state in the 2000s as pursued by the neoliberal aid discourse (and bearing in mind the historical shallowness of the state in the region – Larmour 2003), this now became an important consideration across the Pacific and especially in Melanesia. It reached a zenith in Solomon Islands, where the apparent gradual collapse of the state since a failed state-of-emergency in

1999 seemed to confirm donors' worst fears of a highly unstable region; an "arc of instability" (Fry 1997). The subsequent intervention through RAMSI in 2003 (Box 2.6) was a clear attempt not only to restore order and a government but also to build from the ground up a new set of state institutions and even a sense of nationhood. Continuing our earlier Pacific terminology, we might term this era *vaka matanitū vou* – a new state-led approach but one that was more global in economic orientation.

Thus, the poverty agenda, which had been emerging and given considerable impetus with the MDGs, became intertwined with the new security concerns. Neostructuralism provided a suitable unifying philosophy in this regard: supporting state capacity and legitimacy allowed donors not only to see that poverty-related measures were being implemented but also, they could be reassured that stronger states delivering basic services for their populations would act as a bulwark against terrorism and anarchy. Furthermore, businesses both local and global, would thrive in this environment as consumers would hopefully achieve higher standards of living (and consumption) and property rights would be protected. Neostructuralism, then, was both a political and economic project in the Pacific as elsewhere. Simultaneously it would strive to get more children in schools and more people to health centres whilst keeping governments stable, democratic and responsive. Aid was the cornerstone: donors provided long-term funding to governments and there was supposedly shared agreement on the principles of poverty alleviation, welfare provision, democracy and market-led globalised economic growth. The core elements of this neostructural approach can be summarised as follows (based on Murray and Overton 2011a):

1 A move from orthodox neoliberalism, though with a continued emphasis on globalisation and market-led economic growth.
2 Sovereignty and ownership were 'handed back' to the nation-state – an ostensible reversal of the power relations hitherto.
3 Inclusiveness, harmony and participation became central components in political economy.
4 More holistic approaches to development and social consensus became important recommendations. A new culture of consultation both across countries and within countries was sought.
5 A move away from pure comparative advantage to innovation and value added and the concept of competitive advantage

However, was neostructuralism really a significant change? Some commentators (such as Craig and Porter 2006) have argued that such an approach with its focus on PRSPs was little more than neoliberalism in a different guise; fundamentally the essential role of the market lay at the centre. This has led some commentators to term this period neoliberalism 3.0 (Hendrikse and Sidaway 2010). However, with regard to aid and development in the Pacific we argue there was a clear shift in focus and commitment to lowering poverty and governments in the region, with New Zealand in particular under Helen Clark's Labour government, making a particular stride towards this (see Banks et al. 2012). The evidence is still unclear as to whether the anti-poverty focus successfully and sustainably reduced poverty and the ravages of the global financial crisis of 2008 would soon obscure progress and lead to the rise of a new regime in the 2010s.

We need to pose a key question with regard to the neostructural aid regime nonetheless. It is true that there was a noticeable turn towards ownership and partnership, both laudable goals. Yet our research in the Pacific and elsewhere has shown that this shift to less top-down modes of aid, whilst attractive on paper, actually led to an increase in dependence and compliance costs on the recipients – what we examine in this book as the 'inverse sovereignty' effect. To an extent it is this regime in aid that our research reports on in this volume in Chapters 5 and 6 (together with an overlap into the subsequent retroliberal regime).

Box 2.6 RAMSI and rebuilding the state in Solomon Islands

Finbar Kiddle

On 24 July 2003, the Regional Assistance Mission to Solomon Islands (RAMSI) was mobilised after a request from the Solomon Islands Government (SIG) and a fear that Solomon Islands could become a failed state and destabilise the region. From the beginning, RAMSI was conscious of maintaining in-country legitimacy and avoiding any neo-colonial connotations. The decision to make it a regional operation with the endorsement of the Pacific Islands Forum, rather than a purely Australian operation, was the first step in this direction. RAMSI staff were 'friends from around the Pacific', there to restore law and order and rebuild the nation (Warner 2003). The legislative framework that established RAMSI also set out that SIG maintained executive, judicial and legislative authority (Fullilove 2006). This reinforced that RAMSI was not an international imposition on the country, but there at the invitation of SIG.

The initial deployment of RAMSI focused on the display of overwhelming military force to intimidate militants. This approach was successful and saw over 3700 firearms recovered and 340 arrests by the end of 2003 (PIFS 2004). RAMSI then moved into a long-term project of what it called state and nation building (Hameiri 2007). This was done through the deployment of advisors into capacity building and in-line positions. However, this new role for RAMSI proved not as popular.

Goldsmith and Dinnen (2007) describe how the volume of advisors in key positions led to RAMSI acting as a shadow government and usurping sovereignty. Braithwaite et al. (2010) describe the 'Lime Lounge Syndrome' (a local café), where RAMSI staff only congregated together and did not mix with Solomon Islanders. A piece of research in 2006 also described that amongst RAMSI staff there was a belief that 'culture gets in the way' (Morgan and McLeod 2006: 423). The public relations efforts of RAMSI also caused issues. A legal expert interviewed in Solomon Islands described how when RAMSI first arrived their public relations team was in overdrive, holding community events, funding musical events, printing t-shirts, etc. However, they realised that "this is really unhelpful because it creates a perception about a completely parallel police force, and a parallel civilian component that was superior to the SIG systems" (Kiddle 2014: 91). These issues came to a head in the 2006 riots and the subsequent attacks on RAMSI by then Prime Minister Manasseh Sogvare, threatening RAMSI's image as the benevolent state builder.

In order to address the damage done to RAMSI's in-country legitimacy, significant shifts were made. The Participating Police Force was gradually withdrawn from frontline roles. Work was done to establish jointly agreed targets with SIG and a joint consultative forum was created (Richmond 2011). A community outreach programme was implemented to obtain the voices of Solomon Islanders. Two specific discourses also became prominent in RAMSI public relations. The first emphasised that Solomon Islanders are the ones working to move the country forward, while the second called on Solomon Islanders to step up into the space that RAMSI had created to fight corruption from the top-down and the bottom-up. The withdrawal of military personnel in 2013 and the transfer of the state-building component of RAMSI to more traditional bilateral programmes represents a shift to a more 'business as usual' approach.

Through these changes, RAMSI was able to respond to public and political opinion. It shows a consciousness of the dangers of taking too much of a public role, and a desire to retreat from the spotlight so as to not cast a neo-colonial shadow.

Retroliberalism

In the late 2000s there was another marked shift in the dominant aid regime. We have characterised this shift as a move towards 'retroliberalism' (Murray and Overton 2016a).[4] The abandonment of the Paris principles and budget support practices that embodied neostructuralism was underpinned by several factors, most critically the Global Financial Crisis (GFC), as well as the ensuing trend among governments in the Global North to reorient their overall policies in support of 'exporting stimulus', and an increasing desire to respond to the rise of China as a competitor power (Mawdsley et al. 2018). These factors, among others, saw the further reconstruction of the state and public policies in order to facilitate accumulation in, and international expansion of, the private sector. The geo-economic interests of the private sector and geopolitical objectives of the state are increasingly intertwined. Many of the welfare, education, health and sustainability oriented development aid programmes of the neostructural period have been rapidly undone during this new, retroliberal, aid regime as donors moved to a much more explicitly self-interested stance in the political sphere created by the ravages of the GFC. One of the surprising aspects of this regime has been increased aid expenditure – at least in the early 2010s. A closer analysis, however, reveals aid expenditure is increasingly spent in support of 'exporting stimulus' and other aligned objectives as explored below largely explain this (see Mawdsley et al. 2018). The five main pillars of the retroliberal regime can be outlined as follows:

1 *From poverty alleviation to sustainable economic growth*: donor aid agencies have shifted from poverty to the promotion of economic growth as the core concern. This has returned to the neoliberal belief that economic development is the key to welfare improvements.
2 *From welfare to infrastructure*: aid budgets shifted from funding education and health and poverty-related goals as central concerns (though expenditure was kept up) to supporting economic infrastructure (energy, roads, ports etc.) and key economic sectors in which the donor often has an interest (agriculture in the case New Zealand, financial services in the case of the UK, etc.).
3 *Exporting stimulus*: Keynesian stimulation models were applied in most of the West to tackle the GFC that emanated from over-speculative neoliberal economies. State expenditure was increased, not only, but in no small part, to bail-out the private sector (especially financial institutions). We argue that this has been applied in the aid sector: 'public-private partnerships' and the promotion of private sector activities are increasingly part of aid policy. In addition, aid budgets have been used to pay for domestic expenses related to dealing with the refugee crisis (housing, language classes, welfare, but also detention). 'Exporting stimulus' often favours large companies from donor countries rather than the small and medium enterprises that are the ostensibly target of popular capitalist rhetoric associated with this regime (Murray and Overton 2016a).
4 *'Shared prosperity'*: The World Bank in particular has pushed this new motto broadly. Economic growth in both donor and recipient economies is seen as a laudable target. Therefore aid has been recast as only one of a range of foreign policy instruments that should deliver international outcomes in the national interest (see Mawdsley et al. 2015).
5 *From State Security to Securing the State?* Building state capacity has remained central in this regime. States must be able to deliver aid

programmes and, crucially, provide security. The aim appears to be a securing of those institutions and not a growth in them. As when exporting stimulus, these heightened security concerns have also had an impact with the militarisation of aid: aid budgets have increasingly included the costs of long-term uniformed peace-keeping missions in fragile states (Murray and Overton 2016b).

The shifts outlined in the five pillars above were remarkably similar across most of the major donors (see Table 2.2) and also led to an extent to a convergence with China (Overton 2016). This is despite the fact that there is no overarching global aid framework such as that defined by the Paris Declaration in the previous neostructural period, though the financing for development conference convened by the United Nations in Addis Ababa in 2015 (United Nations 2015) does articulate some of these elements. The retroliberal approach was underpinned in the international arena by the agreements made at Busan in 2011 which saw effectiveness more stringently targeted and a more 'inclusive' approach to aid established. The latter effectively meant a broadening to include the private sector and non-traditional donors, particularly China. This recognition of South-South cooperation, led Mawdsley et al. (2014) to suggest a movement towards a 'post-aid world'. It was out this concern for effectiveness and inclusion that the idea of 'shared prosperity' (see Table 2.2.) was born. In general there has been a shift away from the multilateral consensus embodied in the MDGs and the Paris Declaration of 2005 towards more explicit self-interest of donors in an economic and security sense (Overton and Murray 2016). China refused to be bound by the Busan agreements which given its economic rise and increased role in international assistance signalled a weakening of the OECD Development Assistance Committee.

At the core of this evolving regime is the conviction that the market is the most efficient allocator of resources and the primary vehicle for economic growth and the role of the state is to facilitate, and where possible support, private sector activity. In this sense the central goal has once again become economic growth, and a re-ignited belief in trickle-down. As such, aid donor policy across the world has shifted away from a concern with poverty to a concern with economic growth. Often the term 'sustainable' is appended to 'economic growth', referring to an economy that can 'sustain' its rate of growth over time, not an economy that evolves in an ecologically sustainable way.

The retroliberal aid regime differs from the neoliberal regime in two key aspects. A first point of difference with the neoliberal regime is that it explicitly acknowledges that markets are considered to fail more often – there is less blind faith in the market. In a retroliberal regime, the government actively provides where market failures prevent the creation of public goods such as essential infrastructure. Within the same framework of an active state, a second point of difference lies in the fact that governments will seek

Table 2.2 The retroliberal aid regime and selected donors

	Government change	*Central Mission*	*Institutional change*	*Private Sector*	*Total Budget*
Australia	Liberal Govt (Abbot/ Bishop) 2013 (Turnbull replaced Abbot 2015)	Poverty focus diluted: "promoting prosperity, reducing poverty, enhancing stability" 'Aid for trade' 'Australia's national interest'	AusAid (stand-alone) folded into DFAT and disestablished 2013	Move to infrastructure projects	Cuts to aid budget (12% in 2013, more in 2014) Capped at $5 bill for 5 years
Canada	Conservative Govt (Harper/ Fantino) 2011 (until Trudeau Govt election 2015)	Poverty reduction enshrined in law but . . . 'sustainable economic growth' given prominence 'economic diplomacy'	CIDA amalgamated with FATDC (alongside trade and foreign affairs) 2013	Involvement of Canada's private sector Interest in countries with mineral resources	Aid budget cuts then stabilisation beyond 2015 at $4.62 bill (0.3% GNI)
New Zealand	National Govt (Key/ McCully) 2008–17	'Poverty alleviation' changed to "sustainable development in developing countries in order to reduce poverty and contribute to a more secure, equitable and prosperous world"	NZAID (semi-autonomous) reintegrated into MFAT	Direct involvement of NZ companies (Fonterra, Meridian) tying of aid (e.g. tertiary scholarships increase) Infrastructure projects (airports, energy)	Aid budget increased but at lower rate of increase
UK	Conservative/ LibDem Govt (Cameron) 2010	Long-term programmes to help tackle the underlying causes of poverty . . . 'economic development for shared prosperity'	DFID retained but rebranded (UKAid)	Involvement with UK companies in partnerships e.g. financial services, energy	Aid budget increased (30% in 2013 – to 0.7% GNI)

Source: adapted from Murray and Overton (2016a) and Mawdsley et al. (2018)

to promote, create and protect business opportunities for their domestic private sector. In an historic context, we see this state support for the international expansion of private businesses harking back as to the classic liberal era of the early 19th century where Britain, Germany, the USA and France provided diplomatic and military support to enable their private trading houses establish themselves in the Pacific (see Chapter 1).

These two distinguishing features of a retroliberal aid regime are conveniently converging in infrastructure projects in the Global South that are (co-)funded or subsidised with development aid on the condition that the recipient government accept that companies domiciled in the donor country are employed to build roads, electricity generation facilities or telecommunications systems. These are then handed over to recipient governments to manage or privatise and the resultant revenue streams may be used to repay soft loans used to fund this process. The earlier user-pays free market approach of neoliberalism 1.0 to develop infrastructure – rarely very effective – has now become a (donor) state-brokered and subsidised project that involves private sector construction and ultimately, in some case, private sector ownership. In addition, whereas the neostructural discourse on aid underscored that this tying of a donor's development aid to the companies from the donor country was ineffective and more expensive and the OECD had made untying of aid an official objective in 2001, the emerging retroliberal discourse on aid is gradually overturning this view and accepts or promotes, implicitly or otherwise, that recipients will use donor companies as suppliers and consultants.

In addition to using development aid to support domestic businesses entering the markets in the Global South, the aid budgets are also increasingly redirected to meet expenses in two other areas: paying for the welfare expenses within donor countries related to settling refugees and paying for expenses related to peace-keeping operations. In 2015, the number of refugees arriving in Europe to escape war and conflict (most coming from in Syria, Afghanistan and Iraq) was about 1.5 million, more than double the number arriving in 2014. Governments in many European countries used some of the development aid budget to cover the costs of settling these migrants within the EU. Ten of the OECD members used between 10% and 34% of their ODA budgets to assist the settling in of refugees (OECD n.d.). Norway, for example, spent about 10% of its ODA in 2015 to meet refugee related expenses and planned to spend up to 21% of ODA on this in 2016. Inevitably, this is having a significant negative impact on the main objective of ODA: promoting development in the poorest countries of the Global South.

Retroliberalism may also explain the increasing use of development aid budgets for peace-keeping operations. In a high level meeting of early 2016, the OECD-DAC agreed to start working on 'the modernisation agenda' to redefine what would be counted as Official Development Assistance (ODA). The meeting noted that 2% of its members' current ODA was used for military expenses related to peace-keeping operations and agreed to investigate how this may need to evolve in the future. Anticipating changes, the OECD already agreed that from 2016 onwards three specific types of expenditure would qualify as ODA: the training of military personnel in fragile stages, expenses associated with the delivery of humanitarian aid by the military of the Global North, and training, education and research programmes "to prevent violent extremism" (DAC Secretariat 2016a: 2). The OECD-DAC also agreed to accept expenses related to the settlements of refugees as ODA, "during the first twelve months of their stay" (DAC Secretariat 2016b: 2). Such tampering with the definitions of aid have allowed donors to conveniently shift many domestic costs of the state, such as settling refugees or supporting military forces, into the aid budgets. This has had the effect of seeming to maintain or even inflate some aid budgets at a time when the underlying aid expenditure on established development aid activities, such as poverty alleviation, have been substantially reduced. In fact, the percentage of ODA spent by OECD countries on settling refugees in the OECD countries rose from 2.7% in 2010 to 9.1% in 2015 (DAC Secretariat 2016b: 1).

In some senses then this represents a return to the state developmentalist projects of the post-World War Two era. But there is an essential difference: if the state was in that earlier period conceived as the facilitator and guardian of economic development, in the new regime it is seen merely as the former. In the retroliberal regime the state exists to facilitate the private sector which is the ultimate engine of economic growth. Furthermore, aid is seen as something which ought to *explicitly* bring benefits to both donor and recipient. Whilst it could be argued that aid has never been anything but this, in the retroliberal regime this is made explicitly clear through the concept of 'shared prosperity'.

Retroliberalism and aid in the Pacific

We can discern elements of the retroliberal turn across the full range of aid donors in the Pacific. We will focus this section on the clear changes in in the aid programmes of Australia and New Zealand (Table 2.2) but it is also evident in the reforms within the Japanese aid programme (Box 2.7).

Box 2.7 Japan: JICA and new JICA

John Overton

Japan is a major aid donor in the Pacific and has been one of the largest bilateral aid donors globally over the past 30 years at least, although its ODA has fallen in relative terms over the past decade. Japan has seen aid as a key element of its foreign policy: it has used aid to build relationships and promote the image of itself as a donor that does not interfere in internal affairs and is interested in building mutually beneficially economic growth. It has been part of the OECD-DAC group and a signatory to major international agreements regarding ODA. However, it has tended to follow a quietly independent line. Historically, its aid programme has put much emphasis on technical cooperation, particularly with regard to infrastructural development, it has used soft loans extensively and it has often tied its aid to the use of Japanese contractors and suppliers (Rocha Menocal et al. 2011: 5). In Oceania (the destination a very small proportion of total Japanese aid) Japan has operated rather differently from the main Western donors and, indeed, has tried to position itself as part of the wider region, as in its '*We are Islanders*' publication in 2009 (JICA 2009).

For many years Japan's ODA was conducted through a rather complex institutional arrangement. Many of Japan's aid activities were overseen by JICA (Japan International Cooperation Agency) but this tended to focus on technical cooperation projects. Significant funding was also channelled through its Ministry of Foreign Affairs which controlled most grant aid, and the Japan Bank for International Cooperation (JBIC) which disbursed the largest ODA funds in the form of loans. This structure lacked coordination and reflected the different objectives of the three institutions: technical cooperation promoting the use of Japanese expertise and equipment; grant aid being guided by Japan's wider diplomatic priorities and loans being guided by development bank policies and procedures.

This changed in 2008 with the launch of 'New JICA'. This reform put all three sets of operations within the enlarged JICA organisation (though the Ministry of Foreign Affairs retained some control over a portion of the grant aid "as dictated by specific diplomatic policy" (JICA 2010)). The new JICA aimed to promote greater coordination of Japan's ODA and, in principle, give prominence to the development objectives of the aid programme. However, as noted in a DFID review of Japanese aid in 2011, a degree of institutional confusion remains and, whilst adopting the rhetoric of the MDGs and the Paris Declaration, Japan has been slow to fully engage with poverty reduction strategies and to incorporate the alignment and harmonisation principles of the OECD (Rocha Menocal et al. 2011). In this sense, we might interpret changes (or lack of change) in Japan's aid programme not as being a laggard in terms of the neostructural aid regime but rather as offering a lead towards the retroliberal regime of the 2010s.

Echoing this retroliberal turn, a recent review of JICA by the Abe government led to the launch of the country's 'Development Cooperation Charter' and this brought the aid programme more closely in line with Japan's wider strategic and economic interests. There is a strong emphasis on promoting economic growth, with poverty alleviation given less explicit mention, and there is even a controversial clause to allow for Japan's aid funds to be used for foreign troops in 'non-military operations' (Dugay 2015).

Finally, we can also see some interesting similarities between the way Japan has operated as an aid donor (the use of loans, the focus on infrastructure and the use of Japanese suppliers and contractors) and the way the Peoples Republic of China conducts its development cooperation activities in recent years.

In the case of the role of the state in the post 9/11 world, it is clearly not in the strategic interests of the donor to have client states with hollowed-out and weakened public administrations – this does not deliver security and the case of Solomon Islands is perceived as a clear warning by the region's donors. Thus, the focus in the retroliberal period has returned to ensuring efficient and effective states. In part, this is a return to the good governance principles, but with an explicit concern that states and their effective administrations sponsor and facilitate the private sector. This has been particularly evident in the Pacific where aid policy in the so called 'arc of instability' that stretches from formerly war-torn East Timor, through unstable PNG and across to the Solomon Islands and Vanuatu (Fry 1997). Aid in recent decades has involved a significant continued emphasis on good governance together with the shift to the private sector that has occurred elsewhere in the region.

After Murray McCully became New Zealand's Minister of Foreign Affairs in 2008, the New Zealand aid programme shifted a considerable proportion of its development aid towards the private sector and infrastructural development. There may also be an element of competition with Chinese aid which has been principally focused on infrastructure – and to this we turn later. A good example was the role of New Zealand aid in subsidising the flights of Air New Zealand to Tonga and Samoa until 2010. A subsidy was provided by the governments of Samoa and Tonga to sustain an Air New Zealand flight from there to Los Angeles and "New Zealand government aid has helped Samoa and Tonga pay the Air New Zealand bills" (Field 2010, np). The nature of using development aid in infrastructure has shifted to an extent as revealed in NZAID's strategic plan of 2015, which emphases energy and ICT infrastructure rather than the construction of roads, airstrips and harbours (New Zealand Foreign Affairs and Trade Aid Programme 2015a, 2015b, see also Box 2.8).

By instigating a shift to infrastructural work, it is clearly only another step to suggest that the donor has expertise in the area that is deemed to be wanting. This may also be the case in terms of security interventions and aid programmes. Examples of this include programmes in the Pacific that have involved New Zealand and Australian Defence forces, deployed first as peacekeepers and observers and eventually undertaking delivery of aid functions in places such as the Solomon Islands. The implicit return to the 'tied' aid regime has been discernible especially in the case of the post-war reconstruction in Afghanistan, Libya and Iraq and as such can be seen as a legacy of the War on Terror. Given the recent expansion of conflict in the Middle East in the wake of the Syrian crisis and the rise of ISIS, this is likely to rise. Indeed, there is likely to be a significant upturn in the militarisation of aid more generally over the coming years – embodied in a phrase used by the UK Secretary of State in 2013 on the need to consider the role of 'non-lethal' aid in the case of Syria.

Box 2.8 Supply-led aid: New Zealand's strategic plan 2015–2019

John Overton

In 2015 New Zealand released its new strategic plan for the aid programme (New Zealand Foreign Affairs and Trade Aid Programme 2015a). It mapped out a strategy for the following five years and a commitment for increased funding. It set a clear direction for the programme – its 'investment priorities' – and it constitutes a good example of a retroliberal approach to aid. Its overall mission repeated its earlier sustainable development focus and its stated purpose was explicit in promoting the shared prosperity mantra:

> *The purpose of New Zealand's aid is to develop shared prosperity and stability in our region and beyond, drawing on the best of New Zealand's knowledge and skills.*
>
> *(p. 3)*

Central to the plan are its 'Pacific focus, global reach' motto and twelve 'investment priorities', two of which are highlighted as 'flagship' activities:

1 Renewable energy (flagship)	Expand access to affordable, reliable and clean energy
2 Agriculture (flagship)	Increase economic and food security benefits from agriculture
3 ICT	Expand ICT connectivity, access, and use in the Pacific
4 Fisheries	Increase economic and food security benefits from sustainable fisheries and aquaculture in the Pacific

5 Tourism	Increase economic benefits from tourism in the Pacific
6 Trade & labour mobility	Increase economic benefits from trade and labour mobility in the Pacific
7 Economic governance	Strengthen economic governance in the Pacific
8 Law & justice	Strengthen law and justice systems in the Pacific
9 Health	Improve the health of people in the Pacific
10 Education	Improve knowledge, skills, and basic education
11 Resilience	Strengthen resilience (to disasters)
12 Humanitarian response	Respond to humanitarian emergencies

Some of these (fisheries, ICT, tourism etc) are focused on the Pacific but others (agriculture, energy) seem to have a more global ambition. Reading of the plan reveals some other interesting aspects of the agenda. A simple word count analysis (how many times a certain word appears in the document – see Alfini and Chambers 2007) of an early draft of the plan shows the contrast between the language of the former neo-structural era of NZAID and the new retroliberal approach after 2008:

'Neostructural' Aid Regime Terms		*'Retroliberal' Aid Terms*	
Education	10	Investment	53
Health	10	Economic	32
Climate change	8	Agriculture	19
Poverty	5	Results	16
Environment	2	Energy	14
Sustainability	2	Sustainable development	12
Maternal health	1	Market	11
Gender	1	Trade (excl MFAT)	7
Equality	1	Evidence	5
Social justice	0	Labour	5
Inequality	0	Shared prosperity	4
Ownership	0	Private sector	3

With these discursive shifts and the prioritisation of sectors in which New Zealand – and New Zealand companies – have particular interests and strengths (agriculture, energy, tourism, fisheries) we can discern what might be called a 'supply-driven approach to aid' (Wood 2015).

Yet, despite these shifts to a much more market-oriented and explicitly self-interested agenda than before 2008, there are still important elements of a programme that retains a concern for the well-being and development aspirations of the Pacific region. Although there are signals to increase aid to ASEAN countries – which would be hard to justify if addressing poverty were a criterion but would make sense if opening up huge markets of consumers for New Zealand businesses and their product – there is also a stated intent to increase aid to Melanesia (especially PNG and Fiji) and education and health expenditure remains a major element of the aid budget.

The re-entangling of the diplomatic, commercial and developmental roles of aid has been solidified by restructuring the aid administration in both Australia and New Zealand. As part of the neostructural aid regime AusAid was established as a semi-autonomous department in 1995, and the New Zealand aid programme was partly taken out the Ministry of Foreign Affairs and Trade in 2001 and given a similarly semi-autonomous identity. The separation of the aid division (NZAID) into this 'semi-autonomous' agency, within MFAT but with a direct voice in Cabinet and a separation from diplomatic and trade functions of MFAT had been seen by international observers as a role model example of the separation of national interest and developmentalism. In doing this, Australia and New Zealand followed patterns across the OECD, where development aid programmes were partially or largely insulated from a country's diplomatic and commercial interests by giving them a semi- or full autonomous status. The Netherlands, for example, had a Minister for International Development who participated as an equal to the Minister of Foreign Affairs in Cabinet meetings. Retroliberalism rolled this back. In Australia, following the election of Tony Abbot's liberal government in 2012, the AusAid agency was absorbed into the Department of Foreign Affairs and Trade (DFAT). This mirrored what happened in New Zealand following the election of the John Key-led National Party Government in 2008 (see Banks et al. 2012).

The re-orientation of aid programmes to also – or perhaps much more explicitly – serve commercial and geopolitical interests proved to have a particularly contentious effect on the relations between governments and NGOs in the Global North. The New Zealand Minister of Foreign Affairs Murray McCully has been quite explicit on the role of foreign aid in promoting the business interests of New Zealand. In one of his first public speeches as Minister he promoted the concept of 'NZ Inc' and stated, "Most important in this respect will be our ability to align aid policy with trade policy" (McCully 2009). Inevitably, the retroliberal dismantling of the former aid regime meant the slashing of funding to New Zealand's development NGOs who did have a clear primary focus on the challenges of poverty and inequality. New Zealand NGOs were informed, subtly or otherwise, to align with the more business-oriented new aid policies. Similarly, the new course at the aid programme of the Ministry of Foreign Affairs and Trade led to awkward disagreements with the academic sector engaged in monitoring and evaluation of aid programmes: 'Pointy headed academics' in the words of the Minister (See Banks et al. 2012; McGregor et al. 2013).

Given the explicit shift of aid to suit geopolitical and geoeconomic interests we have witnessed a re-orientation in the range of recipient countries and territories with which any given donor engages. Initially, in the New Zealand case, this involved a reduction in the number of recipients with a focus on the Pacific region at the expense of Asia, Africa and Latin America. In some ways this represents an extension of the neostructural regime where guided by the Paris Declaration principles of effectiveness and harmonisation donors were encouraged to focus on the regions in which they had a particular expertise. This has manifested in the recent past with effectively principal aid recipients often having old colonial ties with the donor. As a consequence, significant geographical concentrations of influence have evolved. This is most notable in the Pacific region where there are clear zones of engagement among the various principal aid donors, Australia, the USA, France and New Zealand (see chapter 3 on aid patterns). In the case of the latter this represents an explicit policy of the 2008 National (conservative) government to re-engage with what it terms its 'realm' territories (see Box 2.1).

However, as the retroliberal era has evolved, new geographies of aid have begun to emerge, and the early re-concentration has become more dispersed again. In the New Zealand case, new aid recipients have been developed often coinciding with the country's wider diplomatic and economic interests. New Zealand's push for a seat on the UN Security Council seemed to spur a rash of diplomatic activity – and aid – in regions such as Africa and the Caribbean as it sought countries to vote in its favour in 2014. It is also notable that some of these new recipient countries, such as Sri Lanka and Ethiopia have been countries where New Zealand's largest company, the dairy enterprise Fonterra, has sought to establish new operations. And in the Caribbean, New Zealand's new aid interests have centred on renewable energy: an initiative involving its energy company Meridian (half state-owned) which has been active with solar power projects in the Pacific.

Seeking a Pacific term for this new development and aid era in the Pacific perhaps leads us to the term '*vaka matanitū tani*' (again in Fijian).[5] *Matanitū tani* here refers to 'foreign governments' or 'foreign relations' and external links and the term implies that this type of development is primarily, and more explicitly than ever before, driven by the interests of foreign powers. It employs rhetorical devices such as 'shared prosperity' and 'sustainable economic development' to disguise the way aid is being used to promote the growth and prosperity of donor economic interests by co-opting the discourse (and budgets) of development aid.

The rise of China and the retroliberal convergence

One of the central trends that has had a major impact on the Global South is the rise of the People's Republic of China (PRC) as an aid donor in the post-Cold War world. This has been especially marked throughout the 2000s as China's government reformed the economy along market lines, making the sourcing of natural resources and locations for investment a priority. China has become an increasingly major player across Asia, Africa, Latin America and the Pacific. Underlying the expansion into these areas in an economic sense is a range of factors which, while they vary from place to place, are essentially rooted in the rise of China as a major industrial power and increasingly a major geopolitical power. In the case of the former China requires inputs for its industrial processes and in the case of the latter it requires governments that are aligned with its interests and likely to provide support for its intentions in forums such as the United Nations. Yet, despite becoming a major global player in the aid world, the PRC can still portray itself as being part of the 'South' and its aid is styled more as 'South-South cooperation' than convention North-South 'aid' (Mawdsley 2010).

In many cases the types of projects involving Chinese capital are infrastructure-based. They are often large-scale modernist projects and almost exclusively involve the use of Chinese expertise, often Chinese construction companies and in some cases machinery and labour. Major examples include the current move towards the building of a second canal through Central America that will facility the export of natural resource from Brazil, Venezuela and other parts of South America without accessible Pacific coasts. Elsewhere, and especially in Africa, China has been extensively involved in the construction of roads and highways and in constructing government and public buildings, such as sports facilities. China has also been extensively involved in the purchase of primary products, including minerals, agricultural and forestry products. It is, for example, a major investor in Peruvian gold and silver mines. Such Chinese-backed large-scale projects are also visible across the Pacific and include road building, proposals for harbour improvements, government offices and sports stadiums in locations as diverse as Fiji, Samoa, Vanuatu and Tonga (Prasad 2016).

This money is often disbursed by China without interference in domestic politics or policies of the recipient country and with only the implicit understanding that the recipient favours Chinese goals. Sometimes it comes in the form of direct tied grants, sometimes it is in the form of soft loans. In some regions, then, the Chinese government is replacing, or at least complementing, the World Bank and other Bretton Woods institutions as the funders of modernisation. This has been particularly the case in countries that have fallen out of favour in the west, for example in Fiji where as a result of a number of coups, most recently in 2006, it was expelled from the Commonwealth and cut out of some forms of aid until the restoration of democratic elections in 2014. This gap in Fiji's development aid has been filled – although nobody is sure to exactly what extent – by Chinese investment. PRC's 'aid' is not listed officially as such and much of what it has exported is highly opaque. The diffusion is facilitated at a state-to-state level and the 'aid' has no explicit social or other political objectives. It largely bypasses civil society and the local business sector in the recipient countries. In this sense we are seeing something of a convergence in terms of the nature and to an extent the intentions of western and Chinese aid towards a state-led modernist regime, the difference being that in the case of Chinese aid the corporations are not fully private but Chinese state-owned. We suggest that there are some interesting parallels between the way the PRC is disbursing development funds and the emerging retroliberal approach of Western donors (perhaps even Western donors may be learning from the apparent success of PRC in infrastructure development). Despite Western claims that China is acting simply to further its own interests, to buy diplomatic support and to support the activities of its state-owned companies, we see a discursive convergence between terms such as 'South-South cooperation' of PRC and 'shared prosperity' of the OECD. Similarly, we see little difference in principle (though more in practice) between the PRC seeking support for its stand in the South China Sea by using its development cooperation activities and New Zealand seeking favour for its seat on the UN Security Council.

Summary and perspective

Table 2.3 reproduces an earlier table concerning aid regimes and their evolution globally but changes the focus to the Pacific region. As we have seen trends in the Pacific Islands have indeed been influenced by broader shifts in the global aid regime, but there have also been significant differences in terms of timing, emphasis and indeed outcome. The historical forces of change have been complex and variable in the case of the Pacific Islands. Decolonisation began later than elsewhere in the colonial territories, due in part to the continued geopolitical comparative strategic importance of the islands during the Cold War together with the desire among some territories to retain constitutional ties with their colonial metropoles. The current state of decolonisation in the islands is thus highly variable.

Following the Second World War and through the incomplete decolonisation era a period of modernist-informed aid policies evolved. Relatively large states, which would play a role in development objectives, were envisaged for the region. This combined with continued strategic relevance of the islands and their small size led to large aid budgets per capita that were administered largely through projects. Following decolonisation there began a large-scale migration from the Pacific Island region to the rim – principally to Australia, New Zealand and to a lesser proportional extent, the USA. This gave rise to the increasing importance of remittances that were sent home from migrants to their home countries across generations. In some ways this process can be seen as the transnational urbanisation of the Pacific peoples, adept as always in carving out pathways and niches for propagation and survival (Hau'ofa 2000). This set of historic contingencies gave rise to what some have termed the MIRAB period – where migration, aid, remittances and bureaucracy were paramount for a group of smaller Pacific territories maintain close constitutional ties with the metropole (Bertram 2004). To a

Table 2.3 Aid regimes in the Pacific – selected events, principles, goals and policies

	Modernisation	*Neoliberalism*	*Neostructuralism*	*Retroliberalism*
	1950–1980	*1980–2000*	*2000–2010*	*c. 2010 to present*
Pacific Events	Nation-building, preparation, independence and other sovereignty arrangements, migration to the Pacific Rim, the rise of MIRAB economies	Consolidation of independent and semi-sovereign territories. Coups in Fiji and instability in Melanesia. Attempts to export resources, withdrawal of the US from the Pacific	Rise of environmental agenda, continued instability, moves towards regionalism, rise of tourism as main economic sector TOURAB, the Pacific Plan	Rise of China as a power, reduced instability, splitting of Pacific geopolitics following Fiji influence, Australia and the Pacific Solution
International events influencing the Pacific	Cold War politics, Kennedy Alliance for Progress, Formation of the EEC	End of the Cold War, Debt crisis, Thatcherism, Rogernomics, consolidation of the EU	MDGs, Labour governments (NZ, UK and Australia) and 9/11	GFC, swing back to the right – Liberal (Australia), National (NZ), Conservatives (UK)
Pacific Principles	*Vaka matanitū*	*Vaka vuravura*	*Vaka matanitū vou*	*Vaka matanitū tani*
Pacific donor development goals	Towards a decolonised Pacific, self-representation politically, in the context of continued Cold War alignment with the West	Economic self-sufficiency in an era of globalisation, outward orientation and export-led development, more efficient economies and reduced government expenditure	Poverty targets following MDGs, diversification of economy and global linkages, environmental sustainability and role in UN climate change, Solving instability in Melanesia through state-building, RAMSI,	Sustainable economic development, shared prosperity and exporting stimulus, securing the state
Aid policies and modalities in the Pacific	Budget transfers, infrastructure physical and political readying for independence and economic self-sufficiency.	Structural adjustment policies, conditional aid, project specific market-based goals	Aid effectiveness, Poverty reduction strategy papers, SWAps, GBS, microfinance and popular capitalism	Infrastructure projects, sectorally-aligned aid priorities, Chinese soft loans

large extent the MIRAB hypothesis has overgeneralised the reality in the Pacific and it was not intended to describe the development situation in the larger independent Melanesian states. As such it should be seen as originally intended – a hypothesis against which the empirical reality of any given country of territory can be measured. To an extent elements of the MIRAB structure remain today – although the emphasis has shifted significantly particularly to include tourism and other service activities (see Bertram 2006).

Until the mid-1980s at least, the donors saw no reason to alter that fundamental development dynamic that MIRAB described for the smaller states and modernisation theory seemed to prescribe for the larger countries. However, by the 1990s, with the rise of the neoliberal idea at the global scale and the influence this had on the aid regime, the continued role of remittances and high levels of per capita aid were called into question, as were trading preferences that sustained forms of nascent industrialisation elsewhere in the region. Export-oriented development models were actively diffused during this period in territories and nations that could never hope to achieve the economies of scale or the political leverage to access overseas markets that was required to prosper in the neoliberal global marketplace. Even so, the downsizing of governments and aid policies that promoted this restructuring were rolled-out together with conditionalities for aid that mirrored trends elsewhere in the global economy. The idea of endogenous economic growth (in effect globalisation and export-led growth) became all important during the neoliberal aid regime and in the Pacific, this translated into abrupt and deeply damaging policies that forced inequalities wider and threatened environments substantially. Although neoliberalism came later than elsewhere to the Pacific region, it resulted in abrupt shifts with major consequences in small island states (Murray 2001).

With the rise of the MDGs and the focus on poverty as the main target together with aid effectiveness in the 2000s a new consensus on changes to the development aid regime seemed to wash across the Pacific. This was against the backdrop of serious instability especially in Melanesia (Solomon Islands and PNG in particular). As such the need to rebuild or strengthen state administrations and to place them at the centre of managing effective aid policy was a central objective of aid in the Pacific. This period of neo-structuralism led ostensibly to the return of ownership of the aid agenda to the Pacific nations and territories through various forms of budget support. As we argue at length in this book, it was during this period the process of inverse sovereignty was compounded.

Finally, more recently we can discern the rise of a retroliberal aid regime across the region which has seen an explicit return to the self-interest of donor governments and the private sector in the Global North. The GFC led to the eclipsing of poverty reduction as a goal and the return to economic growth. At the same time as we see aid being used to further wedge open Pacific economies for investment and capital flows, we see no corresponding attempt to liberalise substantially labour markets and migration beyond a few tightly managed seasonal labour migration initiatives. In some ways the region has been an experimental site in this regard, as New Zealand was one of the first, if not the first to move towards the retroliberal regime in its aid strategy. The principles of 'exporting stimulus' and 'shared prosperity' were embraced by most donors and diffused rapidly across the Pacific donors and the world more generally. This coincided with the rise of the People's Republic of China and its policies of soft loans with few strings attached, spreading across the Pacific and resulting in a significant competition in terms of geopolitical influence with the traditional Western donors – especially in increasingly non-aligned nations such as Fiji. As an unintended consequence there has been something of a convergence between the practices of Western and Chinese 'aid' policy, which has seen the increasing prominence of infrastructural investments and the role of donor-based companies or institutions in building advising and implementing development plans. In some way the retroliberal phase echoes earlier colonial systems, with colonial trading houses holding sway over colonies.

The shift towards the current retroliberal aid regime does not – yet – represent an explicit international consensus. There is no global agreement or declaration that captures the heart or principles of retroliberal aid in the same sense that the MDGs or the Paris Declaration captured neostructural aid. However, the new direction is visible in the remarkably similar shifts in donor approaches in the Global North and debates in the OECD-DAC since 2015 reflect this. The clarity of the current regime is perhaps less obvious than previous ones, as they represented to lesser and greater degrees internationally agreed consensus. The current regime is – still – messier and involves complex and non-aligned geopolitical players. It is these unexplored and to an extent undefined spaces that we are most interested in engaging in the reminder of this book. In a sense, aid regimes or aid discourses are artificial constructs, and perhaps hide more than they reveal. Principles and modalities are as dynamic and polymorphous as the geopolitical and economic trends that give rise to them. Furthermore, in any given locality they represent the unique combination of local and global trends, as these crystallise in that particular location. In this sense we do no subscribe to a deterministic view of the impacts as determined by a set of processes from above. There is always room for agency to 'speak back' and reconfigure regimes and influence the manner in which they unfold in specific geographies comprised of unique configurations of historic, cultural, environmental and social conditions.

Notwithstanding these particularities, there can be no doubt that the history of aid in the Pacific shows that global aid discourses and development agendas of the donors are often very influential on what happens in the Pacific Islands when it comes to factors that shape the livelihoods of Pacific families: public services such as health, education, justice, poverty, inequality, employment, infrastructure, mobility and security. The promise of local, Pacific ownership over development policies of the earlier neostructural regime was undermined by the various inversions of sovereignty that we explore further into this book. The current aid regime sees donor countries explicitly taking back control in ways that we have not seen perhaps since the colonial era. We believe that this is not healthy for holistic, appropriate and sustainable development in the Pacific and that analysing this asymmetry in aid power relations is a first step in recalibrating it to a genuinely inclusive and socially progressive system that delivers significant and meaningful economic, cultural and environmental goals. With this in mind in the following chapter we look in detail at aid flows, bearing in mind the regime chronology as outlined above, we then move on to consider notions of sovereignty and how these have been impacted by the shift between various regimes.

Notes

1 *Matanitū* refers in Fijian to large confederations occasionally mobilised for military functions. It has been used since colonial times as a word for 'government' (Routledge 1985).
2 This remains the case in sectors in the economies of the Global North that are particularly important from the Global South's point of view, such as agriculture. WTO-negotiations to remove protection for agricultural production in the Global North and enable farmers in the Global South to access these markets – the so-called Doha Agenda – have gone nowhere since they started in 2001.
3 We are indebted to Randy Thaman for alerting us to the possibilities of this term, referring to life beyond the horizon.
4 Other terms used for such a new phase might include 'postneoliberalism' (Peck et al. 2010) and 'zombie neoliberalism' (Peck 2010) and we could continue and adapt Hendrikse and Sidaway's (2010) metaphor of updating and relaunching the neoliberal project as 'neoliberalism 4.0'.
5 We wish to thank Maciu Raivoka for suggesting and advising us on the use of this term.

Chinese funded Aquatic Centre, Apia, Samoa

3 Aid in the Pacific Islands

An overview

Introduction: what is aid?

Development aid has been a prominent feature of Pacific Island economies for the past fifty years, though there has been considerable spatial and temporal variation in how this aid unfolds across the region. In this chapter we describe, analyse and map changes in aid flows in the region, noting diversity and dynamism in the sources of aid, its dispersal and its uses. We identify particular concentrations of aid and inequalities in its distribution. We focus on evolving and emerging trends with regard to the way aid is applied to development objectives in the Pacific.

The task of mapping-out aid flows is beset with problems of definition. Although there may appear to be clear specification of what aid is and should be, and reasonably clear data collection procedures, in practice the measurement of aid has been much more problematic. We will see some examples of where these official guidelines have not worked well. The key international organisation for the definition of aid and collection of data about it is the Development Assistance Committee (DAC) of the OECD. The DAC – comprising twenty-nine of the thirty-five OECD members who provide significant development aid – defines aid as 'official development assistance' (ODA) thus:

> [T]hose flows to countries and territories on the DAC List of ODA Recipients and to multilateral institutions which are:
>
> i. provided by official agencies, including state and local governments, or by their executive agencies; and
> ii. each transaction of which:
>
>   a. is administered with the promotion of the economic development and welfare of developing countries as its main objective; and
>   b. is concessional in character and conveys a grant element of at least 25 per cent (calculated at a rate of discount of 10 per cent).
>
> (DAC n.d.)

Such a definition sets out some key elements of aid. Centrally, under this definition ODA should originate as government funding in source countries and flow to institutions and communities in designated recipient states. It is funding that must have strong development and welfare objectives (rather than, say, assistance to buy military equipment) but it may be in the form of concessional loans (as well as grants) which does lead to a return flow of resources to donors. The DAC definition is detailed beyond the above definition and there are strict criteria about what might be included or not as ODA. This has given rise to the term 'DAC-able'; vernacular within the aid world for describing what might be counted as development aid within the DAC's terms and thus be legitimate as a form of 'aid'. As noted in Chapter 2, the OECD is currently reviewing its definition of, and criteria for inclusion in, ODA. It is likely to include additional types of spending on peace and security and incentives for private sector development (Anders 2016).

Given the role of the DAC in collecting and verifying the ODA data, there would seem to be a reasonably robust data set dating back to the early 1960s when the OECD first began such information gathering. Indeed, it is this source that we rely on in this book: we use both current and constant (adjusted for inflation) measures, the standardised $US amounts and actual ODA disbursements in a given year (rather than commitments). For the Pacific Islands region, the DAC ODA data recognises the following sixteen countries as recipients for the 2014–2016 period:

Least developed countries: Kiribati, Solomon Islands, Tuvalu, Vanuatu
Lower middle income countries and territories: Federated States of Micronesia, Papua New Guinea, Samoa, Tokelau
Upper middle income countries and territories: Cook Islands, Fiji, Marshall Islands, Nauru, Niue, Palau, Tonga, Wallis and Futuna

This list however, does not encompass the full actual spectrum of Pacific Island states and territories that receive development assistance. Firstly, there are territories that were previously listed as recipients by the DAC (i.e., New Caledonia, French Polynesia, Northern Mariana Islands) but which are now no longer included. Secondly, there are Pacific Island territories whose close constitutional association or full integration with a metropolitan country (American Samoa, Rapa Nui, Guam, Pitcairn – one might even include Hawai'i) has led to their omission from analysis. Indeed, there is a degree of political consideration in some of these definitions. For example, the decision by France to exclude New Caledonia and French Polynesia from the DAC system after 1999 appeared to be driven by a desire to see these territories not as developing states-to-be but as integral parts of France with levels of welfare and development roughly equivalent to, and maintained directly by, metropolitan France.

Another complicating factor in the definition of aid has to do with donors not covered by the OECD-DAC membership. Membership of the

DAC certainly includes nearly all traditional North American and European donors who have been active in the Pacific, together with Japan and regional donors such as Australia and New Zealand. The OECD aid database also now includes some 'new' donors (Chinese Taipei [Republic of China], South Korea, Russia, Saudi Arabia and Thailand) who are not DAC members but who, in theory, provide data to DAC and are becoming more active as aid donors. But it does not include new non-traditional donors, particularly China (PRC – the People's Republic of China) and India, who are emerging as new major global aid donors. China, in particular, is now significant in the Pacific region but reliable data on its development assistance/ development cooperation funding is extraordinarily difficult to obtain. We examine China as a separate case later in this chapter. However, PRC is not alone with regard to a lack of data. It seems as much of the financial assistance given by the Republic of China to Pacific countries such as the Solomon Islands is not captured by the DAC statistics and aid spending by Taiwan may be almost as opaque as the People's Republic of China.

Finally, the difficulty in quantifying aid flows also has to do with forms of assistance that are not covered by the ODA definition, but which can have significant development implications for Pacific Island countries. For the Pacific region, the most significant of these is migration policies. Some Pacific countries and territories have relationships with metropolitan states (the USA, France, New Zealand) that allow their citizens freedom of entry to the latter. For example, residents of the New Zealand 'realm states' of Cook Islands, Niue and Tokelau are eligible to hold New Zealand passports and enjoy the full rights of citizenship there. The ability to travel, migrate and work in the metropole brings significant benefits – and arguably some costs – for these Pacific Island peoples. Remittances allow Pacific Island people to contribute significant resources back home either via relatives or through their own return migration. Significantly, such liberalisation and widening of migration and labour markets remains one of the key policies that many Pacific Island leaders see as vital for their future economic development.

A second key development assistance aspect not covered by ODA has been the special trade and other concessions offered by some metropolitan countries to selected Pacific countries and territories. This has been significant in the past, particularly before neoliberal global trade reform in the last twenty years lowered or eliminated many trade barriers. For example, perhaps the most successful Pacific Island agricultural industry in the past century – sugar cane cultivation in Fiji – was maintained and subsidised for many years by preferential access to the European sugar market under a succession of agreements such as the Lomé Conventions of 1975–1990 and its successor the Cotonou agreement (Taylor 1987). Other agreements, such as PACER also helped support the garment industry of Fiji (Storey 2004) and the Japanese owned Yazaki automotive wiring plant in Samoa for example. Despite the considerable development impacts that these forms of assistance, whether preferential migration or trade policies, have had on

Pacific economies and societies, their accurate quantification is not possible and are therefore only referred to in broad terms in this study.

This chapter begins with an overview of ODA flows in the region, showing variations over time and clustering of aid in the form of various 'spheres of influence' of donors. It then moves to consider some of the forms of aid not covered by ODA and finishes by identifying several key issues such as relationship between aid, poverty, dependence, security, political status and scale.

Mapping aid flows in the Pacific

Aggregate data on ODA for the Pacific region reveal some marked trends over the past fifty years that warrant analysis. Not surprisingly, the trend for ODA in current prices was subject to a major rise over the time period, from a mere $US130 million in 1965 to a peak of $US 1.98 billion in 2011. However, as Figure 3.1 illustrates, when real values are calculated using constant (2014) prices, aid flows are revealed as surprisingly stable – hovering around $US 1.5 billion in 2014 prices until the mid-2000s.[1] It should be noted here that the precipitous drop of ODA to the Pacific region in 1999 is not as dramatic as it seems. The drop is simply explained by the fact that in 1999 the OECD decided that the large volumes of aid to the French territories New Caledonia and French Polynesia and the US territory of the Northern Mariana Islands – which made them three of the largest four recipients of aid in the region at the time – could no longer officially be counted as 'DAC-able' because they were deemed to be domestic fiscal transfers.

Figure 3.2 depicts these aid flows in terms of recipients. We must first note that these data sets are not consistent. Some ODA flows were counted prior

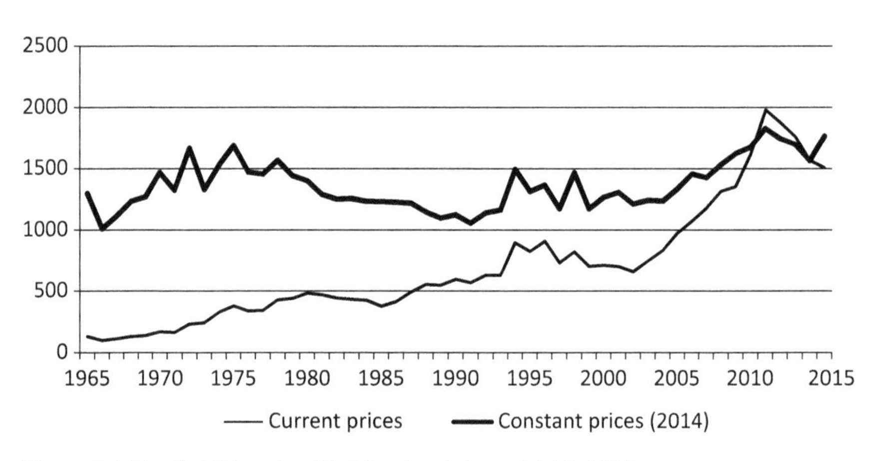

Figure 3.1 Total ODA to Pacific Island recipients 1965–2015 ($US million)

Source: http://stats.oecd.org/ accessed 3 March 2017

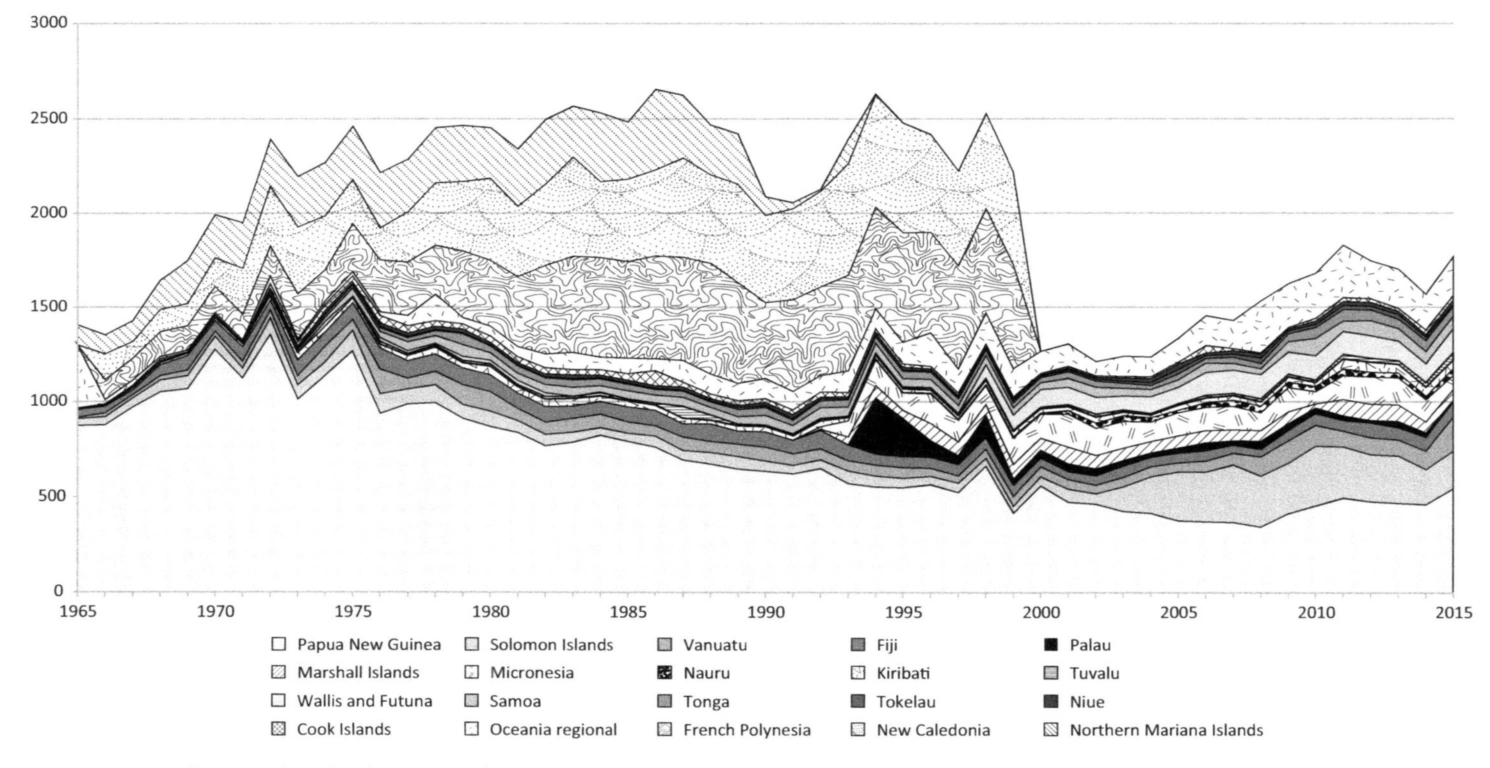

Figure 3.2 ODA to Pacific Island recipients by recipient 1965–2015 ($US million constant prices 2014)

Source: http://stats.oecd.org/ accessed 3 March 2017

to countries becoming independent (such as Papua New Guinea before 1975) whilst other colonial era flows (as to the American territories in Micronesia) were not included.[2] To deal with some of the inconsistencies of data and the definition of aid flows to some countries, we can depict flows to twelve 'core' Pacific recipients (Figure 3.3).[3] Here the trends are clearer, with an early peak in the mid-1970s followed by a long decline in real aid volumes until a sustained increase after 2000. To complement these data depicting recipients of development aid, we can add an analysis of ODA by major donors (Figure 3.4). Again, the removal of French Polynesia and New Caledonia in 2000 reveals a dramatic drop in French ODA to the region at the time.

Waves of aid in the Pacific

Firstly, we can discern some general phases or waves of aid to the Pacific over time. These are the shadows of the various aid 'regimes' we suggested and analysed in the previous chapter:

1 *Modernisation and decolonisation 1960-c.1975*: This period was dominated by the very large flows from Australia to Papua New Guinea, reflecting that country's preparation for independence in 1975. Aid volumes were relatively high and concentrated (colonial donors giving directly to their colonies or recently independent territories). As well as PNG (and the French territories), the major recipients were Fiji, Vanuatu and Solomon Islands, all on the cusp of political independence.

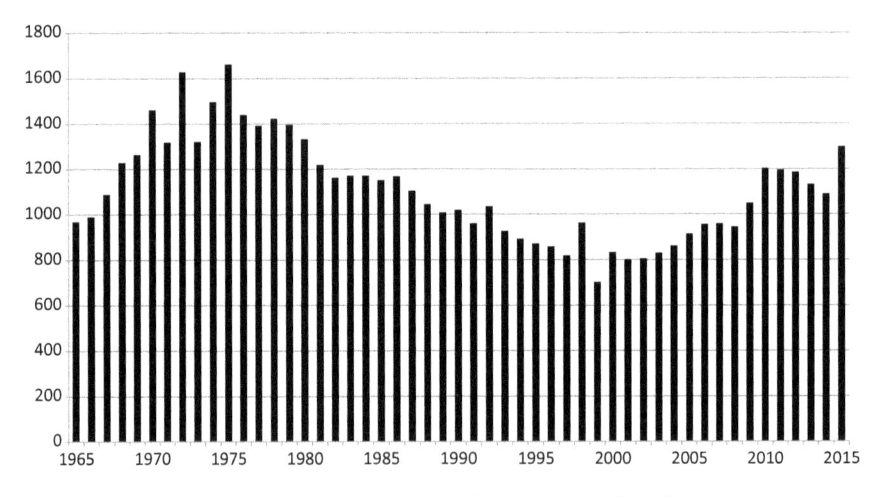

Figure 3.3 ODA to twelve core Pacific Island recipients 1965–2015 ($US million constant prices 2014)

Source: http://stats.oecd.org/ accessed 3 March 2017

Table 3.1 ODA from all donors to Oceania 2010–2014 ($US million constant prices 2014)

	ODA yearly average 2010–2014 ($US million)	*Percent*
Total OECD bilateral	**1704.6**	**84.91**
Australia	984.07	49.05
Austria	0.7	0.03
Canada	3.1	0.16
Finland	0.8	0.04
France	136.3	6.79
Germany	10.2	0.51
Italy	0.7	0.03
Japan	117.0	5.83
Korea	5.0	0.25
New Zealand	235.8	11.74
Norway	1.8	0.09
Spain	1.1	0.05
Sweden	0.4	0.02
UK	8.7	0.43
USA	197.1	9.82
Other OECD	0.6	0.03
Total Multilateral	**299.0**	**14.89**
Adaptation Fund	7.7	0.38
ADB	61.3	3.05
Climate Investment Funds	0.7	0.04
EU	103.7	5.17
FAO	2.0	0.10
Global Alliance for Vaccines and Immunisation	3.6	0.18
Global Environment Facility	12.9	0.64
Global Fund	31.5	1.57
International Development Association	49.3	2.45
IMF	5.6	0.28
OPEC Fund for International Development	1.7	0.08
UNAIDS	1.5	0.08
UNDP	5.8	0.29
UNFPA	2.0	0.10
UNHCR	0.7	0.03
UNICEF	7.2	0.36
WHO	8.9	0.44
Other multilateral	1.7	0.08
Non-DAC Donors Total	**4.8**	**0.24**
Russia	1.9	0.10
Turkey	0.4	0.02
UAE	3.7	0.19
Other non-DAC	–0.1	0.00
TOTAL	**2007.6**	**100.0**

Source: OECD. Stat 1 June 2016

Notes: This list only includes donors enumerated by the DAC of OECD. Donors with an average of less than $US200,000 are aggregated into the various 'others' categories. Furthermore, as mentioned previously, it does not include transfers within France from the metropole to the Pacific territories.

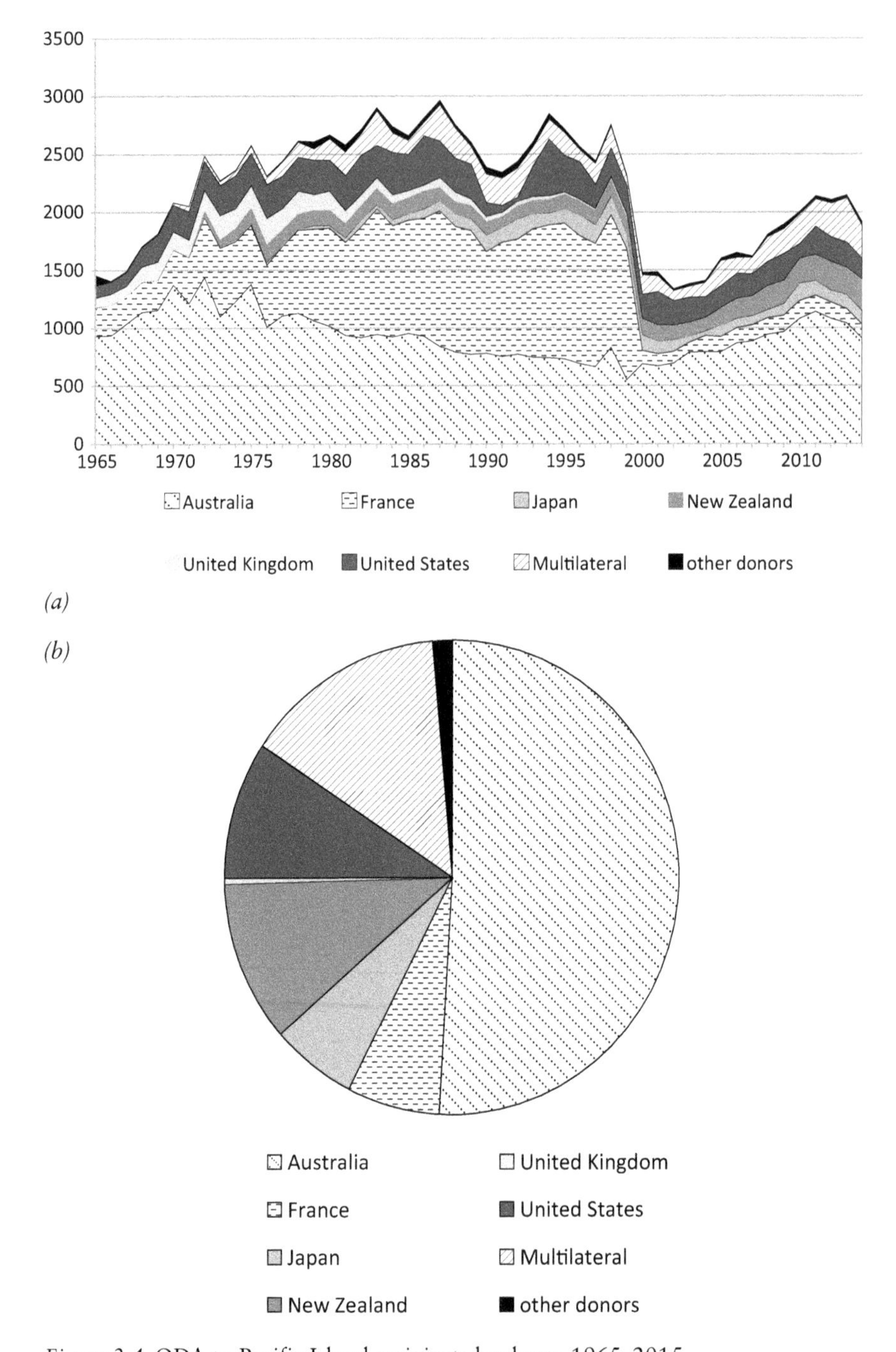

Figure 3.4 ODA to Pacific Island recipients by donor 1965–2015
(a) major donors 1965–2015 ($US million constant prices 2014)
(b) share by major donors 2015

Source: http://stats.oecd.org/ accessed 3 March 2017

2 *Developmentalism (c.1975–1990)*: To a large extent this was a continuation of the decolonisation process, with aid flows again dominated by links between colonial powers and their former territories. Much of this ODA went to both the maintenance of new political and bureaucratic institutions (the daily business of the state) and to new development projects, particularly those relating to infrastructure (roads, harbours, airports, electricity generation). During this period, however, we can see a gradual decline in real aid volumes (i.e. at constant prices) particularly to Papua New Guinea. There was some diversification in the dispersal of aid: Tonga, Samoa and Cook Islands received more, and regional institutions began to get more attention.

3 *Neoliberalism (c.1990–2000)*: This period of neoliberal reform was perhaps felt later in the Pacific Island region than the rest of the world and, on the surface, appears to have been less severe. Yet its imprint is still apparent. Although Figure 3.1 appears to show no major decline in aid, despite some variability, during the 1990s, if we discount the addition of the Micronesian recipients during the period, there was a clear reduction in real aid volumes to the established Pacific countries during the decade. Aid to PNG and Fiji was squeezed and countries such as Cook Islands experienced significant reductions. If we examine twelve 'core' Pacific countries (Figure 3.3) we can see a drop in aid volumes from just over $US 1 billion in 1990 to about $US 700 million in 1999 (in constant 2014 prices).

4 *Neostructuralism 2000–2010*: This period was dominated by the first ten years of the Millennium Development Goals and a focus on both poverty alleviation and security and state-building in the region. There was a marked increase in real aid volumes, though a slight reduction to some (such as PNG). A wider range of recipients received more assistance and regional institutions in particular appeared to gain. The Solomon Islands, following RAMSI, received the most notable increase during the period. ODA was spent on poverty alleviation programmes (in education and health) but there was also significant attention given to building state capacity and governance.

5 *Retroliberalism 2010–*: In the wake of the global financial crisis of 2007–2008, major donors to the region responded in intriguing ways. Although reductions in aid have been experienced in particular cases, in aggregate since 2010 and especially with Australia's change of policy in 2014–2015, there was a surprising lack of aid cuts in the immediate wake of the financial crisis. Yet behind the continuing relatively high ODA volumes, there was a major change in the nature of aid, as we have seen, with more focus on economic growth, trade and explicitly supporting donors' own interests. The poverty focus was diluted but there has been a return to the former developmentalist concern for infrastructure. This has also been a period when the emergence of China as a significant donor has led to a response to, in some cases a mimicking of, Chinese development cooperation strategies in the region.

The evolution of the donors

We learn a good deal when we examine shared and differentiated trends among the donors in the Pacific, which has evolved considerably over time. Figure 3.4 shows data for various donors. Despite differences between donors, the broad patterns illustrate elements of similarity. Below we discern a number of major trends with respect to the donor group in the region.

Firstly, there has been a major change in the composition of donors in the region in the past fifty years. Aside from the issues surrounding the inclusion or not of French ODA to its territories (and US aid to the Northern Mariana Islands), the most notable feature is the major withdrawal of United Kingdom as a donor to the region – and its replacement, to a large degree, by Australia. Thus, former British territories – Fiji, Solomon Islands, Kiribati, Tuvalu, Vanuatu – saw Australia become the surrogate donor, taking over Britain's position as a major source of aid. This trend underlined the emergence of Australia as the major donor in the region, providing as much ODA as all other bilateral and multilateral donors combined – or at least the non-Francophone Pacific and excluding aid from China and Taiwan. Although the UK has made some steps to re-engage in the region as a donor, its presence today is relatively small (less than 1%of total ODA to the region in 2014). Secondly, we can see the growth of Japanese aid to the region, especially after the 1970s, though this may be waning recently. Within these donor trends, there were marked patterns of concentration – a point we turn to below when we consider spheres of interaction.

Secondly, it is apparent that all the major donors have conformed largely to the features noted above with regard to global aid regimes, albeit with some small variations. Australia and New Zealand were closely involved in the colonial and developmentalist eras and both also participated in neoliberal cutbacks in the 1990s. And although New Zealand was rather slower in adopting the 'new' poverty agenda in the late 1990s, it did so enthusiastically following the launch of the MDGs in 2000 (Storey et al. 2005). Australia was a key driver of these changing policies in the region. It cut its aid in 1989 and again in the mid-1990s but more than doubled its aid in real terms during the neostructural and early retroliberal phases in the years 2005–2012. Its major aid cuts in 2014 may signal a general reduction across other donors in the coming years. Japan seems to have been rather less of an enthusiastic follower of global aid trends, its dispersals in the region seemingly more conditioned by its own economic fortunes and political priorities. Thus, its volumes increased during the 1980s and even 1990s when other donors were cutting back but its increases during the 2000s were much less marked than others.

France and the USA as aid donors to the region are more difficult to track in terms of DAC data and their patterns of aid in the Pacific are strongly conditioned by their former colonial ties (see below): both were strongly in evidence during the colonial and developmentalist phases. France, indeed, rivalled then outstripped Australia as the region's main aid benefactor until it reclassified its ODA in 1999. Both also participated in neoliberal reductions in the 1990s and both, in various ways, engaged in neostructural programmes of poverty reduction in the early 2000s. Yet neither has greatly extended its ODA donations in the region outside of its former colonial relationships or engaged in wider regional initiatives such as RAMSI.

Thirdly and finally new donors have proliferated in the region in the past twenty years. Multilateral agencies – the Asian Development Bank (ADB) and the World Bank, and the various United Nations institutions – have a relatively long history in the region but, the ADB aside, few have figured prominently in terms of large aid volumes when compared to the major bilateral donors. The ADB has become more visible as a donor in the past two decades: it was a major donor in the early 1990s, it provided aid to support neoliberal reforms in the mid-1990s and it has backed several large infrastructure projects since. European donors have combined more recently in the form of European Union ODA programmes in the region and new multilateral players, such as the largely privately-backed Global Fund, have appeared in some countries. With regard to other bilateral donors and, with the exception of China and Taiwan (see below), there has been an increase in the number of new and smaller donors, such as Indonesia and Malaysia, though only the United Arab Emirates amongst this group provides any substantial financial assistance (over $US 9 million to the region in 2014).

Despite these trends in aid to the region and the increase in the number of donors, there is a surprising degree of concentration of aid donation as revealed by the OECD data. Figure 3.4b presents a summary of the relative sources of aid to the region in 2015 (including regional organisations but excluding New Caledonia, French Polynesia and Northern Marianas[4] and donors not included in the OECD database). Australia provided half (50.9%) of all ODA to the region, multilateral donors contributed just over 14%; New Zealand 11%; USA 9%; and Japan and France just over 6%each. All other bilateral donors enumerated (including UK, Canada and non-DAC and private donors) together gave a relatively paltry 1.7%.[5] If we count multilateral agencies as one (and the Asian Development Bank is the dominant multilateral agency in the region), then we can conclude that just six major donors dominate the aid landscape in the Pacific and, of these, Australia is by far the largest.

Changes in the sectoral distribution of aid in the Pacific

These broad patterns of ODA flows disguise to a large extent some important changes in the nature of aid. Although the DAC data do not allow for a microscopic examination of what aid is spent on over a long time frame, we can draw some broad points from the extant data on sectoral distribution of aid. We do have a reasonable time series for the example of Australia (Figure 3.5) but only since 2002 can we can analyse the sectoral distribution of ODA for all DAC donors (Figure 3.6). These data use the DAC sector categories for ODA expenditure and several of these categories cover the funding of quite a wide variety of activities. What is labelled as 'commodity aid', for example, includes aid-in-kind (such as food or oil) but often recipient governments are allowed to stock these goods in strategic reserves or sell them and use the cash for the public budget. Similarly, categories such as 'government and civil society' cover many more items because here it is not the nature of the aid but the purpose or recipient that matters. In general, however, some changes are clearly apparent. The earlier phases of aid, covering the colonial and developmentalist phases, included expenditure on a number of sectors that have since become less prominent – or largely disappeared. For example, Australia supported 'commodity aid' to a large extent up until the late 1980s (this may have involved subsidies on inputs such as fertilisers or even forms of price support) but most of the category was probably forms of direct support for government, especially in Papua New Guinea.[6]

It is also important to note that much of the aid from donors to their extant or recently independent colonies was in the form of direct budget support to government coffers. This was particularly the case for expenditure on education, health and infrastructure. Arguably, Australia merely resumed its own colonial practices in PNG when it embraced General Budget Support (GBS) explicitly in the early 2000s in line with the objectives of the Paris Declaration on aid effectiveness. New Zealand could be seen as doing the same when it offered to increase its budget support to the Government of Samoa after the Paris Declaration.

Education and health have been significant items of expenditure throughout the various aid regimes, even though the detailed disbursement (as between hospitals or primary health care; or primary education versus tertiary scholarships overseas) may have varied over time. There was notable growth in these two categories in the 2000s, closely linked to the MDG-inspired aid regimes of the time. The preamble of New Zealand's partnership agreement with Samoa of 2011, for example, explicitly noted "a shared vision . . . [of] achievement of the Millennium Development Goals" and committed New Zealand to "an increasing proportion of its aid spending through higher level aid modalities such as sector wide approaches and . . . financing at . . . general budget level" (Government of Samoa and MFAT 2011: 1).

In the years before about 1985, assistance for agriculture, energy and transport infrastructure was significant, aligned to the developmentalist philosophy of promoting key economic sectors. Japan has been prominent throughout in funding these forms of ODA. In general, these three categories (agriculture, infrastructure and energy) fell from favour for many years after the 1980s but they have re-emerged since about the late 2000s and the retroliberal approach to promoting economic growth.

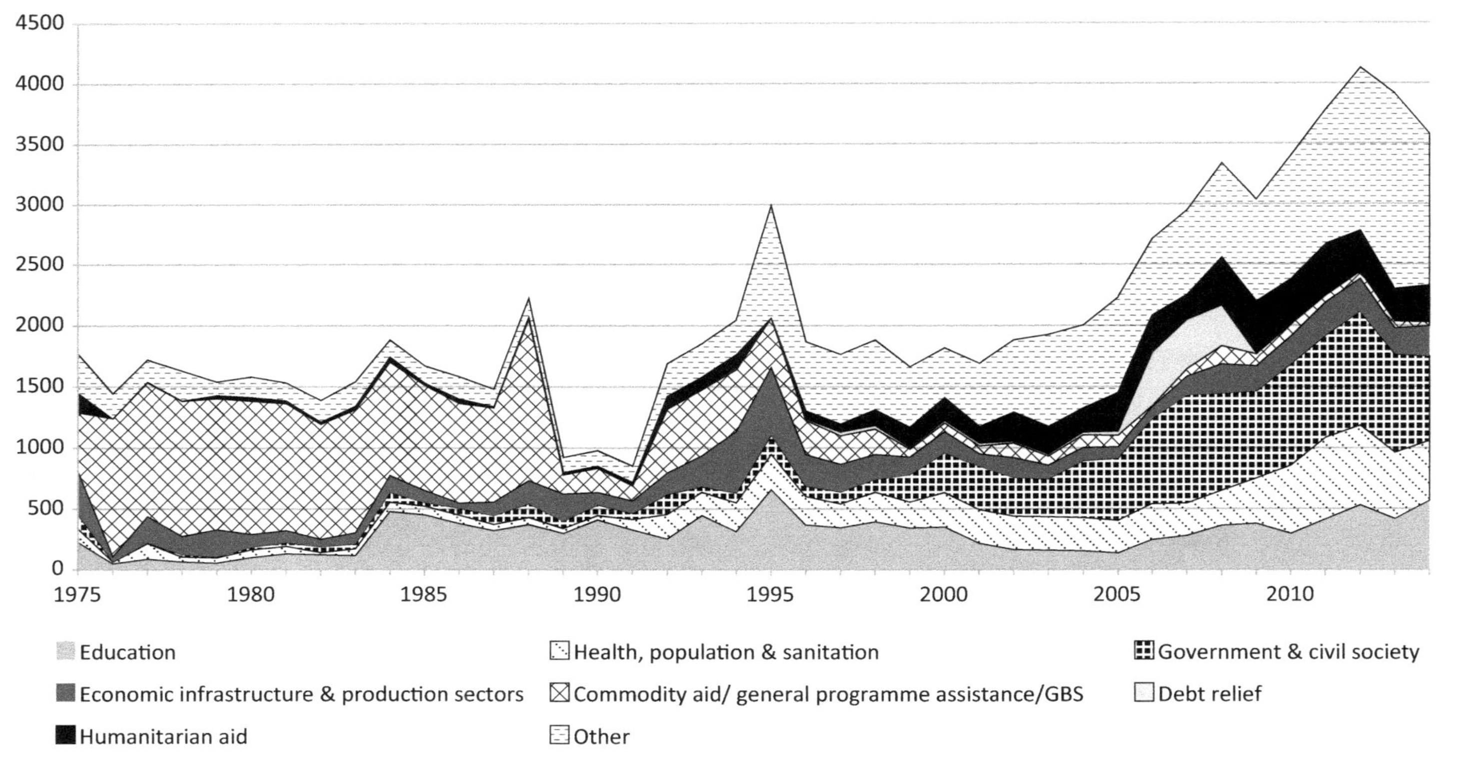

Figure 3.5 Sectoral allocation of Australian ODA to Pacific Island recipients 1975–2014 ($US million constant prices 2013)

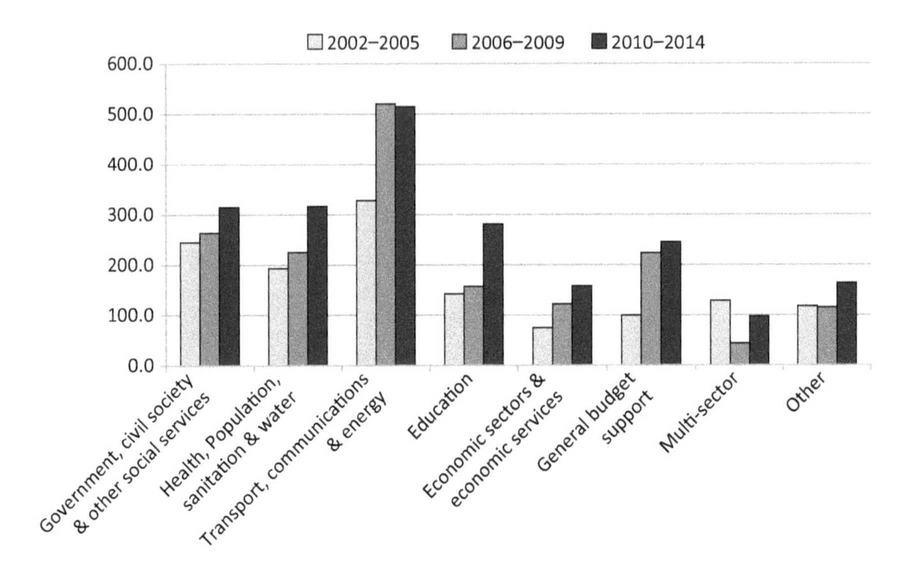

Figure 3.6 Sectoral ODA allocations to Oceania 2002–2014 ($US million constant prices 2014)

The other significant feature of changing sectoral priorities for ODA in the region has been the emergence of expenditure of 'government and civil society' since the late 1990s. Though encompassing a diverse range of activities, this category has been largely associated with the neostructural project of building the capacity and capability of Pacific states to deliver key welfare and economic services. It reached a zenith with the RAMSI intervention in Solomon Islands and the particular state-building project in that country following the civil conflict (Dinnen 2004; Tabutaulaka 2005; Dinnen and Firth 2008; Anderson 2008). Whereas in its earlier years it involved support for civil society, for much of the 2000s the emphasis has been on state institutions. Finally, we can see the rise of expenditure on 'humanitarian aid' in recent years. Arguably, this has occurred partly to deal with the humanitarian consequences and increasing costs of natural disasters and because of increasing concerns since the 2000s that fragile states may both cause humanitarian disasters or be unable to deal with crises. Donors have now identified such assistance as something to be planned and budgeted for over the long-term.

When we analyse aggregate data since 2002 (Figure 3.6) we can discern some trends between the neostructural and retroliberal periods in Oceania. Aid disbursements on education and health have continued to increase (though there may have been shifts within these categories, as with the marked increases in expenditure on scholarships for education in donor institutions for New Zealand for example) but there seems to have been a retreat from the state-building project of the neostructural era with falls in assistance for 'government and civil society' since 2010). The highest rate of increase since 2010 has been in the economic sectors (transport and communication and energy) and there have been increases elsewhere (such as in climate change-related funding).

Geographies of aid and spheres of interaction in the Pacific

Moving beyond these broad time series data for aid in the Pacific, it is now possible to move to cross-section analyses to reveal in more detail the spatial distribution of aid and the particular concentrations of aid that reveal underlying political relationships between donors and recipients. Here we see that different donors have clear areas and countries of interest in the Pacific and their ODA disbursements are closely concentrated. These are revealed in the 'spheres of interaction' maps of Figures 3.7 to 3.11.[7] These spheres of interaction display a remarkable degree of ODA spatial concentration in the region. There is crossover: some donors such as Japan spread their ODA more widely than others and some recipients have a much more diverse set of aid sources than others. Yet the most striking feature of aid in the Pacific is the degree to which most aid recipients rely on a single donor for a very high proportion of their ODA and this dependence is mirrored across several distinct spheres of influence. An overall summary (Figure 3.7) shows a varied landscape of aid: proportional circles relative to the total volume of aid show that the large Melanesian states (PNG, Solomon Islands and Vanuatu) together with Oceania regional institutions attract most total aid. PNG is the largest recipient, averaging $US 608 million in ODA per year over 2010–2014 (in 2013 constant values). At the other end of the spectrum, Niue, Tokelau, Cook islands and Tuvalu each received ODA in the range $US 18 million – $US 26 million. These differences are important in our later discussion of inverse sovereignty for whilst the larger recipients deal with aid volumes and institutions comparable to many other developing countries, the small recipients (the microstates) receive relatively small volumes but have to accommodate this with usually tiny bureaucracies and limited capacity to comply with 'global' aid modalities.

The Australian *sphere of interaction* (Figure 3.8) is drawn more by its geography than its colonial history. Australia provides over a half (and often over two-thirds) of the total ODA to the Pacific countries it lies closest to. The most populous countries of the region – Papua New Guinea, Solomon Islands, Vanuatu and Fiji – compromise this group which draw a broad arc around Australia's northeast margins. The smaller countries of Kiribati, Tuvalu and Nauru are also added to this sphere. Of all these, only PNG had a direct colonial association with Australia; the rest were 'inherited' as after Britain's withdrawal from the Pacific.

The USA, on the other hand has a sphere that is very closely prescribed by its colonial past in the region (Figure 3.9). Its former Micronesian territories,

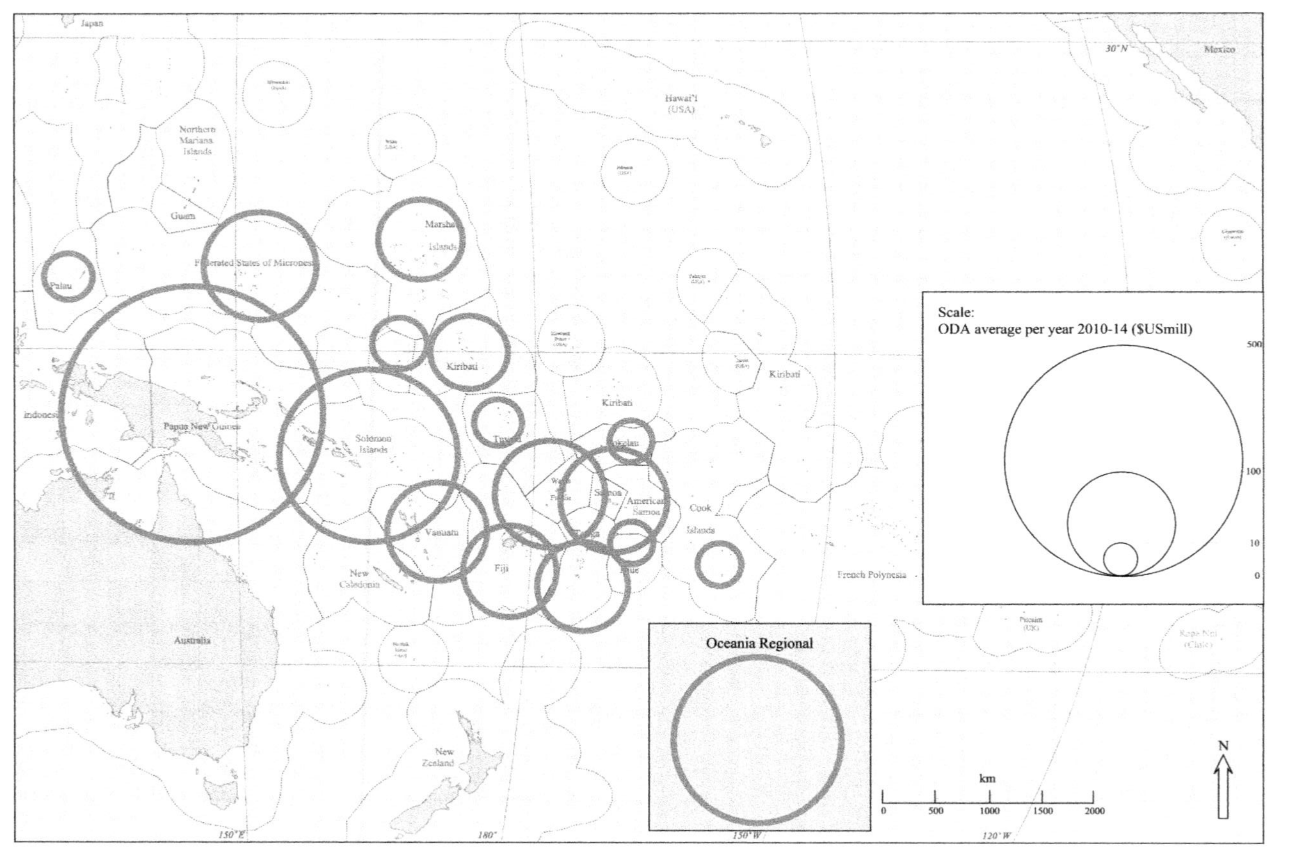

Figure 3.7 ODA to Pacific Island recipients 2010–2014

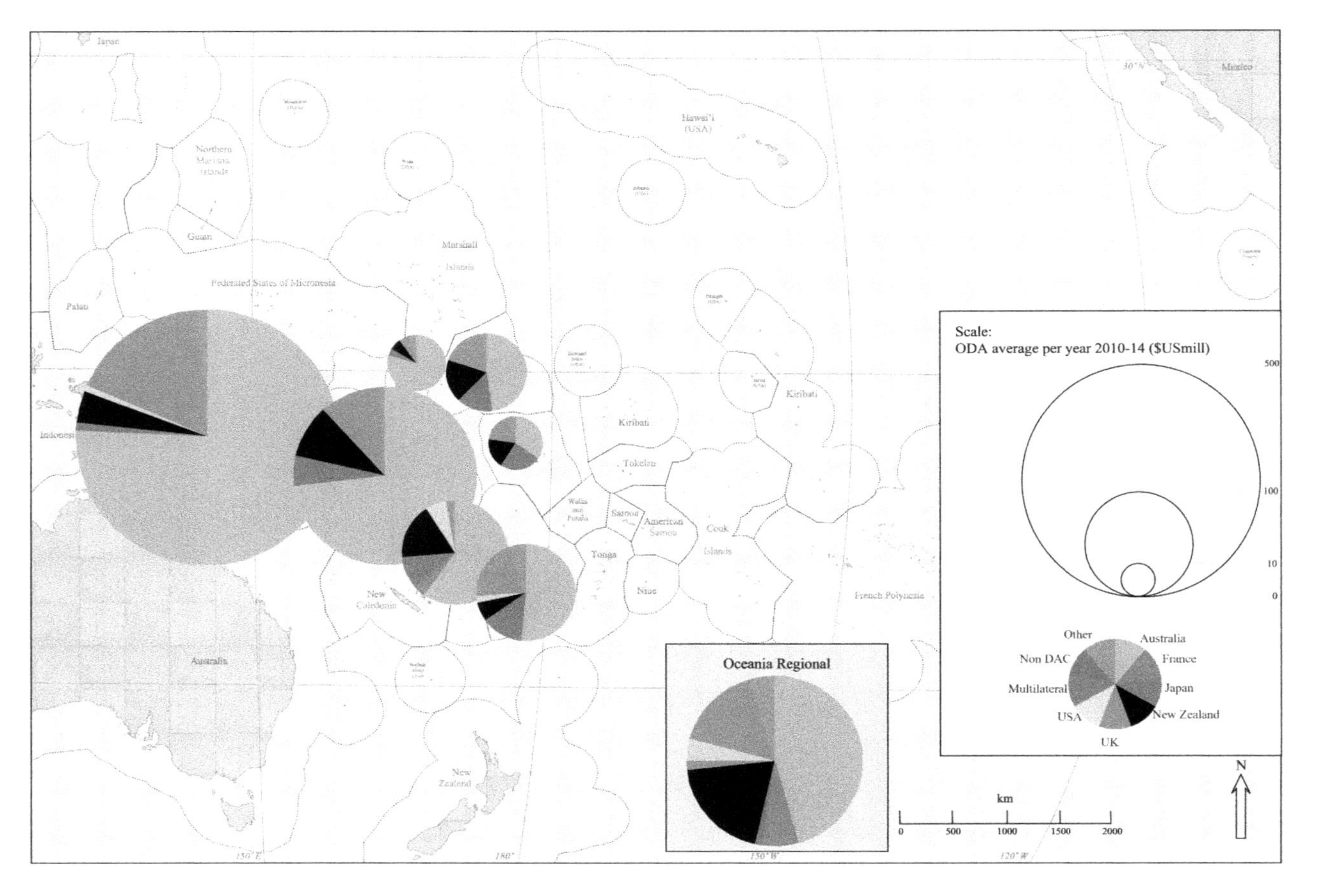

Figure 3.8 The Australian sphere 2010–2014

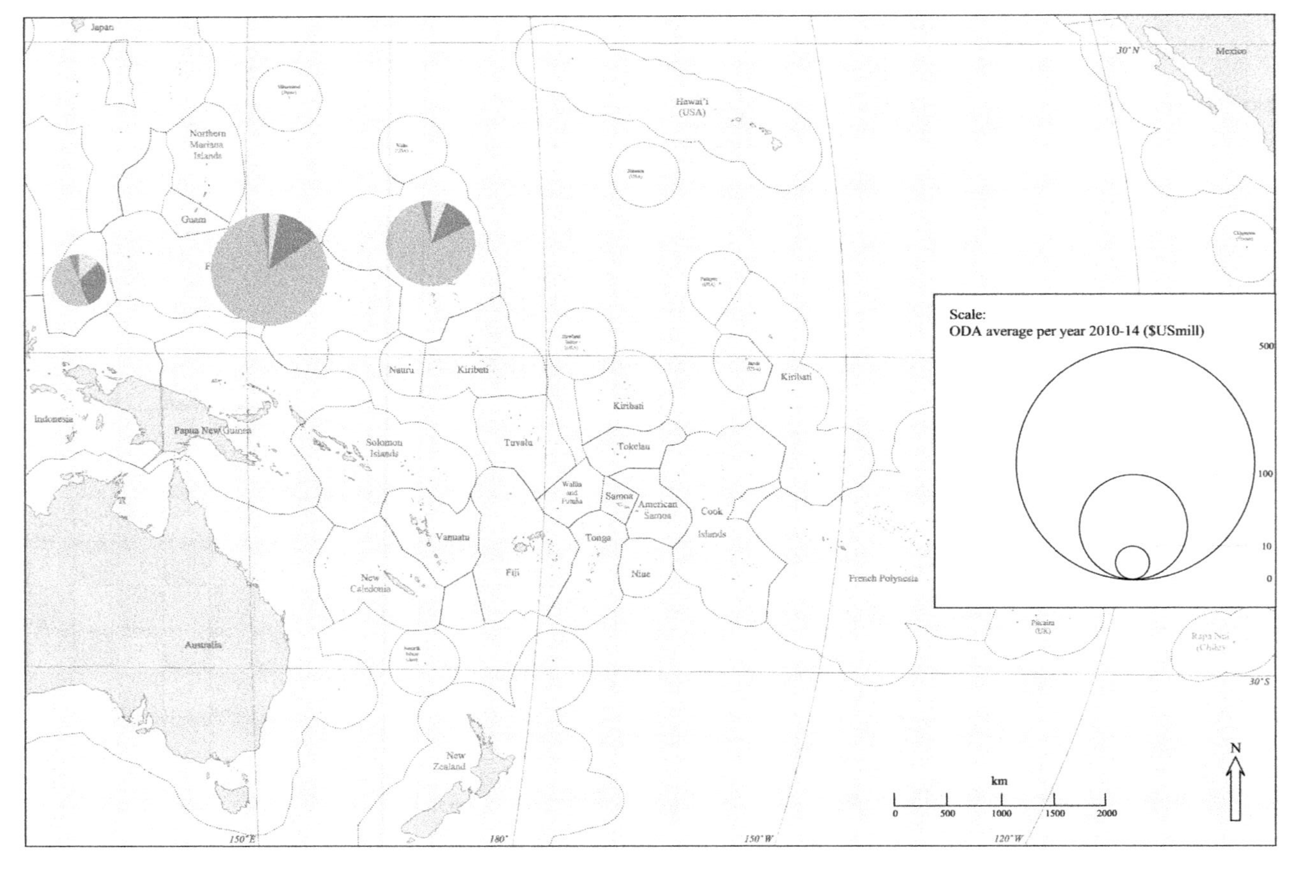

Figure 3.9 The USA sphere 2010–2014

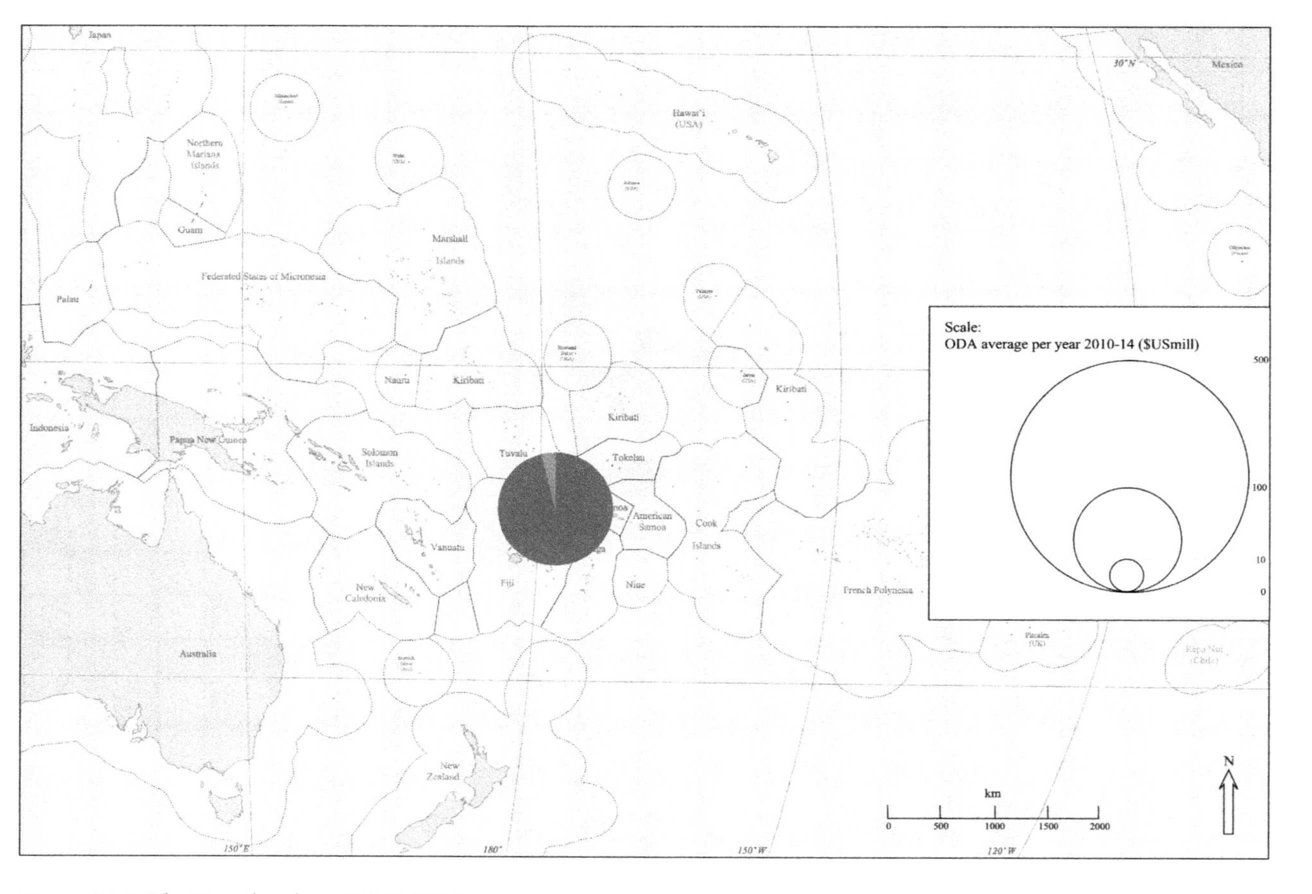

Figure 3.10 The French sphere 2010–2014

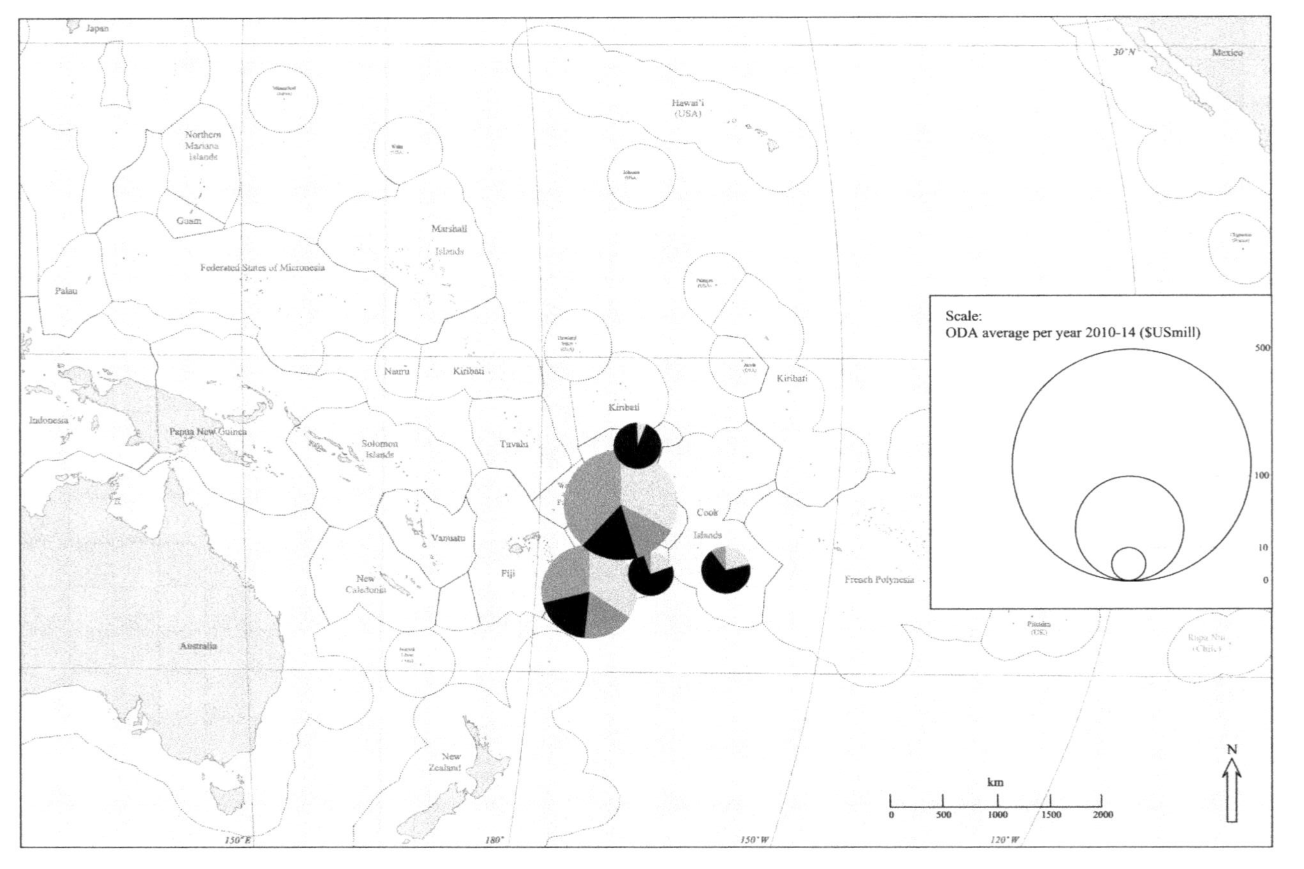

Figure 3.11 The New Zealand sphere 2010–2014

some obtained after Japan relinquished them after the Second World War, have had a strong strategic value for the USA. Marshall Islands, for example, contains islands that were used for American nuclear and missile testing for several decades. After these territories gained varying degrees of independence, the USA has maintained strong ODA flows. The American sphere is thus very much a north-western cluster of countries, close to Asia and with close political association with the USA (Crocombe 1995). Another component of the American sphere is American Samoa (Faleomavaega 1994). This territory, its value again reinforced during and after the Second World War, has a close political association with the USA (see Chapter Two). American expenditure on welfare and development there is not included in the DAC statistics, being considered the 'internal' maintenance of an American territory. American aid also is disbursed to PNG, Solomon Islands and Vanuatu – the latter two with some historical ties resulting from military operations based there during WWII – though such American ODA there is dwarfed by Australian aid in particular.

The French sphere of interaction (Figure 3.10) has perhaps the highest degree of ODA concentration in the region. Almost all French ODA is concentrated into just three territories, all of which rely on metropolitan France for well over 95%of their assistance. Elsewhere, we have estimated what these levels of assistance might be to French Polynesia and New Caledonia (see Box 3.1) given that the flows are no longer captured by DAC so only Wallis and Futuna are depicted here. Nonetheless, the relatively large flows that are counted to Wallis and Futuna and the historical record of flows to the two larger territories indicate that French assistance to the Pacific, though highly concentrated, is also very substantial. Again, some French aid goes to other countries in the Pacific, some under the banner of European Union aid, but it is tiny both compared to other donors there and to French aid to its three focal territories. Thus, the French sphere of interaction in the Pacific both reflects and reinforces the Francophone character of these countries and maintains a very strong French presence in the region.

Box 3.1 The French territories: how non-official aid can be helpful if it is done the French way

Séverine Blaise, Julien Migozzi & Gerard Prinsen

Most of the policies and literature concerning development aid is predicated on the principle of aid as an arrangement to facilitate a flow of resources from sovereign donor states to sovereign recipient states. The OECD's Development Assistance Committee also takes this as a basis for its research policy advice. However, there are several Pacific Islands that are formally overseas territories of France. Legally, the three island groups of New Caledonia, French Polynesia, and Wallis and Futuna are three overseas territories of France and in the framework of the OECD of Official Development Assistance (ODA), the resources flowing from Paris to its three Pacific territories are intra-state transfers from a central government to a local government body. Such transfers do not qualify as development aid.

However, with a French context this transfer of resources from Paris to the Pacific is often seen quite differently. A closer look at the transfers to New Caledonia may explain this. First, Kanak (the indigenous people of New Caledonia) and the French State have had a complex, often uncomfortable, and at times violent relationship when it comes to the question of sovereignty. New Caledonia officially features on the United Nations' list of colonial territories that still need to be decolonised. Vociferous and sometimes violent clashes between Kanak and the French State led to two peace agreements: the Matignon Accord of 1988 and the Nouméa Accord of 1998.

Central to both peace agreements is the idea that as part of the decolonisation process, the French State will transfer resources to the Kanak-dominated areas in New Caledonia to compensate for colonial injustices and to assist in a re-balancing of the inequalities between the South Province (where 91% of the descendants of white settlers live) on one hand and the North Province and the Province of the Loyalty Islands, whose populations are, respectively, 80% and 97% Kanak, on the other. These inequalities are stark. For example, the average monthly household income in the North Province is about $US 2,166, whereas the household income in the South Province is double that at $US 4,276. Similarly, about 54% of New Caledonians from European descent complete tertiary education, but only 12% of Kanak do (ISEE 2013).

The French State's financial transfers to the Kanak-dominated provinces are a clear attempt to redress today's inequalities that are consequences of historic injustices. The Nouméa Accord officially aims to support a 're-balancing' in New Caledonia. This can lead to decisions to transfer resources from Paris to New Caledonia where political arguments weigh as much or more as economic arguments – not dissimilar to a lot of Official Development Assistance. For example, virtually all industrial investments in New Caledonia are made in the South Province. New Caledonia possesses about a quarter of the world's nickel reserves and until recently, all processing plants were located in the South Province. As part of the re-balancing agreed in the Nouméa Accord, the French State has co-invested in the establishment of a nickel metallurgic plant in the North Province worth over $US 2 billion.

Generally speaking, regular transfers between Paris and New Caledonia arrive via two distinct channels. The first channel comprises the funding from the French Treasury for recurrent public budgets in New Caledonia. Most of this entails paying for salaries of teachers and health personnel, as well as expenses in education material and medicines. It is, however, rather difficult to trace these transfers in the public records because rather than being recorded as a transfer from Paris, many of these payments are booked as 'income' by the New Caledonia government and the total volume can vary in the course of time because they are tied to wages, allowances, prices of pharmaceuticals, etc. Nonetheless, the lowest percentage of this type of 'income' for the New Caledonia government was 34% in 2007, while it also rose to 47% in the year 2000 (ISEE 2015).

The second channel for transfers includes financial support and tax exemptions for projects for the 'rebalancing' of New Caledonia's economic structures. This not only refers to big-tickets items such as the nickel plant, but also an array of more minor funding arrangements from which local authorities finance projects in social housing, public works, transport, education and health infrastructures. Between 1990 and 2013, the three provinces and the twenty-three municipalities secured about $US 2.2 billion, averaging $US 86 million per year (ISEE 2015). It is worth noting there is also a third channel which is a transfer of resources whose actual benefit in New Caledonia is most difficult to calculate: tax benefits awarded by the French revenue authorities for investments made in New Caledonia by private companies based in France. (This can be a fiscalist's nightmare – or dream, depending on who is involved – because New Caledonia is, of course, still France.)

In summary, while calculating the net transfer of resources from Paris to New Caledonia is complex, one official report puts the figure at about $US 1.72 billion in the year 2013. With New Caledonia's total population at about 250,000, this comes down to a net transfer of approximately $US 6,800 per capita (IEOM 2014: 67). It may not be labelled Official Development Aid, but for New Caledonians it certainly represents help and assistance in redressing some of the persisting inequalities.

The fourth major sphere, that of *New Zealand* (Figure 3.11) is not as large or strong as the other three, mostly because New Zealand is a much smaller donor in absolute terms and in only three places (Cook Islands, Niue and Tokelau) does it account for over a half of the aid receipts. In addition, New Zealand directs much of its aid to Samoa and Tonga (although in both Samoa and Tonga it is not even the largest donor – that being Australia). Yet these five countries might well be considered to be within New Zealand's 'sphere' given the strong historical ties and durable current political attachments in evidence, not to mention the strong orientation of Samoan and Tongan relationships towards New Zealand rather than elsewhere.[8] This is very much a central Polynesian sphere and reflects New Zealand's political perceptions and priorities in the Pacific. New Zealand is also a donor to other parts of the region, particularly in Melanesia where Australia dominates. Indeed, although New Zealand is very much a junior partner to Australia in PNG and Solomon Islands, its aid budgets there in the past decade have been greater than in its defined Polynesian sphere of interaction. In addition, New Zealand is a relatively generous donor to Pacific regional institutions: it contributed just under 20% of ODA to this sector, relatively sizeable compared to Australia's 45% and far surpassing donors such as Japan, France and the USA.

The final major bilateral aid player in the region is *Japan*. Because of its political history in the region and its defeat in the Second World War, Japan has no formal political ties and its aid disbursal patterns are more diffuse than other donors. It has no clearly defined sphere of interaction. Yet there are still some obvious patterns. There is some correlation with its colonial history in the region and its geography – perhaps seen in its substantial aid donations to the Micronesian states. There also appears to be an interest in what might be called 'middle Pacific' states: Kiribati, Tuvalu, Fiji, Samoa and Tonga (it is the second largest bilateral donor behind Australia in Kiribati and Fiji).

These spheres of interaction, then, paint a picture of the Pacific region where the imprint of colonialism and immediate postcolonialism together with important historic episodes such as the Second World War are still visible in aid flows. Although the spheres have evolved over time, they have remained remarkably constant. More recent policy turns which have seen donors in general focus on a smaller range of countries where they have ties and are seen to be able to make a difference, have accentuated the geographic concentrations. Former colonial relationships – and some continuing ones – are mirrored in heavy dependence by many countries on a single donor, usually the colonial power or, its surrogate (in the case of Australia's replacement of the United Kingdom). Colonialism, it might be argued, continues through present aid relationships. Yet, as we suggested in Chapter Two, this might now be not simply the continuance of colonial dominance by an assertive metropolitan power but also a more subtle maintenance of such close relationships by willing Pacific countries that see benefits in preserving some elements of colonial relationships. These are, however, colonial relationships that have been redrawn and negotiated in a postcolonial world where power is exercised in different forms and directions in complex ways to lever political and economic benefits in multiple directions. This opens up possibilities for peripheral manipulation of the metropolitan centre simultaneously with metropolitan exploitation of political and economic benefits from the periphery.

Recipient profiles

Although these aggregate and donor data sets reveal diverse patterns of aid in the Pacific over time and space, when we consider the individual experiences of different recipient countries and territories, an even more complex picture of aid is revealed (Figure 3.12). There is a remarkable lack of

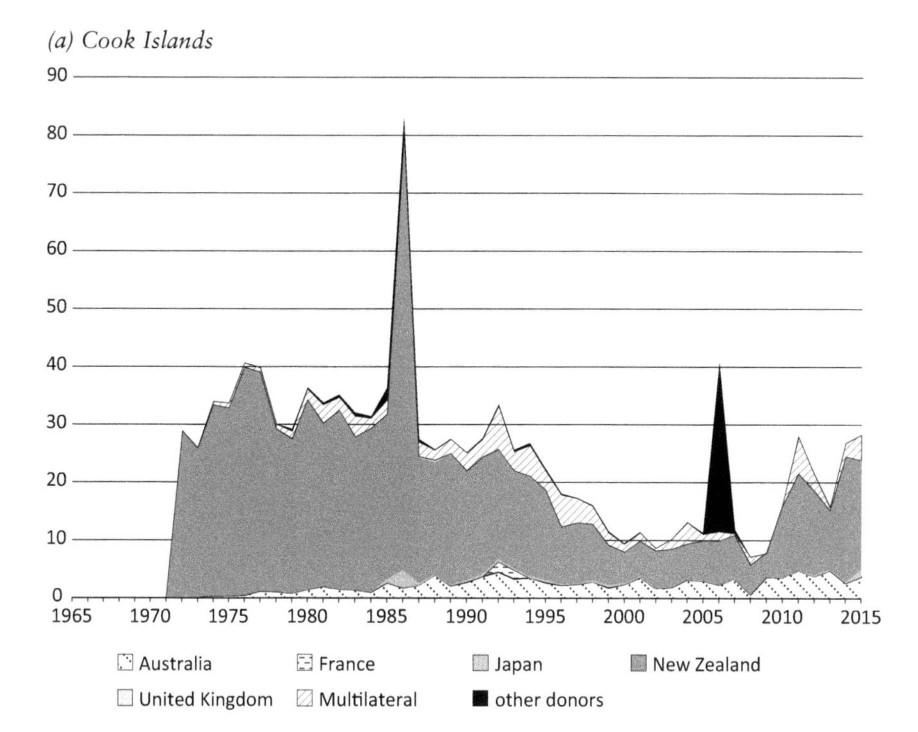

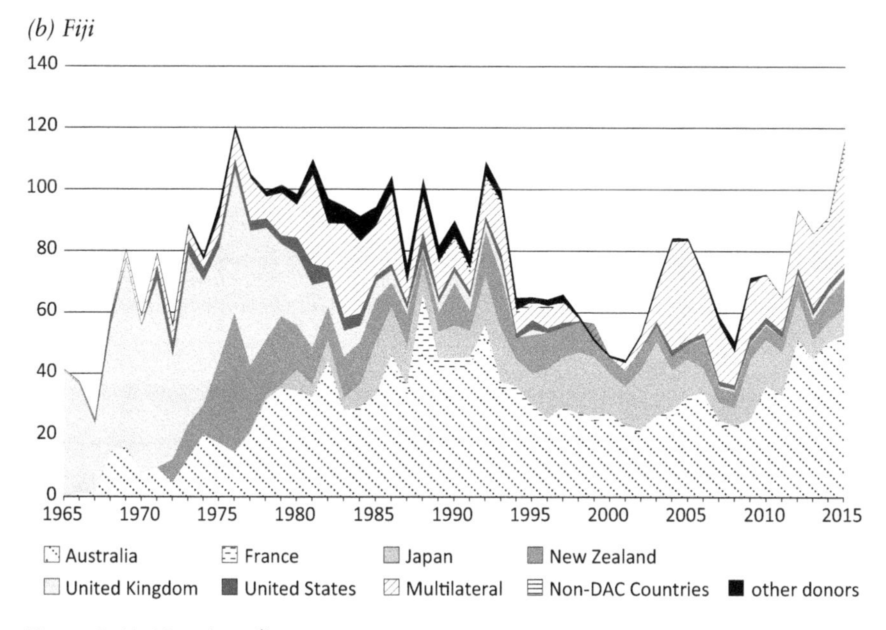

Figure 3.12 (Continued)

(c) Federated States of Micronesia

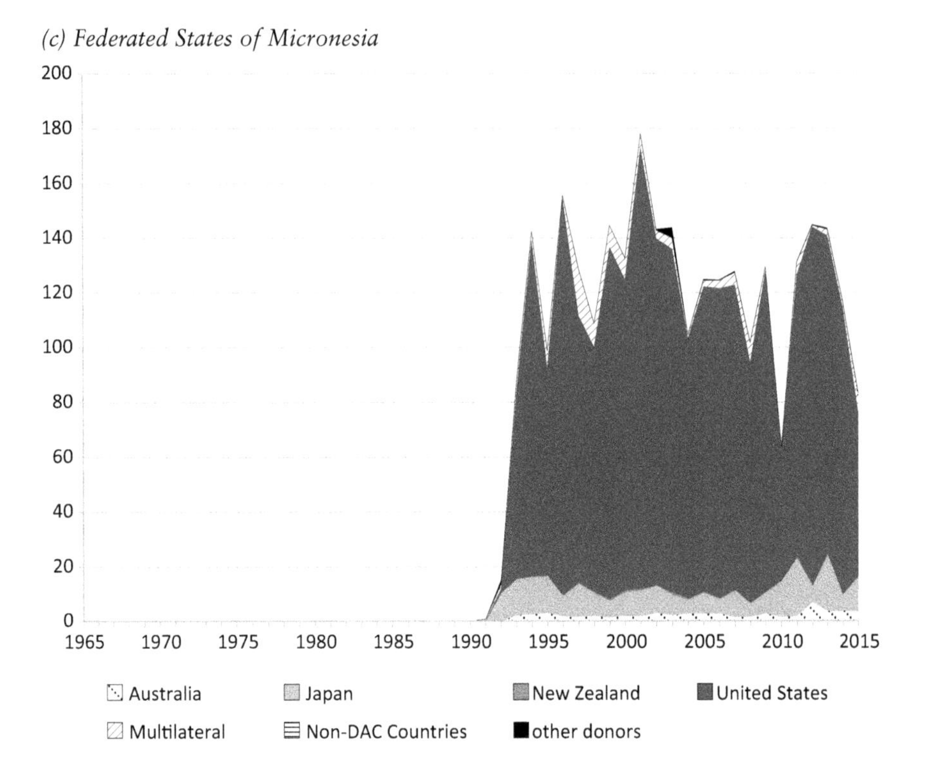

(d) Kiribati

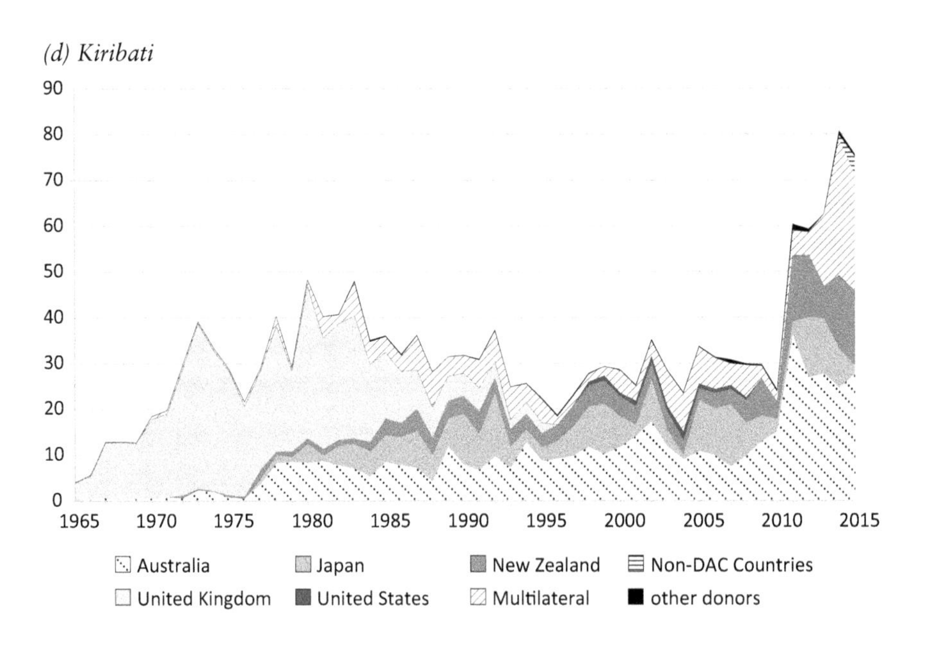

(e) Republic of the Marshall Islands

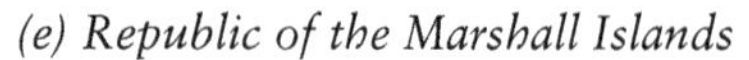

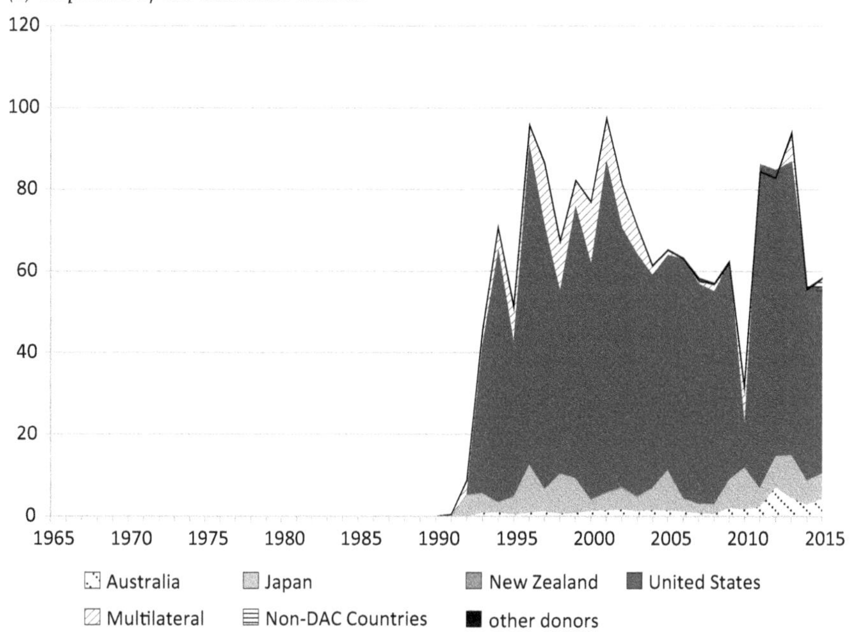

(f) Nauru

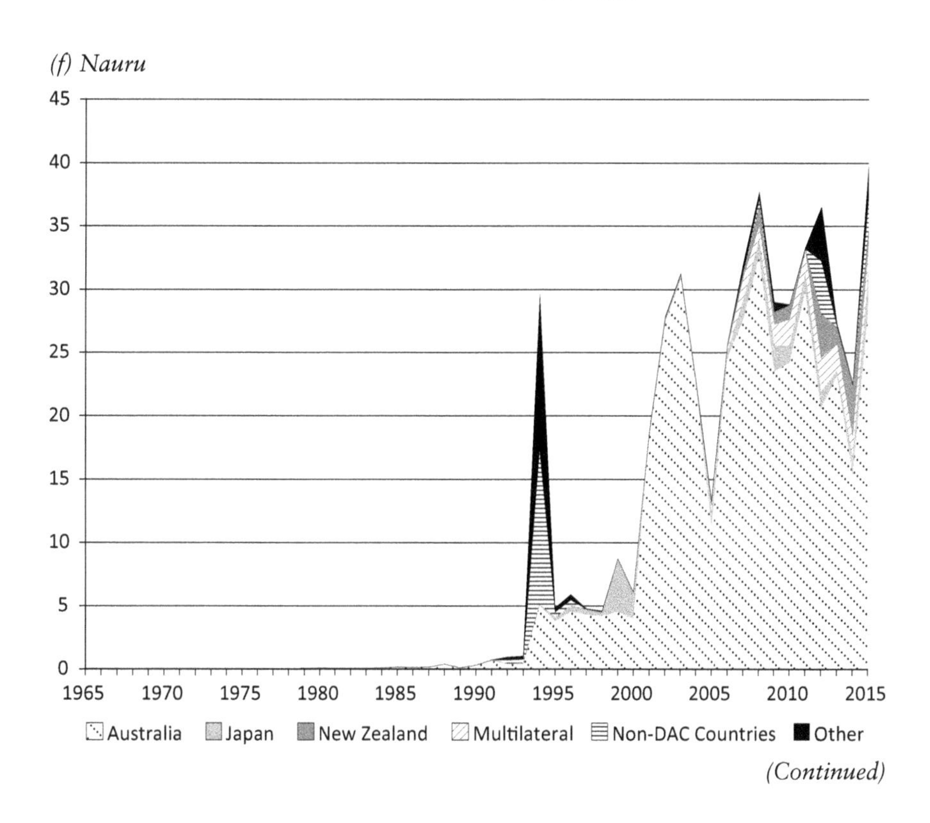

(Continued)

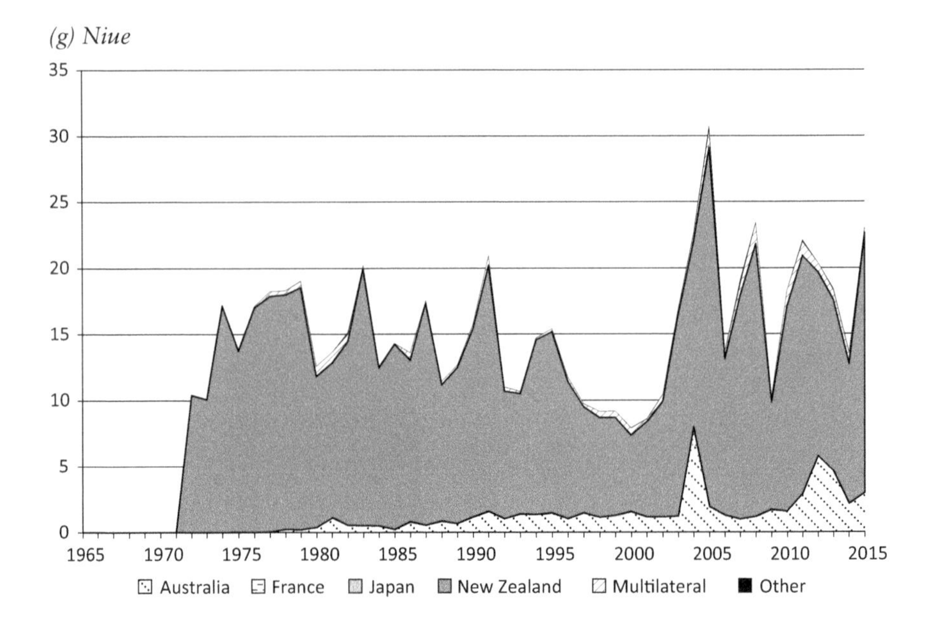
(g) Niue
35
30
25
20
15
10
5
0
1965
1970
1975
1980
1985
1990
1995
2000
2005
2010
2015
Australia
France
Japan
New Zealand
Multilateral
Other

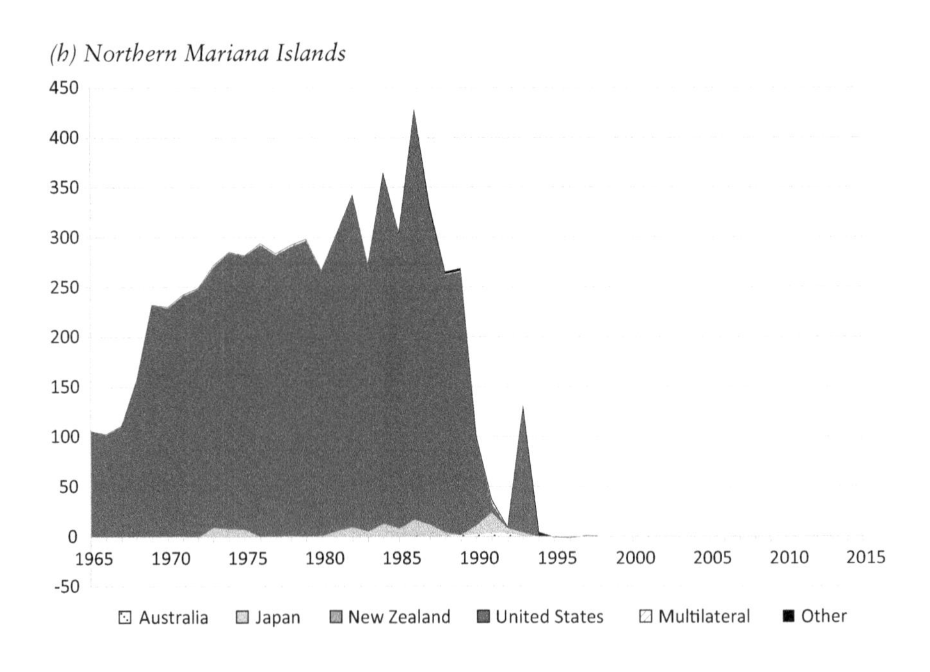
(h) Northern Mariana Islands
450
400
350
300
250
200
150
100
50
0
-50
1965
1970
1975
1980
1985
1990
1995
2000
2005
2010
2015
Australia
Japan
New Zealand
United States
Multilateral
Other

(i) Palau

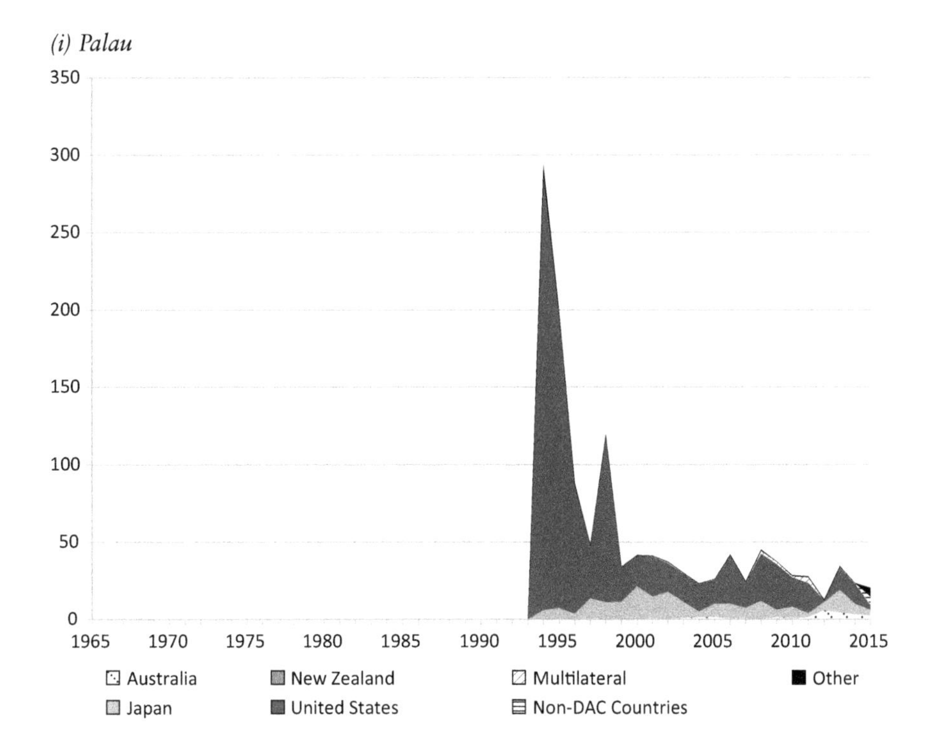

(j) Papua New Guinea

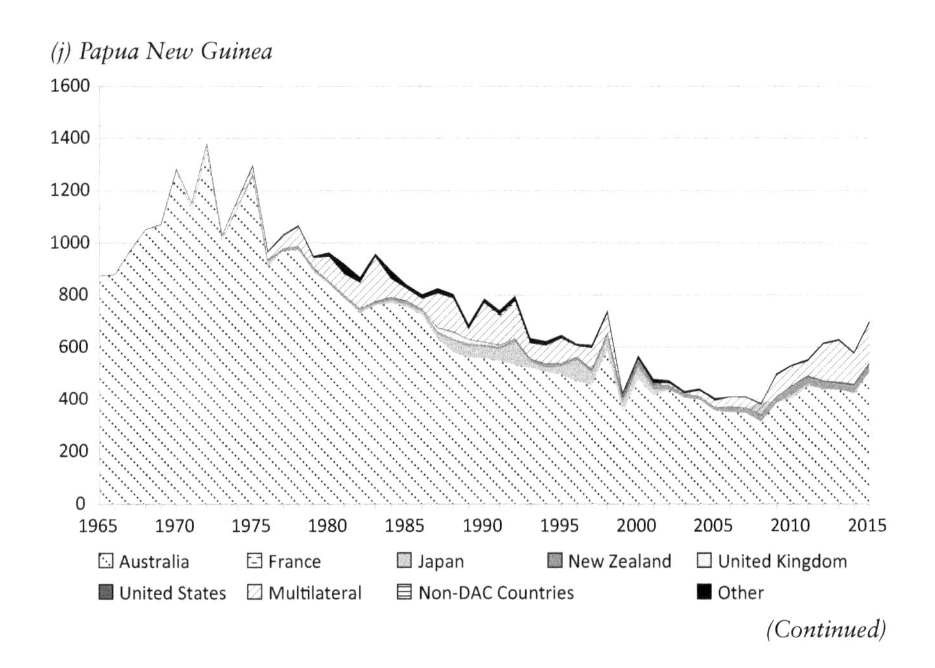

(Continued)

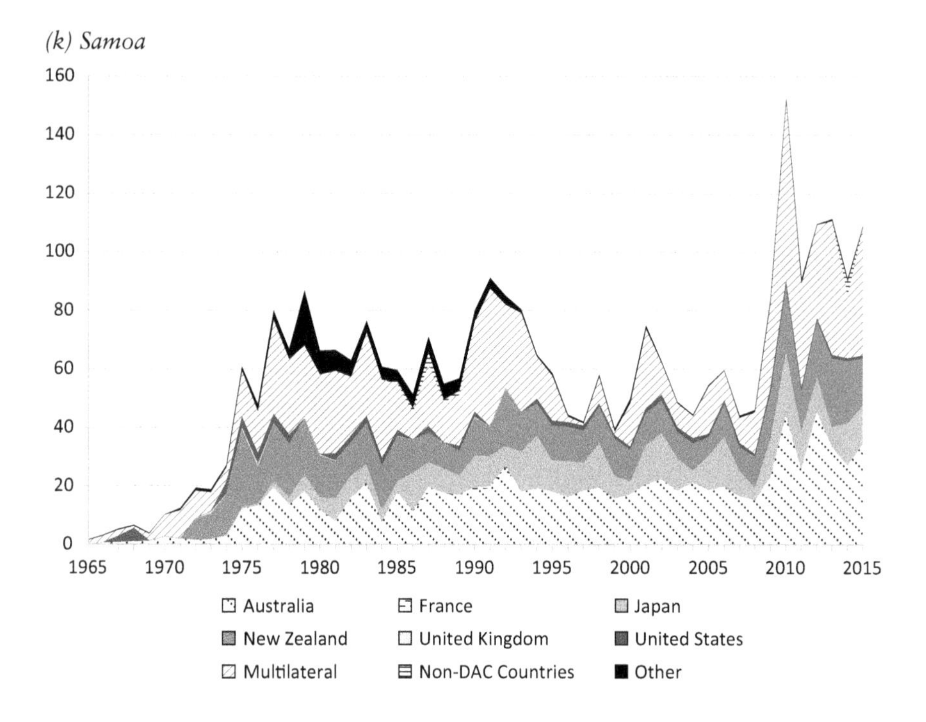
(k) Samoa
160
140
120
100
80
60
40
20
0
1965
1970
1975
1980
1985
1990
1995
2000
2005
2010
2015
Australia
France
Japan
New Zealand
United Kingdom
United States
Multilateral
Non-DAC Countries
Other

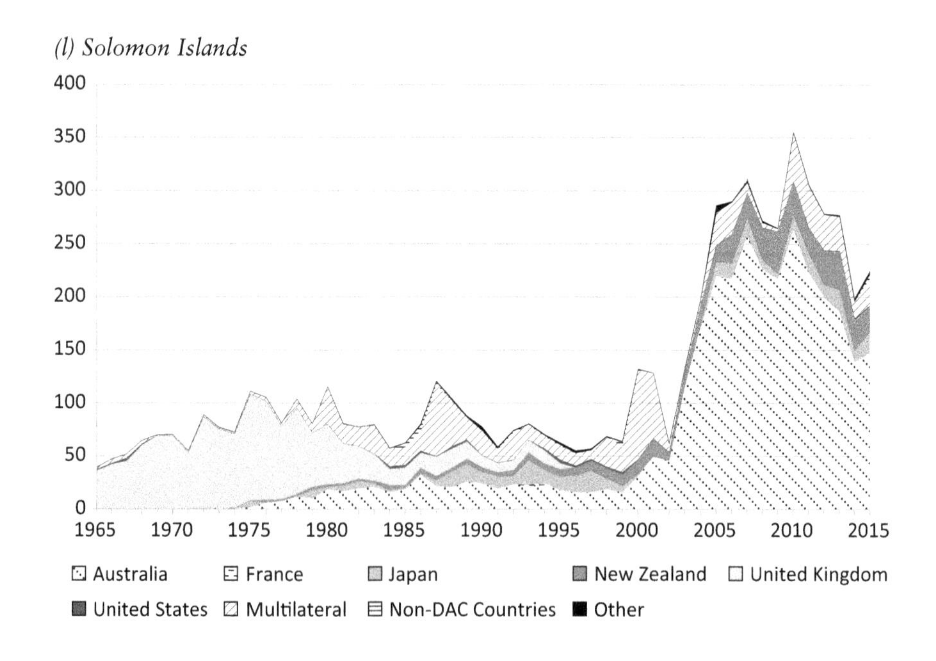
(l) Solomon Islands
400
350
300
250
200
150
100
50
0
1965
1970
1975
1980
1985
1990
1995
2000
2005
2010
2015
Australia
France
Japan
New Zealand
United Kingdom
United States
Multilateral
Non-DAC Countries
Other

(m) Tokelau

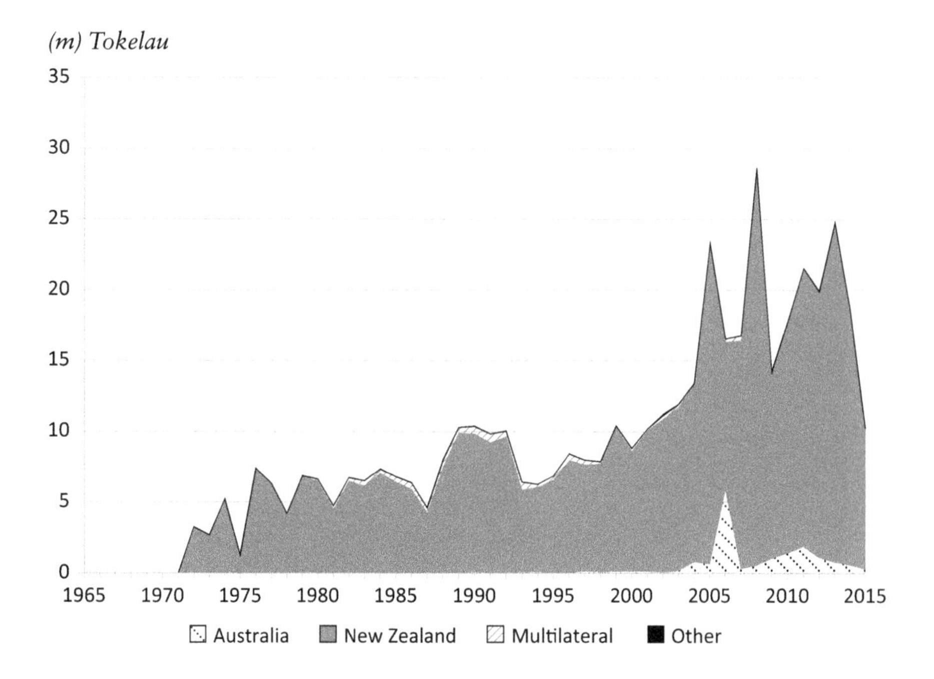

(n) Tonga

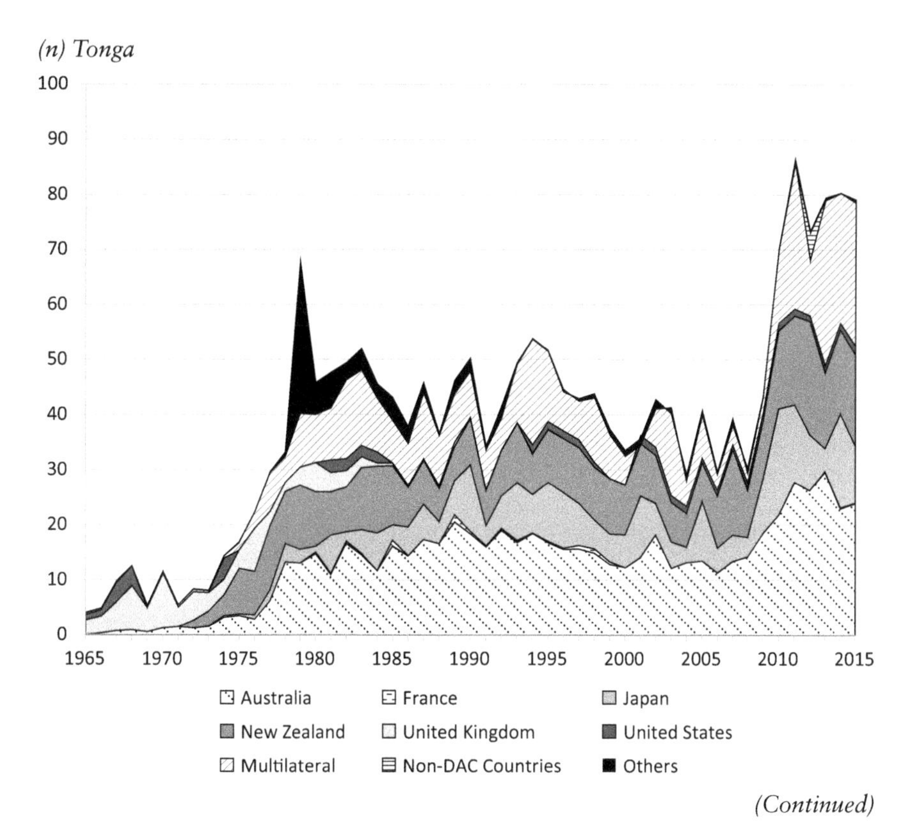

(Continued)

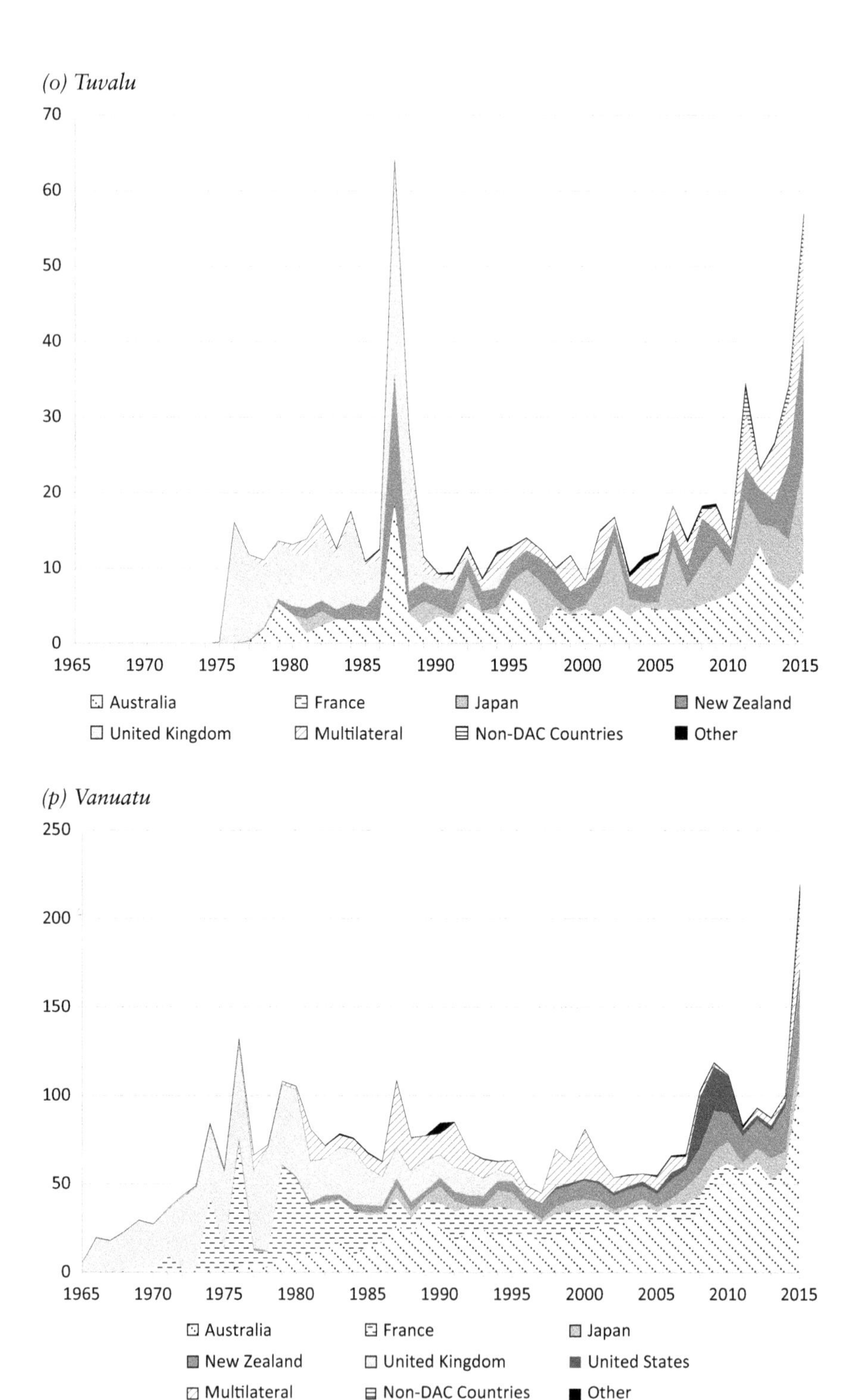

(o) Tuvalu
70
60
50
40
30
20
10
0
1965
1970
1975
1980
1985
1990
1995
2000
2005
2010
2015
Australia
France
Japan
New Zealand
United Kingdom
Multilateral
Non-DAC Countries
Other
(p) Vanuatu
250
200
150
100
50
0
1965
1970
1975
1980
1985
1990
1995
2000
2005
2010
2015
Australia
France
Japan
New Zealand
United Kingdom
United States
Multilateral
Non-DAC Countries
Other

(q) Wallis and Futuna

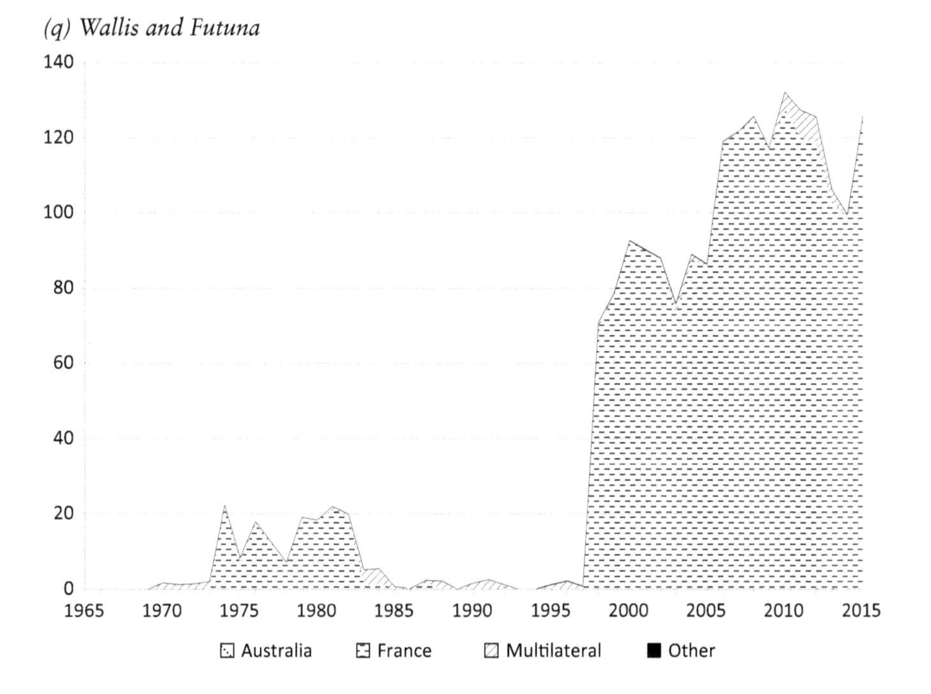

(r) French Polynesia

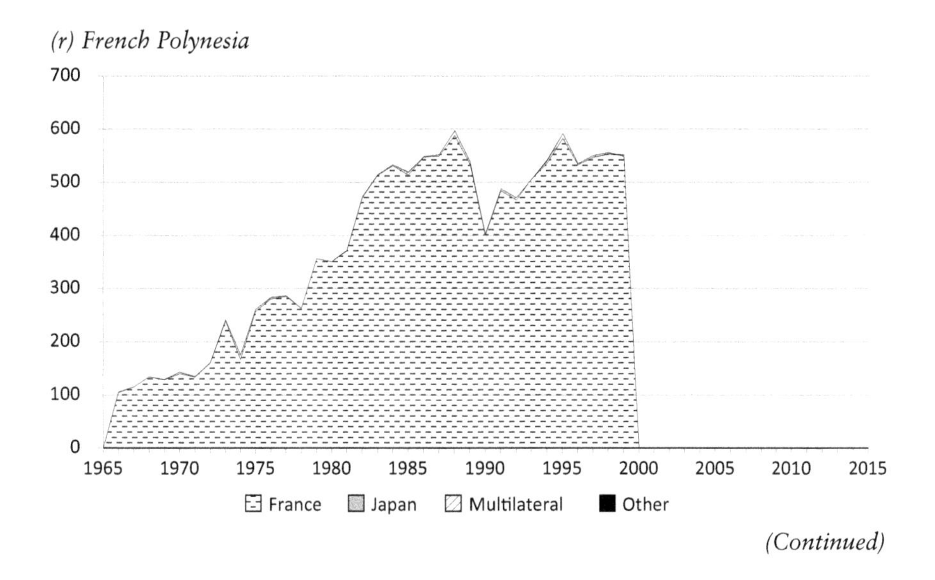

(Continued)

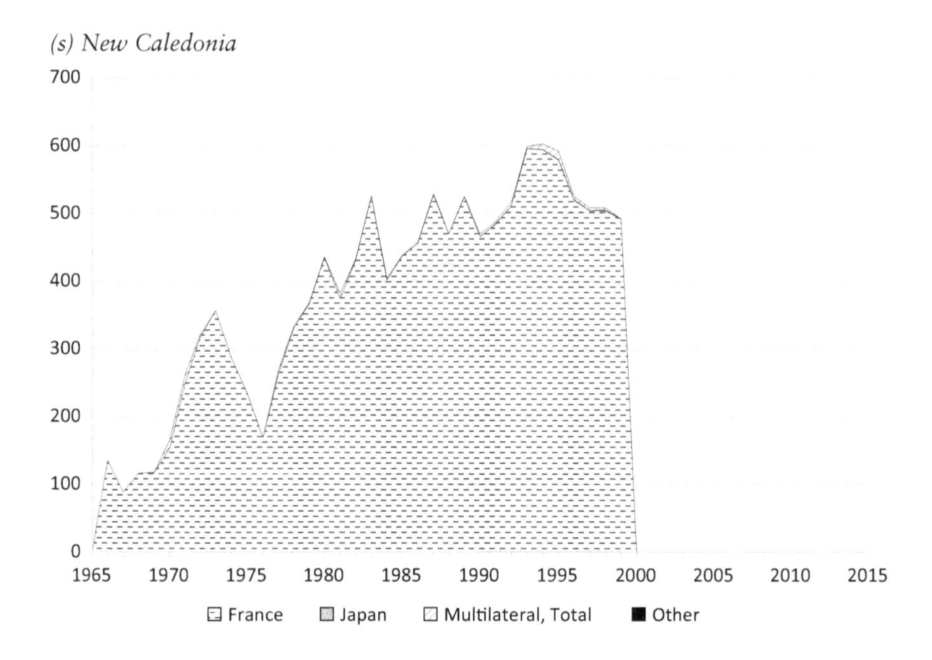

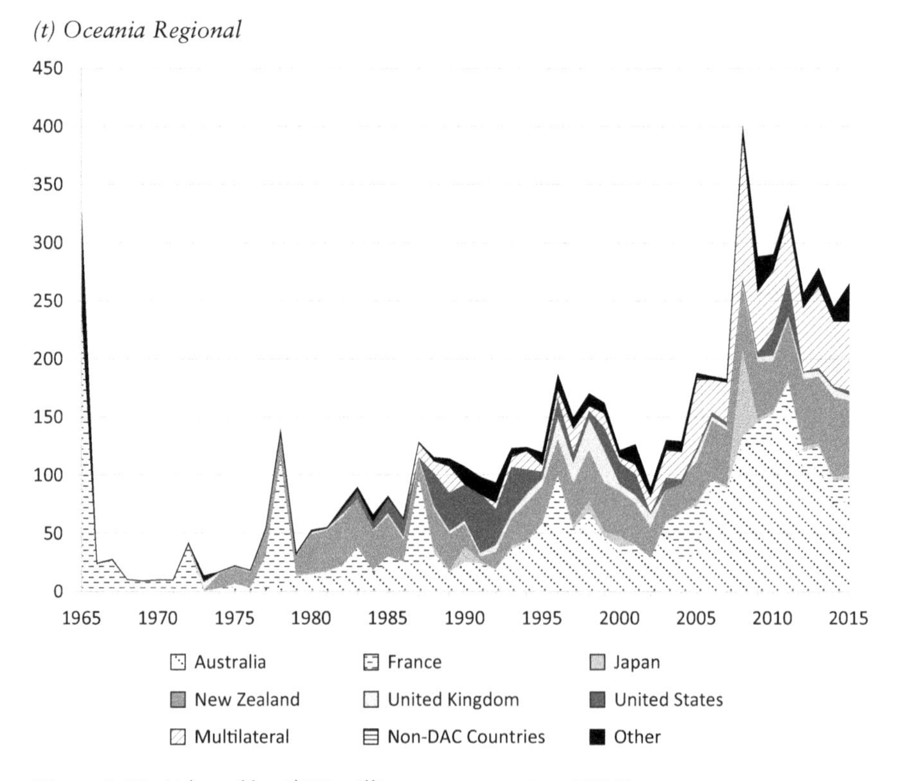

Figure 3.12 Aid profiles ($US million constant prices 2014)

similarity across the profiles of twenty countries depicted in Figure 3.12. Each has experienced different relationships with donors; many have had periodic disasters which have spiked the aid volumes in certain years; and nearly all have experienced the effects of major shifts in the aid paradigms of donors.

Whilst much of the variability across Pacific aid recipients has to do with changes in donor policies (and some reflect arbitrary changes in ODA accounting (as in French Polynesia and New Caledonia after 2000), it is possible to suggest that varied strategies have been adopted by Pacific countries and territories. Some (Wallis and Futuna, Tokelau, FSM) depend very heavily (over 80%) on a single donor. This may have significant benefits in terms of high per capita aid volumes and other special relationships with the metropole, but it also creates a high level of dependence and vulnerability if that donor shifts their policy sharply. If we examine another group (PNG, Niue, Marshall Islands, Nauru) that also rely on a major donor (for between 70% and 80%of ODA receipts) we can see that dependence does not automatically result in certainty. Here we see examples of major fluctuations from year to year (as with Niue), long-term decline (PNG) or periodic shifts related to specific donor requirements (Nauru).

Relatively few Pacific countries have been able to develop a diverse portfolio of donors. This group (Samoa, Tonga and Tuvalu – and less so Fiji) do not have close ties to a single metropolitan patron (though there may be some special relationships as with Samoa and New Zealand) and their per capita receipts are lower than those who do. Furthermore, such diversity can result in higher costs of managing and harmonising aid portfolios across several significant donors. Yet, as we will explore in Chapter Six, such a relative lack of dependence may provide more independence and more space to assert local development objectives.

Overall then, Pacific recipients have experienced quite different trajectories and sources of aid. There has been interplay between changing global and donor aid regimes and policies and various local responses and strategies. Some are more constrained than others in their opportunities to negotiate favourably with a range of donors, but this may come with the benefit of greater ODA volumes. Others, as a result of different colonial histories, political regimes or geopolitical strategies, have been caught with fewer options to link with a single generous patron but this has allowed arguably for a more assertive and independent set of relationships with the outside world. Yet for no recipient in the Pacific has there been an easy, stable and reliable stream of aid, despite the fact that some are amongst the highest recipients of aid in the world.

Beyond total flows – per capita aid trends in the Pacific

Hitherto, our analysis of aid flows in the Pacific region has focused on total flows. This reveals a particular geography of aid in the region that centres on the dominance of several large donors (Australia, France, the USA, Japan and New Zealand) and a few large recipients (PNG, Solomon Islands, Vanuatu, FSM and Marshall Islands as well as French Polynesia and New Caledonia if we accept estimates of actual flows of development and welfare assistance). Yet these data present a somewhat misleading view of aid in the region, especially if we are interested in relative levels of aid receipts and aid dependence. Given the vast contrasts in population sizes amongst the countries and territories of the region, from perhaps around seven million people in PNG to barely one thousand people in Tokelau, total flows of aid do not uncover the subtleties of the geographies and politics of development aid. An analysis of aid receipts per capita is an important additional indicator of the impact of development assistance.

Table 3.2 and Figure 3.13 present data on per capita ODA. The resultant map of aid in the Pacific is strikingly different to that depicted by aggregate aid flows. In the latter, the largest country in the region, Papua New Guinea, dominates and the smallest (Tokelau) barely figures. Yet their relative positions are all but reversed in the per capita map. PNG receives the smallest volume of ODA per capita (less than $US 100 per head per year) whilst Tokelau receives the largest (nearly $US 17,000). There appears to be a strong inverse correlation between size (population) and aid per capita.

Thus, we can sustain the argument that, in general, smaller territories as sovereign or semi-sovereign entities receive much higher levels of aid whilst those who depend on a single large donor appear to be better off than those who rely on several sources of aid. Furthermore, the data shown in Figure 3.13 also indicate how this link between aid per capita and political status is closely drawn. Those states similar to Tokelau – small dependent states incorporated or closely integrated with metropolitan patrons (Wallis and Futuna, American Samoa, and Niue[9]) – all receive levels of aid per capita that are very high (around or above $US 10,000 per year). There is another category of countries (Palau, FSM, Marshalls, Cook Islands) that have a similar if slightly looser close association with a metropolitan power (usually in the form of a compact of free association) that also have high levels of aid per capita (all above $US 1,000 per head).

The only real anomaly in this sense (countries with aid per capita receipts over $US 1,000 per year but with a relatively high level of political independence) are Tuvalu (with a small population but able to attract a range of donors) and Nauru (high levels of Australian aid there arguably have resulted directly from Nauru's decision to host refugee seekers for Australia). Elsewhere, independent states receive less aid per capita. Samoa and Tonga appear to be relatively successful in having a diversified range of donors and moderate aid levels, but the large independent states (PNG, Fiji, Vanuatu – even Solomon Islands despite the heavy RAMSI attention) are the smallest recipients in per capita terms.

Thus, in analysing aid flows in the Pacific, whether in aggregate or per capita terms, we see a strong correlation with the themes regarding sovereignty that were raised in Chapter Two. States that maintain a higher

Table 3.2 ODA and ODA per capita by recipient
($US million constant 2013 prices, average per annum 2010–2014)

	Population (estim 2013)	*ODA per capita ($US)*	*% from largest donor*
American Samoa	56,500	0	100
Cook Islands	15,200	1386	69
Federated States of Micronesia	103,000	1158	83
Fiji	859,200	102	51
French Polynesia	261,400	0	100
Guam	174,900	0	100
Kiribati	108,800	536	58
Marshall Islands	54,200	1282	79
Nauru	10,500	2901	79
New Caledonia	259,000	0	100
Niue	1,500	12193	75
Northern Mariana Islands	55,700	0	100
Palau	17,800	1453	49
Papua New Guinea	7,398,500	82	76
Samoa	187,400	615	36
Solomon Islands	610,800	483	73
Tokelau	1,200	16657	93
Tonga	103,300	772	34
Tuvalu	10,900	2472	35
Vanuatu	264,700	372	60
Wallis and Futuna	12,200	9636	96

Source: OECD. Stat 1 June 2016

degree of political association (as dependent territories or degrees of special political status) clearly receive higher levels of per capita aid than those who have acquired full political independence. There is also a correlation with size – smaller states receive more aid per capita than larger ones. However, this association is not as strong as that revealed for sovereignty and per capita aid for we can see with larger and mid-sized countries such as French Polynesia, New Caledonia, FSM or Samoa that higher levels of dependence do compensate to some degree in the form of higher levels of aid per person.

In summary, then, although the picture of aid flows in the region has become more complex, with more donors in evidence and the distribution of aid being less dominated by a few major recipients (PNG and the French territories), there has been a fairly consistent pattern of aid to the region. With the exception of increases in the first decade of the new millennium, volumes in real terms have been relatively even, albeit with occasional episodes of cutbacks. Australia and France have been the major donors, with significant contributions from the USA, Japan and New Zealand. The clearly drawn geographies of ODA in the Pacific show that financial flows of aid are channelled through well-worn paths constructed over decades of political and economic interaction. They both reflect colonial pasts and postcolonial

Figure 3.13 Total aid per capita in the Pacific, 2010–2014

renegotiations, yet they also help construct and strengthen a political geography of the region criss-crossed by complex political relationships and continually reworked understandings of sovereignty.

South-South cooperation: the example of China

The well-established patterns of aid flows in the region, particularly as represented in DAC data, have begun to change and there is increasing awareness of the presence of new donors (or 'development partners') operating outside of the parameters of the established aid landscape. In recent years, the growing visibility of Chinese investment and aid in the Pacific has received increased critical attention (Ratuva 2011; Powles 2016). A wider global literature has also seen such activity in other parts of the world, such as sub-Saharan Africa, and there has been a tendency to portray China and other non-traditional aid donors such as India, Saudi Arabia, Indonesia, Russia and Brazil, as 'not playing by the rules', pursuing forms of aid which appear to be more motivated by self-interest than altruism (Mawdsley 2010). The fact that many of these donors have decided to stay outside of the OECD-DAC set of agreements, such as the Paris Declaration, has heightened a sense of aid behaviour that is driven by its own political logic not that of the OECD 'donors club'.

In the Pacific region, such a portrayal is not new. In the past, diplomatic manoeuvring by the former Soviet Union (as with apparent offers to build an international airport in places such as Tonga in the 1970s) was met with a swift response from Western donors and efforts to persuade Pacific countries to decline such offers. Similarly, some forms of Japanese or Taiwanese aid have been treated with scepticism by other donors who portrayed some aid as efforts merely to 'buy' the political support of Pacific countries to, say, recognise Taiwan (the Republic of China) in diplomatic circles rather than the People's Republic of China or vice versa. Japanese initiatives were also seen as vote-buying efforts to secure support on international bodies such as the International Whaling Commission. Again though, aid and political deal making has hardly been the preserve of a few non-Western donors. France was accused of trying to improve its tainted post nuclear testing image in the Pacific by stepping in to provide assistance to post-coup Fiji in 1987 when other donors instituted a boycott. New Zealand, too, has often attempted to speak in the name of small Pacific Island states in furthering its own political objectives, whether in opposition to French nuclear testing in the 1970s and 1980s or in attempts to secure a seat at the Security Council of the United Nations in 2014. The use of aid as a political and diplomatic tool, then, is not new and the portrayal of the motives behind it is heavily tainted by the political-economic positioning of these donating it and those making comment about such donations.

What has become increasingly evident however in the case of China – and not exclusively China – is the tying of aid to broader economic objectives or commercial interests of the donor country. Much Chinese aid and investment in the Pacific region and elsewhere has been seen in the last decade as being linked to infrastructure and construction projects in particular. The sight of Chinese contractors building new government offices, sports complexes, roads or ports has become common in Fiji, Samoa, Papua New Guinea and elsewhere. Some of these are portrayed as 'aid' with concessional loans or grants from China being used to fund them, even if the materials, plant, expertise and often labour comes from China. They have helped promote the activities of Chinese construction firms: work on such aid projects has allowed them to establish themselves as potential bidders for projects that are on offer on the free market.

It is true that much of this activity is not 'aid' as defined by ODA or delivered through the OECD-DAC principles of established aid practice (e.g. the desirability of untied aid, grants are preferable to loans, etc). Yet, as discussed in Chapter Two, the recent approach by China towards South-South cooperation should not be framed and judged within a DAC setting: it is a new[10] form of interaction between states that does involve some concessional elements and grants for development purposes, but is also unashamedly about mutual benefits and wider economic interests – and this objective is not really that far away from the current retroliberal aid regime as practiced by DAC countries (see Murray and Overton 2016a). Indeed, a Pacific context, the latter is arguably partly in response to the increasingly active role of China in development projects.

Because Chinese 'aid' is not quantified through the DAC system, and much detail is not in the public domain, it is very difficult compare China with other donors or even see this form of cooperation in an aid framework. Nonetheless data collected by the Lowy Institute provides some basis for a depiction of the geography of Chinese aid in the Pacific (Table 3.3 and Figure 3.14- see Brant 2016). Although there are obvious problems in comparing the data with that of DAC donors, we can see a picture of a donor with a much more dispersed strategy and a less marked sphere of influence than other donors (Liu 2016; Wesley-Smith 2016). There are several countries that recognise the Republic of China and do not receive PRC attention in the form of loans and infrastructure projects (Solomon Islands, Tuvalu, Kiribati, Nauru, Marshall Islands and Palau). And while China's aid is not in evidence in the Pacific islands associated with the USA, China is otherwise relatively active across the region – including opening commercial collaboration in the metallurgic industry with the Kanak-dominated Government of New Caledonia's North Province (Prinsen and Blaise 2017).

Box 3.2 Tuvalu, sovereignty and the Republic of China

John Overton

Tuvalu gained full independence in 1978. It had been part of the British protectorate of the Gilbert and Ellice Islands but decided to become a separate country with full sovereign rights. Independence has meant that it has taken on the apparent burden of maintaining the full apparatus and costs of statehood with a population base of fewer than 10,000 people. It has also meant that it does not have access to the patronage of a metropolitan power – in the form of high aid levels and migration access – that free association or compact of association status might have brought (though this was not on offer for Tuvalu).

However, independence has also brought benefits – not just the satisfaction and pride of being fully self-governing. Tuvalu was able to convince donors (initially UK, Australia, New Zealand and the Government of Tuvalu itself – and later others) to contribute to the Tuvalu Trust Fund which began in 1987. This has been wisely managed and returns a dividend to the country to allow it to meet around a quarter of the costs of government in most years. It has also used its status as a country to gain revenue from the sale of stamps to collectors worldwide, its sovereignty over a large marine economic zone has led to income from the sale of fisheries licences, and it has sold rights to its national internet domain name (.tv).

Sovereign status has led to other opportunities. Tuvalu became a member of the United Nations in 2000. Its vote in the UN General Assembly then is theoretically equal to states with populations a thousand and more times the size of Tuvalu. Its vote has value. And independent status has meant the country has been able to lobby effectively on issues such as global climate change.

Tuvalu has had diplomatic relations with the Republic of China (Taiwan) for nearly forty years and has supported Taiwan internationally (such as its bid to join the United Nations). This has put it offside with the People's Republic of China which does not wish to see Taiwan with separate sovereign status and recognition. Taiwan has proved to be a grateful partner. It maintains the only full diplomatic post in Funafuti and Tuvalu's post in Taipei is one of only three it supports overseas. The extent of Taiwan's development assistance to Tuvalu is not clear – it is not counted by the OECD nor is it explicit in Tuvalu's national accounts. Yet the physical manifestation is evident in a number of projects, most notably Tuvalu's prominent and modern three storied government building in the centre of Funafuti. Other projects include contributing to Tuvalu's goal to switch entirely to renewable energy by 2020 (Radio New Zealand 2017).

In sum then, Tuvalu may not reap the material benefits of aid and access to remittance income through migration that countries such as Cook Islands or Wallis and Futuna might – and its per capita incomes are lower perhaps as a result. Yet it has been able to sustain its independence and place in the world by using the value of its sovereign status in careful and strategic ways.

In assessing the impact of China's aid in the region, it is difficult to compare directly with the main DAC donors and statistical comparison is highly problematic, given the problems regarding definition and the likely public under-reporting of Chinese donations. Nonetheless, if we use the Lowy data, based as it is on publicly reported 'investments' by China in the Pacific, we can begin to gain an impression of relative involvement by China. Firstly, it is necessary to differentiate, as the Lowy Institute data allows, between outright grants and concessional loans. Table 3.3 presents data for the DAC major donors and China. Even accounting for the likely under-reporting of data for grants from China, it is probable that China is a relatively small grant donor compared to Australia, New Zealand and Japan. Across the thirteen Pacific recipient countries monitored by the OECD, the volume of Chinese grants was about a quarter of those of Japan and New Zealand and around 4%of Australia's. Even if we discount the Pacific countries China does not give grants to, the picture is similar. For Cook Islands, Fiji, Samoa, Tonga, Vanuatu, PNG and FSM, the amount of Chinese grants per year over the 2010–2014 period was about 40%of the grants from Japan and New Zealand and a mere 6%of Australia's grants.

However, this measure of Chinese involvement in aid in the Pacific is rather misleading, though it does throw into question some of the more alarmist views of the recent role of China in the region. If grants are relatively small, China is much more active as a lender. Its concessional loans are significant,

Table 3.3 China and selected major donors to Oceania 2010–2014 (annual average 2010–2014 $US million current prices 2014)

	China (PRC)	*Australia*	*New Zealand*	*Japan*
Cook Islands	1.9	4.3	13.9	0.2
Fiji	6.3	44.9	5.5	14.1
Kiribati		28.0	10.2	9.3
Marshall Islands		4.0	0.1	9.2
FSM	4.6	3.7	0.1	16.3
Nauru		24.1	1.9	1.4
Palau		3.3	0.0	8.9
PNG	6.1	457.8	23.1	7.0
Samoa	6.7	37.2	18.1	16.5
Solomon Islands		213.8	27.3	18.0
Tonga	6.0	27.3	14.8	15.5
Tuvalu		9.4	4.7	7.4
Vanuatu	3.5	58.5	16.4	11.3
Total	35.1	916.3	136.1	134.9

Source: OECD: Stat for Australia, New Zealand and Japan (DAC data)

Data for China extracted from Lowy www.lowyinstitute.org/chinese-aid-map/# accessed 3/5/16 only includes disbursements for the period 2010–2014 (to be consistent with other donors above). Grants only. Projects excluded where no amount is given.

and it is these that have received most media attention, especially with regard to the possible difficulties Pacific borrowers may have in repaying them. In this case, a suitable comparison can be made with other major lenders in the region, specifically the Asian Development Bank and the World Bank (Figure 3.14). Firstly though, we can note that the ADB, like China, gives grants as well as concessional loans; in both cases these are often linked so that a grant may be used to complete some aspects of a large project which will then be supplemented by a large loan component.

Over the period 2010–2014 (as much as can be ascertained from the Lowy Institute data) China gave $US 35.1 million in grants whilst the ADB gave $US 39.7 million (albeit to nearly double the number of countries).

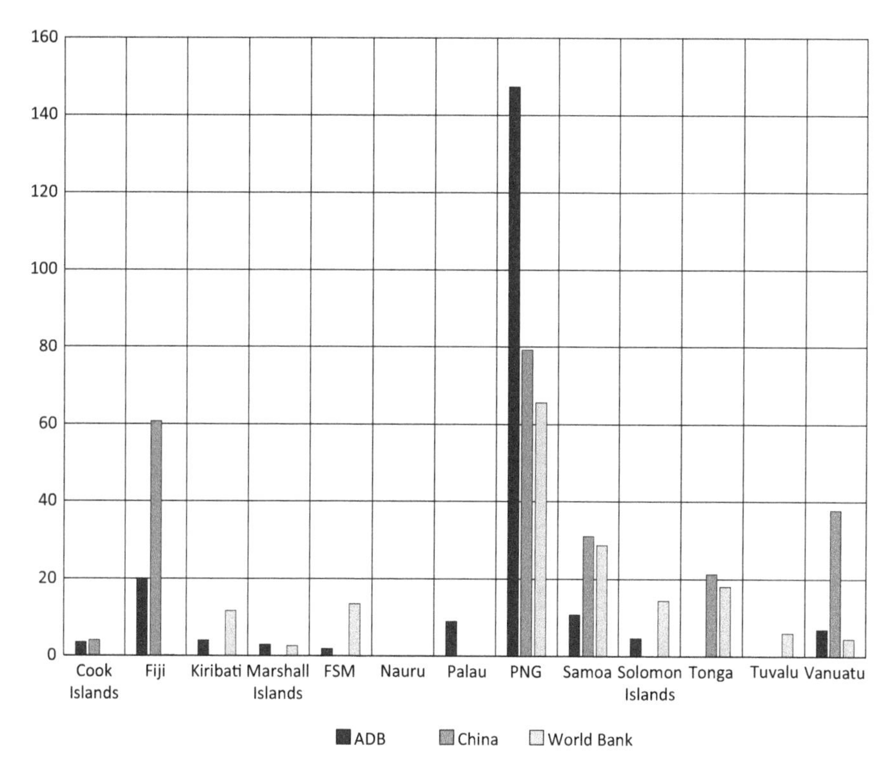

Figure 3.14 Grant assistance and concessionary loans from China and development banks 2010–2014

($US million current prices 2014)

Note: Data will only be partial and not 'DAC-able'. (see Lowy for how data collected)

Data extracted from Lowy www.lowyinstitute.org/chinese-aid-map/# accessed 3/5/16 only includes disbursements for the period 2010–2014 (to be consistent with other donors above). Amounts for concessionary loans includes just the principal amounts – no data on amount of concession – included if loan commenced or was included in 2010–2014 period – for full amount of loan. See ADB 2016.

Projects excluded where no amount is given.

Figure 3.14 shows that China was a significant grant donor compared to the ADB in several countries (Cook Islands, Fiji, FSM, PNG and Vanuatu) but ADB was a larger donor in Samoa and about the same in Tonga. For concessional loans (Figure 3.14) China again is much more significant: its average of $US 234 million exceeds both the ADB ($US 210 million) and the World Bank ($US 165 million). Furthermore, given the smaller number of countries it lends to, it has become by far the largest lender to Fiji and matches the World Bank in Tonga and Samoa and the ADB in Cook Islands. Yet, although its loans to PNG are very large (over $US 80 million per year) this is still less than loans from the ADB ($US 147 million) and it is apparently only present as a lender in five of the twelve Pacific countries recorded as concessional loan recipients.

Thus, China has become a major actor in the Pacific aid landscape. It is far from a traditional 'aid' donor, though its grants are locally important, if relatively minor in aggregate. Instead it operates more as a large commercial assemblage: its government provides very large loans to certain countries at concessional rates, largely for big infrastructure projects, and these are used in large part to pay Chinese corporations to undertake construction. Rather than being greeted with alarm and cynicism, perhaps such operations should be seen (with caution) as a form of development cooperation that provides the region with resources for desired projects that are not always available from traditional donors, especially not through the bilateral programmes of established donors (see Iati 2016 for a more layered perception of Chinese aid in the region). In this sense, we should see China as sitting alongside the ADB or the World Bank as a significant lender, rather than framing it as an 'aid donor'. As such, China may well impose its own (possibly opaque) conditionalities in terms of economic or political concessions but, we might suggest, this may not be that different (in scale if not nature) compared to the loan conditionalities historically imposed by the IMF, World Bank or ADB.

Estimating non-DAC aid flows

Given this discussion on the role of China as a development partner in the region, it is now apposite to broaden our discussion to include a range of aid flows that similarly are not counted by the OECD. In doing so, we move from the seemingly precise (if not always consistent) measurements of ODA by the OECD to the realm of estimation and approximation. There are many potential financial flows which could be considered forms of 'aid' that are not quantified by the OECD such as trade concessions, migration access, superannuation payments, investment flows, private donations, family remittances, etc. These are simply too complex and numerous to quantify and compare in any meaningful way. However, we can begin to incorporate some elements of these other flows that we consider significant in the region in order to gain a better picture of official (government/ multinational agency to government) assistance. Here we embark on an exercise to estimate approximately a wider picture of aid flows. There are several elements to this overall estimate of aid to the region:

- *OECD data on ODA disbursements*
  These data (which we have used throughout this book) cover development assistance in uniform data sets and give an accurate picture of the flows of aid from major bilateral and multilateral donors and some private agencies. This we get a good estimate of the aid donations from Australia, New Zealand, Japan, the European Union, and multilateral agencies such as the Global Fund, the International Development Association and the Global Environmental Facility.
- *Estimates of Chinese (PRC) development cooperation.*
  These were discussed above and the data relies on the excellent data gathering exercise conducted by the Lowy Institute (Brant 2016) mainly dependent on in-country media coverage and government reports. The data can be split between grants and concessional loans and can be roughly annualised. Nonetheless, the data should be used with caution: it may underestimate the total flows on one hand, yet it might also overestimate the 'development' component of the flows on the other.
- *Other forms of South-South cooperation.*
  As well as direct financial assistance, some non-OECD countries do provide forms of development assistance to the Pacific Islands. These may involve small-scale technical assistance or scholarships for tertiary study from countries such as India (see Liu 2016 for China) and they remain generally small in volume and often ephemeral. However, one of the most prominent in the region has been the offering by Cuba of medical training for a number of Pacific countries (see Box 3.2). The financial amounts involved are almost impossible to quantify, especially given the low rates of remuneration in Cuba, yet the contribution of trained medical staff in significant numbers in the region is substantial, especially given the costs of such training in conventional Western institutions.

Box 3.3 South-South cooperation: Cuban medical training and the Pacific

John Overton

One of the features of the Cuban Revolution after 1959 has been its active and widespread programme to develop strong relationships with many non-aligned countries through its use of bilateral health assistance programmes. Asante et al. (2012: 2) claim that Cuba has the world's largest health cooperation programme "with more than 38,000 Cuban health workers in over 74 countries while hosting more than 20,000 students studying medicine in Cuba from 60 countries." Although Cuba has been active in this regard for some time, especially in Sub-Saharan Africa, its efforts in the Pacific have been relatively recent. Nauru received eleven Cuban doctors in 2004 (but these did not last due largely to language problems). Since then, Cuban doctors have been sent to Kiribati, Solomon Islands, Tuvalu and Vanuatu.

Of arguably more significance has been the training of Pacific Island medical students in Cuba and such students have been sent from Fiji, Kiribati, Solomon Islands, Tonga, Tuvalu and Vanuatu. Asante et al. (2012: 5) estimated that in 2010, there were 177 Pacific Island medical students studying in Cuba. The training such students receive is not the same as in medical schools in, say, Australia or New Zealand. Much emphasis is given to primary health care, and relatively 'low-tech' medicine, whilst there are efforts to train large numbers of such health workers so their impact on health care can be 'catalytic' (Negin 2012: 2). For countries such as Kiribati, Papua New Guinea and Solomon Islands, where medical staff are sorely lacking in more remote areas and there are only one or two doctors per 10,000 people in the countries as a whole, the prospect of many new doctors to tackle issues such as infant or maternal mortality, is highly appealing. Moves were made to extend the Cuban programme to Papua New Guinea in 2016, following a meeting in Havana between President Raul Castro and Prime Minister O'Peter Neill (*The National* 2016). This would involve the sending of Cuban doctors, the training of PNG medical students and possibly also the supply of Cuban-produced pharmaceuticals.

Cuban medical assistance has not been without criticism. The language issue has been highlighted: Spanish-trained doctors have to operate in an environment where vernacular languages and English are used in clinics, hospitals and dispensaries. Others have also pointed to the potentially very high costs to government of paying the salaries of many new doctors and one senior PNG specialist even described Cuban doctors as 'useless' in terms of general skills (ABC 2016).

Yet there are promising signs that Cuban medical assistance may become a valuable and continuing feature of health development programmes in the Pacific. Perhaps recognising the low cost of training and the skills Cuban-trained medical staff have (appropriate to primary and rural health care), Australia has indicated a recent willingness to work with and support the Cuban programme (Negin 2012), whilst New Zealand agreed to provide language training for Cuban doctors in the region (Locke 2015). Furthermore, the example of the large Cuban medical programme in Timor-Leste (where assistance is planned to evolve into the establishment of ongoing medical training within the country) could provide a model for Papua New Guinea and the Solomon Islands in particular to train a continuing and large stream of local health professionals to work in remote areas and tackle the significant primary health care issues at a scale that expensive and extensive Western-style medical training could not achieve.

- *Estimates of development assistance from Taiwan (Republic of China).* We can assume these flows are significant, especially to countries that have a diplomatic relationship with RoC, such as Solomon Islands, Kiribati, Tuvalu, Palau, Marshall Islands and Palau. However, despite the fact that Taiwan is listed as a non-DAC donor under the OECD rubric, little or no such data is collected and the amount of development cooperation funding from Taiwan remains extremely opaque. We include Taiwan in this picture just to note its presence and likely significance though we acknowledge that it cannot be adequately quantified.
- *Concessional loans from the World Bank and the Asian Development Bank.*
These are the two largest multilateral lenders to the region. Their development grant funding is captured in the OECD data and counted as ODA, but their loans are not (though there may be an element of ODA counting in the concessional elements of these loan packages – hence a potential of double counting and overestimation here). Data on concessional loans from the two agencies are included here and annualised and averaged for the 2010–2014 period. As with Chinese loans, it should be noted that these loans, of course, have to be repaid and aggregating grants and loans overestimates the net flow of development assistance (compared to those donors who just use ODA). However, the loans from China, ADB and the World Bank are included here to give a picture of the total volume of financing (Figure 3.14).
- *USA federal funding to associated territories.*
Whilst US ODA to several territories (including its compact partners of Marshall Islands, FSM and Palau) are counted in the OECD data, flows to its more closely associated and incorporated territories of American Samoa, Northern Mariana Islands (after 1995) and Guam are not. These territories receive significant financial support from the US Treasury to run the everyday business of government and fund development activities such as infrastructure. American Samoa, for example received 65%of its total revenue from this source in 2013 (Office of Insular Affairs 2016: 2). Such 'aid' is direct and much more substantial (though less directly tagged to 'development') than the ODA to other Pacific countries. The funds are largely administered through the Office of Insular Affairs (OIA) of the Department of the Interior. Whilst the OIA has a development budget, total federal funding is much greater. We use the data for federal grants for these three territories for 2013 (see

also Furlong and Hamano 2014).[11] This is the amount flowing directly from the US Treasury into the territory government budgets but it might not include other US government spending in the countries, such as federal salaries and direct federal expenditure. Despite these problems of estimation, overall, we believe the rough estimate of about $US 600 million of US federal funding flowing into the three territories is reasonable and it can be seen alongside the $US 191 million flowing into the rest of the region.

- *French funding for New Caledonia and French Polynesia.* The French government directly maintains the governments of these two territories and funds a wide range of budget items and development projects. Until 1999, such funds were included in the OECD data and those Wallis and Futuna still are. It is almost impossible to estimate these actual flows today. There are many financial flows to the countries from metropolitan France, but the two main channels involve firstly the direct flow from the French Treasury to the recurrent budgets of the two countries and secondly, a range of financial support and tax exemptions for development activities. One official estimate of these flows suggested a net transfer of resources to New Caledonia of $US 1.72 billion in 2013 (Box 3.1 above). However, given the difficulties of collecting such data and attributing to it a 'development' value, here we resort to assuming that the ODA recorded as going from France to New Caledonia and French Polynesia in 1999 (the last year it was recorded by the OECD) could be extrapolated forward by converting that 1999 amount to current values. Such an assumption is unsatisfactory given the time difference – it is probable for New Caledonia at least that the volumes have grown in real terms – but is does give some indication that nearly $US 1.2 billion goes each year from metropolitan France to the two countries as some form of 'aid'.

Putting these estimates together is problematic. The amounts we record are not precise and they compare some quite different forms of funding and some major assumptions have to be made. Nonetheless, Table 3.4 attempts to bring together these disparate estimates.

Bearing in mind the above cautions regarding the lack of accuracy in much of the data, we can however suggest a rough 'league table' of aid donors in the Pacific:

- The major players are the USA, France and Australia with funding to the region of $US 800 million to about $US 1.2 billion per year. For Australia this is distributed through conventional ODA programmes and is reasonably widespread through the region, though the Australian 'sphere of interaction' in the Western Pacific is quite marked (see above). France and the USA however distribute their assistance in different ways, mostly through non-DAC forms to the budgets of their former and current territories. France has three main recipients in the region (New Caledonia, French Polynesia and Wallis and Futuna) who each receive very large amounts of French funding. The USA has special relationships with closely integrated territories (Guam, American Samoa and Northern Marianas) as well as three associated states (Palau, FSM and Marshall islands). The first group receive very high levels of funding. As with France, aid from the USA is not widely dispersed in the region beyond these closely associated territories.

- The big lenders are the ADB, the World Bank and China. These donors distribute most of their assistance in the form of soft loans, though grants may also be used. The purpose of the loans is predominantly to fund development, particularly infrastructure. In the case of the ADB and the World Bank, these loans are reasonably transparent, and they fund the operations of local governments and both local and international companies. In the case of China, the loans usually are tagged to employing Chinese companies to undertake development work (mainly the construction of roads, buildings and telecommunications). In all cases, loans are given with the expectation of repayment, but concessional elements exist to lower interest rates or extend repayment terms.
- The medium-sized bilateral donors. This group includes New Zealand, Japan and the European Union (and its component bilateral programmes). These aid flows are included within the rubric of the DAC and generally follow conventional (Western) aid practices over time. They are mostly quite widespread throughout the region, although New Zealand does have an apparent 'sphere of interaction'. We might include Taiwan (RoC) in this group. This would be on the basis of what might be assumed to be reasonably comparable total volumes of development assistance (perhaps of the order of $US 50 million per year though we can only guess at this). However, Taiwan, like China (PRC) operates in practice outside the OECD aid system and we know little of how aid is negotiated and distributed – or what is expected in return.
- A large group of relatively small bilateral, multilateral and private donors. These donors generally distribute less than about $US 20 million per year – and often much less. They are widely spread through the region, they have a wide range of priorities and practices and the funds and conditions they bring often impose a new layer of compliances – as well as some often welcomed funds – on stretched local agencies and communities.

Other forms of 'aid'

As indicated earlier in this chapter, aid defined by ODA and the DAC is but a part of the picture of the way development and welfare assistance is conducted in the region, albeit its most transparent and quantifiable form. In this section we outline other forms of 'assistance' that have an important impact on Pacific islands' development choices but are somewhat more diffuse and

Table 3.4 Estimated aid by donor to the Pacific, 2014 (annual $US million current prices c.2014)

Donor	*ODA average annual 2010–2014 (as recorded by OECD)*	*Other financial assistance (estimates)*	*Concessional Loans*	*Estimate of total 'aid'*	*Notes*
France	134	1040		1174	Other assistance is an estimate of funds from metropolitan France based on 1999 aid returns to New Caledonia and French Polynesia adjusted to $US million current 2013
Australia	1045			1045	
USA	191	623		814	Amount of federal grants for Northern Marianas, Guam and American Samoa for 2013 from OIA data etc.
Asian Development Bank	61		211	272	ODA from OECD; concessional loans from ADB (may be overestimate if concession included in ODA)
China (PRC)		35	213	248	Annual average (from Lowy Institute data)
New Zealand	219			219	
World Bank (incl International Development Association)	49		165	214	
Japan	139			139	
European Union	102			102	
Global Fund	31			31	
Global Environment Facility	13			13	
Germany	10			10	
UK	8			8	

Sources: OECD. Stat 1 May 2016, Office of Insular Affairs (2016)

hard to quantify: migration, trade preferences, 'non-DAC-able' financial flows and new forms of investment, and 'South-South' development cooperation. The underlying geographies of such aid are much more opaque and such a discussion is necessarily different in terms of the detail of analysis possible. In order to put the more formal ODA flows into critical perspective we need to at least reflect on the parameters of these less obvious forms of aid.

Migration and remittances

For many economies of the Pacific, the contribution of remittances from migrants working overseas considerably outstrips income from ODA. Yet opportunities for Pacific Island people to exploit opportunities for work and income from overseas are highly uneven and, as with aid, often proscribed by political and historical relationships with larger economies and current or previous colonial metropoles. Furthermore, the importance of remittances has led to the explicit framing of work and temporary migration permits as a form of development assistance in the programmes of a number of donors. It is also important to note that debate continues regarding the long-term benefits and costs of migration and remittances in the region (Gibson and Mckenzie 2012; Connell 2008).

Table 3.5 outlines some of the broad parameters of migration options and remittance flows for various Pacific Island countries and territories, some of which are amongst the most dependent on remittances in the world.[12] For 2009, for example, the World Bank calculated that of the world's top ten remittance recipients (by percentage contribution to GDP), four were in the Pacific region, with three ranking in the top five: Tonga (27.7 % – the highest in the world), Samoa (22.3 % – ranked second), Kiribati (6.3% – fifth) and Fiji (3.4% – seventh).[13] Yet this analysis did not include islands integrated as territories of a metropole, for which data are often not available. Thus, we might suggest that territories such as American Samoa, Guam, Wallis and Futuna, Tokelau and Niue could well have remittance contributions approximating or exceeding these. In Table 3.5 we have noted the migration options available to given territories and countries and provided a normative comment on the speculated proportional role of remittances in the economy. As noted there is no reliable single data set for such flows, and in some cases the data simply do not exist.

Table 3.5 Migration options and estimates of the proportional role of remittances

Country/ territories	*Migration opportunities*	*Speculated Proportional Role Remittances in the economy*
American Samoa	Rights to enter and work in the USA freely. Recruitment to US military is significant	Very high

(*Continued*)

Table 3.5 (continued)

Country/ territories	*Migration opportunities*	*Speculated Proportional Role Remittances in the economy*
Cook Islands	Full rights as New Zealand citizens to enter New Zealand (and Australia)	Very high Significant return migration
Federated States of Micronesia	Rights to enter and work in the USA under Compact agreement as 'non-immigrants'. US military recruitment access.	High to very high
Fiji	Limited formal arrangements but historical migration flows to NZ, Australia, Canada and UK.	Relatively high. Sources include permanent migrants and involvement in military and nursing sectors overseas
French Polynesia	Full rights as French citizens to enter France and some other French territories	High
Guam	Rights to enter and work in the USA under Compact agreement as citizens (unincorporated organised territory). US military recruitment access.	High to very high
Kiribati	Limited options but involvement in seafaring work overseas	High. Some aid-related assistance with seafaring training
Marshall Islands	Free association with the USA, movement and employment and military opportunities as 'non-immigrants'	High
Nauru	Limited but some visa links to Australia	Low to moderate
New Caledonia	Full rights as French citizens to enter France and some other French territories	Moderate
Niue	Full rights as New Zealand citizens to enter New Zealand (and Australia)	Very high Significant return migration
Northern Mariana Islands	Commonwealth in union with the USA, movement and employment and military opportunities through US citizenship	High
Palau	Compact agreement in free association with the USA. Entry to USA and work facilitated as 'non-immigrants'	High
Papua New Guinea	Few migration options, some elite migration to Australia	Very Low
Pitcairn Island	Overseas territory of UK and special arrangements with New Zealand for work	Moderate to high

Country/ territories	*Migration opportunities*	*Speculated Proportional Role Remittances in the economy*
Rapa Nui	Full rights as Chilean citizens – free flow to Chile	Likely high but also probably net flow out of remittances to the mainland to fund education
Samoa	Treaty of Friendship with New Zealand and some special quota for immigration. Historical flows of migrants and spread to Australia	Very high
Solomon Islands	Few migration options, some involvement in seasonal migration schemes	Low
Tokelau	Full rights as New Zealand citizens to enter New Zealand (and Australia)	Very high Significant return migration
Tonga	Some special quotas for immigration to New Zealand. Historical flows of migrants and spread to Australia	Very high
Tuvalu	Limited options but involvement in seafaring work overseas	High. Some aid-related assistance with seafaring training
Vanuatu	Few migration options, some involvement in seasonal migration schemes	Low but increasing in some communities
Wallis and Futuna	Full rights as French citizens to enter France, New Caledonia and New Caledonia	Very high

Box 3.4 COFA: Palau, the Republic of the Marshall Islands and the Federated States of Micronesia

Alexander Mawyer

The Compacts of Free Association (COFA) are complex bilateral agreements between the United States and three north Pacific Island states – Palau, the Republic of the Marshall Islands, and the Federated States of Micronesia (Yap, Chuuk, Kosrae and Pohnpei). Former colonial possessions of Japan, these three islands states were administratively acquired by the US Navy at the end of World War II as the Trust Territory of the Pacific Islands. US governance under the aegis of a United Nations trusteeship obligated the US to promote the territories' development and eventual independence. Important constitutional developments in the late 1970s and 1980s led to the establishment of COFA agreements in 1986 for the RMI and the FSM and in 1994 for Palau. Under the compacts, the US provides economic and social supports and largely unrestricted access to US immigration and markets in return for privileged relationships for American businesses and exceptional access to insular and marine territories for regional strategic defence purposes. Palau's proximity to East Asian markets and strategic waters, US use of Chuuk's extraordinary deep water lagoon, the ongoing importance of Kwajalein Atoll in the RMI as an active military base and missile and advance weapons research site, and the US' enduring responsibilities for those atolls used as nuclear test sites including what might be seen as insufficient ongoing payments to the RMI government, draw attention to some of the uneven dimensions of these relationships. As a central feature of these relatively recently self-governing states' emerging sovereignties, the COFA agreements exemplify some of the legal and practical murkiness of 21st-century postcolonial sovereignty arrangements arguably including potent examples of inverse sovereignty effects.

For instance, the exceptional ease of movements between COFA states and the US insular territories (Guam, CNMI), the state of Hawai'i, and the US 'mainland' has been a magnetic draw for many COFA citizens. Out-migration from this region has been constant and intense since the agreements' signing, particularly from the FSM and RMI. Today, large diasporic communities are well established in Guam, Oahu, Hawai'i, California, Oregon, Washington, Utah, Texas, Arkansas and elsewhere in the United States and various mobilities and flows back and forth between diasporic communities and home (is)lands for a central backdrop to life in this part of the Pacific.

Meanwhile, in COFA homelands a number of distinct political, economic and social phenomena demonstrate the complexity of contemporary trans- and international arrangements. For instance, in Palau, it has become increasingly clear that US aid and investment for superstructure projects is unlikely to be the primary vehicle of development and that the Asia Development Bank, the Japanese Special Fund, and other international partners will play a superordinate role in local economic development. Meanwhile US originating funds are notably entangled in the marine and terrestrial management of natural resources. From a certain point of view, the special arrangement between the US and Palau defined by COFA appears to enhance the likelihood of increased restrictions on local sovereign government for some of their own lands and waters. Other examples of such seemingly paradoxical arrangements can be identified. For instance, COFA citizens from FSM, Palau, or RMI regularly volunteer for service in the US military. US recruitment is a highly visible and active dimension of life for young persons across all of these islands as a path of social mobility, opportunity for individual growth and career, development, for education and travel. Many young men and women serve with distinction in one or more tours of duty in US 'hot' conflicts over the last decades. However, upon death, after disability due to service, or upon honourable discharge at the end of a period of duty, COFA persons are not entitled to the same death, VA, or national benefits of most other US citizen-soldiers. Finally, the COFA relationship appears to greatly diminish RMI agency in (re)negotiating compensation for US nuclear testing despite increasing publically available knowledge about the extent and depth of US malfeasance and damages both ecological and human in the Marshal Islands due to US nuclear testing. Largely atmospheric testing, between 1946–1958, including the infamous 1956 Castle Bravo test, spread a plume of radioactive contamination across a number of inhabited atolls including Bikini, Enewetak, Rongelap and Utirik for which COFA established a 'full and final settlement' of $US 150 million dollars. Compare, for instance, British Petroleum's $US 42.2 million dollars paid into the trust fund to compensate individuals, communities and businesses affected by the Deepwater Horizon oil spill.

For those who wish to explore some of these arrangements and histories in greater depth, several recent documentaries offer excellent starting points. Adam Horowitz's *Nuclear Savage: The Islands of Secret Project 4.1* (2011) documents the experience of Marshalese islanders as unwilling experimental 'test subjects' for the US Dept. of Energy as well as long-term mortality and health impacts from US nuclear testing. Set in the Marshallese and broader Micronesian experience on the Big Island in the US State of Hawai'i, Andres Williamson's *The Land of Eb* (2012) uses narrative fiction to capture the perverse and ambiguous status of COFA citizens who have migrated to Hawai'i where they experience intense and heart wrenching discrimination and racism including by indigenous Hawai'ians and other Pacific Islanders longer resident in these islands. Finally, Nathan Fitch's forthcoming *Island Soldier: Mirconesians Fighting in America's Wars* documents and illuminates the profound role of the US military service in COFA communities even as returning veterans including those who saw active combat discover that their service may not always come with the recognition received by other vets.

The twenty-year renewal for the Compacts with the RMI and FSM in 2003 is now rapidly approaching and it is not at all clear that voters or the governments in these COFA states will seek to maintain the same relationships. Transformations and changing stances seem to be afoot. Observers note the US's unfulfilled commitments and continuing resistance to efforts to see significant increase in remuneration for decades of nuclear testing and the poisoning of lands and RMI peoples (particularly Bikini and Rongelap communities). Similarly, the Compact's economic provisions for Palau ($US 18 million per annum) expired in 2009, and efforts towards renewal agreements negotiated in 2010 have seen numerous hurdles and complications in front of the US Congress and remain un-ratified. Many factors may lie behind any of these complex relationships. However, the role of contingent phenomena is disrupting established agreements is hard to miss – notably the global financial market collapse in 2007 and 2008 and subsequent market fluctuations have seen the US's Compact Trust Fund underperforming. Locally circulating conversations run from the possibility of even closer integration with the US, a devolution of sovereignty, which seems unlikely, the breakup of the FSM with Chuuk state going its own way and seeking a fresh bilateral relationship with the US, to the maintenance of an uncanny status quo.

Similarly, we can see from Figure 3.15 that the importance of remittances relative to other sources of national income (exports, FDI and ODA) varies greatly. We can conceive of different groupings in terms of the relationship between migration/remittances and aid. The classic MIRAB economies (see Box 2.2) are those which have relatively high levels of aid and remittances as a proportion of GNI (Brown and Connell 1993). Samoa and Tonga are illustrated here but we would add smaller territories in particular (Wallis and Futuna, Tokelau, Niue, American Samoa, and possibly also the FSM, Palau and Northern Mariana Islands). These islands have very close ties to a metropolitan patron which supplies the bulk of aid inflows and offers immigration opportunities.

Other territories in the Pacific – here illustrated by Kiribati – have more limited options for migration and are more relatively dependent on aid as a result. This group would include Tuvalu and Nauru. In another group come the demographically larger states (indicated by PNG, Fiji, Vanuatu and Solomon Islands) which tend to have a more diverse economic base. PNG, despite receiving the highest absolute volume of ODA in the region is the least dependent on it, due to its rich resource base and growing income from commodity exports, especially minerals and energy. Fiji, Vanuatu and Solomon Islands have similar opportunities for resource extraction (though with a different mix of exports) but with relatively more reliance of aid and remittances. These three countries may well become more of a target for special migration schemes (such as seasonal worker schemes in Australia and New Zealand). Tourism adds an additional income stream for Fiji and, to a lesser extent, Vanuatu.

Finally, there are the two large Francophone territories which exhibit both MIRAB and diverse economy aspects (Poirine 1994). New Caledonia and French Polynesia have high levels of financial assistance from metropolitan France and ready access to overseas labour markets and their high average material living standards are supplemented by tourism and, for New Caledonia, considerable mineral wealth.

Box 3.5 Niue: 'sweet spot' of sovereignty?

Warwick E. Murray

Niue is an island nation that, unlike many others in the Pacific, occupies only one island of approximately 250 square kilometres with a population of around 1,600. The capital Alofi is home to an international airport and the island's wharf. Settled by Samoan Polynesians approximately 1,100 years ago, and then Tongans in the 17th century, it became a protectorate of Great Britain in 1900 and governance was handed to New Zealand in 1901. Despite attempts to see Niue administratively added to the Cook Islands, the island remained separate. The colonial story of Niue is not devoid of conflict and 1953 the New Zealand High Commissioner Cecil Larsen, who had abused and exploited the local population consistently, was murdered (Scott 1993). Niue was not settled extensively by Europeans and as such has maintained a very strong cultural identity to this day. From 1974 the country moved to the status of 'territory in free association' with New Zealand which saw the latter responsible for its foreign policy and defence, together with some economic functions (the currency for example is the New Zealand dollar). Today Niue is the smallest sovereign nation in the world.

One of the principal challenges facing the country has been population loss which has fallen from a high in 1966 of over 5,000. Today there are many more Niueans living in New Zealand (as well as Australia) and some estimates place the ratio of overseas to domestic components of the Niuean diaspora at 20:1 (see Overton and Murray 2014). This has been a double-edged sword, as remittances have provided much required income for the country, especially in times of natural disaster such as during Cyclone Heta in 2004. The diaspora maintains close ties and mobility between the island and New Zealand especially (where dual citizenship is permitted) is high.

The continued association with New Zealand geopolitically, and the island nation's isolation and challenges with regard to evolving outward-oriented economic activity partly very high levels of aid per capita (Fisk 1980). In some ways Niue is the classic MIRAB economy. At times in the past, aid amounted to over 50% of total GNI, with the remainder being comprised of remittances, with some tourism receipts, supplemented by various, often frustrating, attempts at the development of an agro-export strategy (Murray and Terry 2004).

In 2015 total aid receipts were \$19.3 million, meaning that per capita receipts stood at over \$12,000 – one of the highest in the world (see Chapter 3). As Figure 3.12 (g) illustrates, the trajectory of aid in Niue is subject to undulations, in part due to the impact of large-scale project-based spending in discreet years, in the case of an airport expansion and wharf strengthening for example. A significant rise in aid in 2005 was due to the aftermath of Cyclone Heta. Generally speaking however, the 1990s saw an average decline in aid given the neoliberal agenda of the New Zealand aid agency at the time. Indeed, Niue was subjected to some of the harshest restructuring of all Pacific territories and government spending was slashed by significant proportions at the start of the 1990s and again towards the end of the decade. Attempts were made, mostly ill-fated, during this period to establish agro-exports but this proved very difficult given the lack of economies of scale and transportation costs, among other things. In the late 2000s and early 2010, while subject to fluctuations real aid levels have been consistently higher than in the 1990s, with project and infrastructural and transport projects (see below) being favoured under the retroliberal policies of the main funder – New Zealand

Over the last decade the relationship has been re-imagined by the former colonial power. Niue has been positioned by the New Zealand government as part of its 'Realm' (see Box 2.1), and private sector aid projects such as that which has funded the expansion of the country's main hotel, the Matavai, and the subsidy of Air New Zealand flights have been undertaken. It could be argued that these 'retroliberal' turns have returned significant benefits back to New Zealand, which sees a significant return on its aid grant in terms of private exports to Niue (Overton and Murray 2014), and which gains through the success of Air NZ as the majority owner. What we witness in Niue is the case of a country that has negotiated its sovereignty in return for benefits that flow in terms of monetary streams and access to overseas labour and education markets. Sovereignty is never absolute, and Niue illustrates how marginality might be 'negotiated' through special arrangements, what we call elsewhere searching for the 'sweet spot' of sovereignty.

Since the emergence of the trade-based neoliberal approach to aid the 1980s, aid agencies have taken a growing interest in the way aid programmes can articulate with the promotion of remittances as a development strategy. The recognition of the importance of remittances to Pacific economies (Connell and Brown 2005) – and perhaps glossing over some of the negative aspects such as decline over time, expenditure of remittance income on current consumption or religious donations rather than economic investment, and uneven social impacts – has been championed by agencies in recent years by agencies such as the World Bank and the Asian Development Bank. Some researchers, such as Charlotte and Richard Bedford (Bedford et al. 2017) also see the volume and durability of remittances from schemes such as the Recognised Seasonal Employer (RSE) scheme in New Zealand and the Seasonal Worker programme (SWP) in Australia as a favourable development option for many Pacific communities (also Brickenstein 2015; Bedford et al. 2010). There have been some attempts by aid agencies to explore and instigate schemes with banks and other businesses in the finance sector, such as internet banking, which can lower the transactions costs of remittances (see Gibson et al. 2007). These initiatives have aimed to increase the net economic return of remittances to Pacific households and, arguably, help spread the benefits of remittance income spatially and socially. A recent World Bank report (Curtain et al. 2016) has highlighted the importance and future potential for increased labour mobility in the region as a driver of development in Pacific economies. In arguing that labour migration involves a 'triple win' for migrants and both sending and receiving countries, the report suggested that metropolitan donors "can use their aid budget through 'aid for migration' schemes. This can be considered as 'aid for trade', a form of aid that is at least, if not, more relevant for many Pacific nations than trade facilitation" (Curtain et al. 2016: ii).

As noted, the most notable recent example of an aid policy incorporating migration and remittance strategies has involved the support and promotion of seasonal labour schemes in New Zealand and Australia. These government-backed schemes have sought to institute special immigration policies that allow in seasonal workers (largely in agricultural and horticultural industries where periodic labour shortages arise) from approved Pacific countries. The schemes aim to support to strategic industries in host countries (in the case of Australia and New Zealand viticulture, vegetable production, and dairying) whilst recruiting workers (largely men) from rural communities (in the Pacific that may have few other alternatives for waged labour). Recruitment is regulated and periodic so that migrants have to return home after a set time, though they (or other members of their villages) may return to work in successive and subsequent years. The explicit linking of strategic industries in host countries and communities in the Pacific – and specific oversight by labour inspectors of labour conditions and accommodation costs – is designed to ensure the proportion of earnings that return home is maximised.

The involvement of donor aid agencies in these schemes is not necessarily direct (host official labour and immigration agencies take much of the lead), but such institutions been active promotors of the schemes. Ancillary aid activities, such as training or improvements in financial transfers and services, have helped broker these new development programmes. Furthermore, the (re-) integration of aid agencies into wider foreign affairs ministries in Australia and New Zealand and the adoption of 'whole of government' rhetoric has arguably fostered a more overarching approach to aid in the Pacific in recent years. In addition, this selective management of Pacific migration supported by aid agencies in order to meet labour demands in the metropole is a strong example of the new retroliberal approach to aid: it uses a 'win-win' (shared prosperity) rhetoric that mixes supposed development benefits for the Pacific with the needs of capital in donor economies.

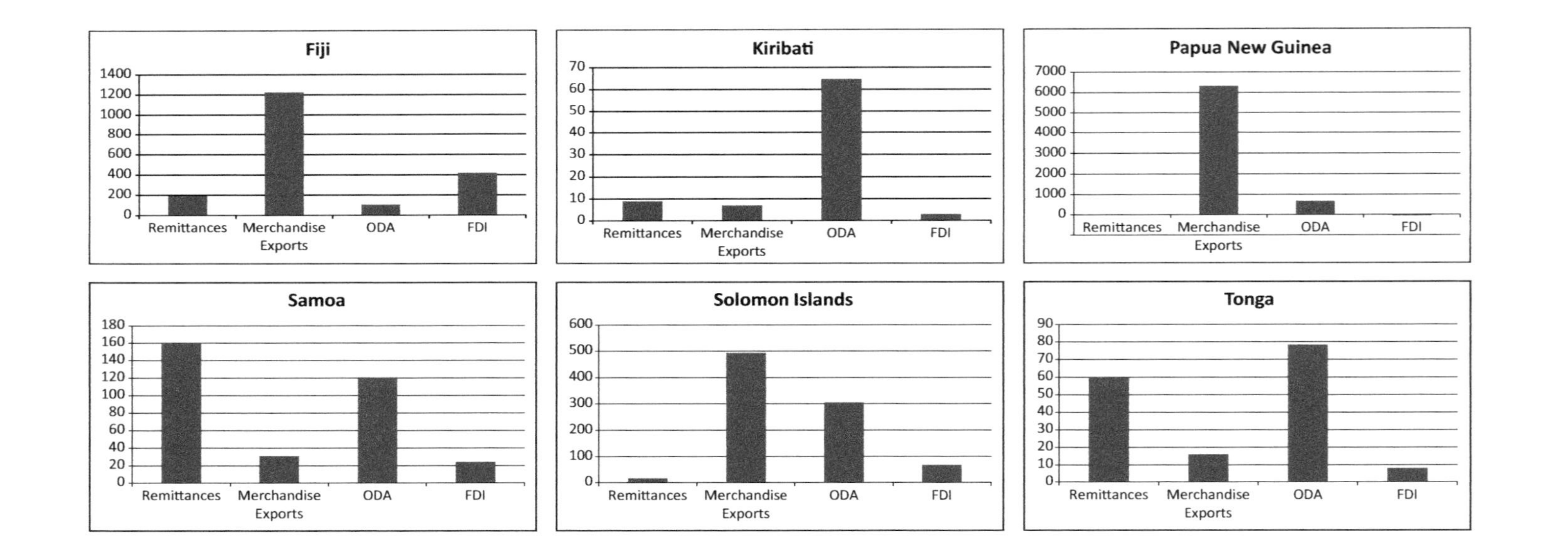

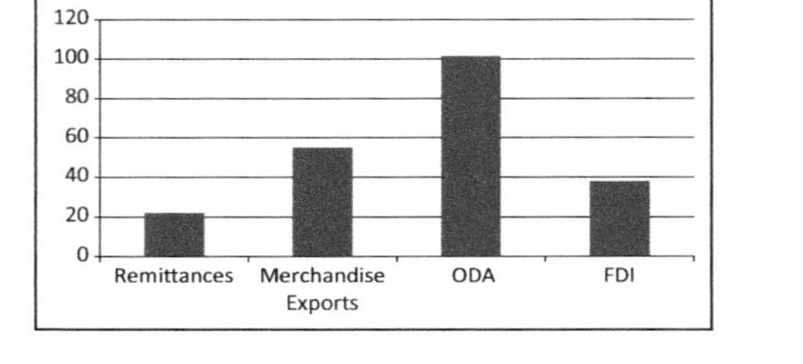

Figure 3.15 Remittances, ODA and exports in selected Pacific Island countries, 2014 ($US million current prices 2012)

Source: ADB (2014: 223). The countries and territories depicted here were the only Pacific countries enumerated for this by the ADB.

Some reports, however, points to mixed benefits for Pacific participants in the scheme (for example Lewis 2017). Due to high deductions for accommodation and travel, it has proven difficult for temporary migrant workers to save sufficient from their wages to meet demands at home (and repeat circuits of involvement in the schemes are necessary). Moreover, the conditions of work can be difficult and unfamiliar, and there are many social adjustments and costs to be faced by returning workers and their families.

Preferential and non-reciprocal trade: 'aid with dignity'?

Preferential and non-reciprocal trade agreements between countries in the top of the OECD list and poorer countries, which we can conceive of as a form of non-ODA aid, has been on a steady decline since the mid-1990s, following the rise of the neoliberal regime. With the exception of the larger and relatively well natural resource-endowed countries of Melanesia (PNG, Solomon Islands, Vanuatu, Fiji), or certain favoured and well-connected tourism destinations in the Pacific (Fiji, Tahiti, New Caledonia), most Pacific Island states are unable to contemplate economic development strategies that are based on resource extraction, commodity production or industrial processing that, under certain conditions, could succeed in distant markets against large-scale competitors. Historically, the absence of economies of scale and the high costs of transport have impeded Pacific commodity sectors such as cocoa, sugar, cotton, coffee or tropical fruit production. Manufacturing industries have faced challenges of diminutive size and friction of distance together with the absence of diversified innovative economies that stimulate and nurture such growth.

In response to the limits that the geography of the region imposes on trade, some metropolitan powers aimed to assist their former Pacific colonies and territories by offering preferential access to their markets, without demanding the former colonies in exchange open their market in equal measure for metropolitan businesses. The most significant such set of preferential and non-reciprocal trade agreements was instituted in the 1970s by the European Community (later European Union or EU) in the form of the 1975 Lomé Convention. The Lomé Convention was a trade agreement between the European Community and more than seventy African, Caribbean and Pacific countries, captured in the acronym 'ACP countries'. In the Pacific, all sovereign island nations – not those retaining constitutional links with the USA, France or New Zealand – are included in the ACP countries. This agreement, adopting various forms over many years, recognised the special relationships between former European colonies and territories and regulated preferential and non-reciprocal trade deals for selected (mostly agriculture) imports from such countries at lower tariff rates than those levied on countries that were not former European colonies. The overall effects fitted the explicitly development-oriented objectives of the Lomé Convention: a better price for agricultural products from the ACP countries, whose revenues enabled the development of domestic (processing) industries that could be protected against direct competition from European industries.

However, as the neoliberal views came to dominate politics, such preferential and non-reciprocal trade agreements came under pressure from various sides. When the World Trade Organisation (WTO) was established in 1995, one of the first cases presented by the USA to the WTO's Dispute Settlement Body concerned the legality of the EU's Lomé Convention. The USA as the plaintiff, supported by several Central American countries, argued that the Lomé Convention distorted global competition and free trade. The WTO ultimately ruled in favour of the plaintiff and the EU was forced to end the Lomé Convention. It was replaced in 2000 by a new trade agreement – the Cotonou Agreement covering the 2000–2020 period – between the EU and the ACP countries. The Cotonou Agreement explicitly acknowledges WTO rules about free trade and the reciprocity of trade and it works out reciprocal trade deals called Economic Partnership Agreements (EPAs) in which the ACP countries are expected to have opened up their domestic markets for competition from EU businesses by 2020.

Whereas the increasing power of the WTO erodes the benefits of preferential and non-reciprocal access by ACP countries to the EU, this erosion did not take place for the overseas territories of EU members. For these twenty-five so-called 'Overseas Countries and Territories' (OCTs) – all islands – the EU established another specific regulatory trade regime. While the people living on OCTs are all EU citizens, their territories are legally not part of the EU. However, a special legal arrangement allows OCTs tariff- and quota-free access to the EU, while under certain conditions allowing them to impose tariffs on EU imports. In this framework, the Pacific OCTs of New Caledonia, Wallis and Futuna, French Polynesia, and the Pitcairn Islands enjoy trade (and migration) benefits with the EU that other Pacific islands and ACP members do not have. However, it may be that at some point in the future a large trading nation or a large corporation may challenge the EU again in the WTO as was done with the Lomé Convention; after all, the OCTs are not fully integrated in the EU's legal frameworks (Prinsen et al. 2017; Prinsen and Blaise 2017).

Although relatively small exporters to Europe in terms of total volume, some Pacific countries have benefited greatly from trade agreements such as the Lomé convention and the Cotonou Agreement – none more so than Fiji and its prominent sugar industry. Sugar cane production was introduced to Fiji during the 19th century. It was overseen eventually by a single large company – the Colonial Sugar Refining Company (CSR) from Australia – and it involved the importation of a large number of manual labourers and their families from the Indian subcontinent (Knapman 1987). Production expanded during the colonial period, although CSR restructured the industry in the 1920s by devolving its directly managed estates to a smallholder system whereby largely Indo-Fijian workers took on, heavily regulated and very limiting, cane growing contracts mostly on land leased from indigenous Fijians (Ellis 1985; Ward

1985). Political independence of Fiji from Britain in 1970 was paralleled by the transfer control from CSR to the parastatal Fiji Sugar Corporation (FSC) which took over CSR mills and management of the industry.

Because Fiji qualified for support under the Lomé Convention, its sugar exports could enter the EU at significant preferential rates and without being required to open up Fiji's domestic markets to competition from the EU. The existence of predetermined price levels stabilised income in an otherwise volatile commodity market – and the industry proved to be particularly profitable. Backed by the Fiji government, a marked expansion of sugar cultivation took place in the two decades following independence; production and exports grew steadily and sugar came to be a pillar of the Fiji economy (Ward 1985). In the mid-1980s sugar dominated the country's economy, approached only by the tourism sector in importance (Taylor 1987). The degree of trade via preferential trade agreements for sugar was significant (Grynberg 1993, 1995), much more at the time than total ODA.

One set of commentators in the 1980s analysed the Fiji sugar industry and concluded that its success (with support from the EU) could be regarded as 'aid with dignity' (Taylor 1987). Not only did this preferential access comprise a substantial form of assistance – in effect the EU was foregoing income from tariffs on Fiji sugar that it did levy on non-ACP sugar exporters – it did so in a way that seemed to offer a high degree of stability to the industry and allow Fiji policy makers, farmers and sugar managers to control and develop their own industry. Support was not in the form of a direct financial handout but rather through the creation of a carefully negotiated and managed trading environment that empowered Fijian sugar producers, made their industry viable and created an income stream that directly employed farmers, harvesters, mill workers and contractors and yielded huge multiplier benefits to the wider economy.[14]

The Fiji sugar industry was the main beneficiary – and Europe the main benefactor – of the preferential trade strategy for development assistance in the Pacific but it was not the only example. There were other exports to Europe – a proposal to export rice from Solomon Islands – and there were also other donor countries which were prepared to offer trade concessions as part of an international development strategy (see Table 3.6). In the 1980s Australia and New Zealand – pushed by Pacific countries – moved to formalise a trade agreement with the Pacific to allow exports to the region to be imported either free of tariffs or at rates well below their non-Pacific competitors. The resultant SPARTECA agreement (and later variants under PACER and PACER Plus) provided a basis for limited industrialisation in parts of the region. SPARTECA was instituted (in 1980) at a time when Australia and New Zealand still had large tariff barriers to protect their own manufacturing industries from low wage economies, especially in Asia. Some Pacific companies, along with companies from further afield, found they could take advantage of lower wages in places such as Fiji and Samoa and then export to Australia and New Zealand successfully in competition with both local

Table 3.6 Summary of trade agreements and bodies relevant for the Pacific

Name	*Non-Pacific partners*	*Pacific partners*	*Features*
Cotonou Agreement (2000–2020)	European Union (EU)	Pacific African, Caribbean, Pacific (ACP) countries: Cook Islands, FSM, Fiji, Kiribati, Marshall Islands, Nauru, Niue, Palau, PNG, Samoa, Solomon Islands, Timor-Leste Tonga, Tuvalu, Vanuatu	Preferential trade (low or no tariffs for Pacific exports into EU) Updated from the Lomé Convention to become reciprocal
Lomé Convention I-IV (1975–1999)	European Community/ European Union (EU)	As above less Timor-Leste	Preferential trade (low or no tariffs for Pacific exports into EU) Non-reciprocal
Melanesian Spearhead Group (MSG) Trade Agreement (MSGTA 1993–2004)	–	PNG, Vanuatu, Solomon Islands, Fiji	Free trade in goods, based on positive trading list only (duty free access)
MSG Trade Agreement (MSGTA 2004–2016)	–	PNG, Vanuatu, Solomon Islands, Fiji	Free trade in goods, based on negative trading list principle (allowing import duties with a phasing-out time frame)
MSG Trade Agreement (MSGTA 2017 – present)	Indonesia	PNG, Vanuatu, Solomon Islands, Fiji	As above and extended beyond goods to include services, labour mobility and investment
OCT-EU	European Union	Pacific EU Overseas Countries and Territories (OCT); New Caledonia, French Polynesia, Wallis & Futuna, Pitcairn Islands	Funding for competitiveness from the EU with an emphasis on trade relations with the EU
PACER Plus (under negotiation)	Australia and New Zealand	PICTA countries plus Australia and New Zealand	Free trade in goods and services, mobility and investment under negotiation

Pacific Agreement on Closer Relations (PACER) (2002-)	New Zealand and Australia as members of the Pacific Island Forum	Pacific Island Forum countries (18, including Australia and New Zealand)	An overarching body that provides facilitation for negotiation of free trade between members
Pacific Islands Countries Trade Agreement (PICTA) (ratified 2001, in force 2006)	–	Cook Islands, Fiji, Kiribati, Nauru, Niue, Papua New Guinea, Samoa, Solomon Islands, Tonga, Tuvalu and Vanuatu are ratified members	Free trade agreement in goods and services, later expanded to include labour mobility and investment
SPARTECA (South Pacific Regional Trade and Economic and Cooperation Agreement) (1981–)	Australia and New Zealand	Forum Island countries	Non-reciprocal preferential access to Australia and new Zealand markets for Forum Island countries
World Trade Organisation (WTO)	164 as of 2017	Fiji, Samoa, Solomon Islands, Tonga, Vanuatu	Membership body; regulation of trade between nations, facilitation of agreement negotiation and resolution of dispute.

and overseas producers. Furthermore, some Australian and New Zealand companies found they could establish (or relocate) operations offshore in Fiji and supply their existing markets with lower production costs.

Two examples of seemingly successfully industrialisation in the Pacific as a result of these preferential and non-reciprocal agreement trade agreements were the Fiji garment industry (Box 3.6)[15] and manufacturing in Samoa. In Samoa, the Japanese company Yazaki established a large automotive wiring plant in Apia. It employed a local workforce of up to 750 people – being the largest private sector employer in the country – and exported to the Australian market in particular. However, in 2016 Yazaki announced it would close down its Samoan plant as a knock-on effect of increasing globalisation. Its Samoan plant was providing parts to the Australian car manufacturing industries and as these plants were closing down in Australia, so would the plant in Samoa.

Although it could be argued that such operations represented a 'race-to-the-bottom' strategy, being based on cheap labour (and there were allegations of very low wages and poor working conditions in the Fiji garment industry, for example), it is clear that the particular combination of preferential trade, (largely foreign) investment, and local manual labour provided a new source of livelihood for hundreds of people and their families in the Pacific. And, again allowing for criticisms of low wages, it was argued that such employment reached some groups in society (women, urban workers) who had few other opportunities for secure employment. Indeed a study of the Fiji garment industry commissioned by Oxfam (Storey 2004 – see Box 3.6) concluded that, on balance, this model of development brought more benefits that harm.

However, despite the putative success of the development model of preferential non-reciprocal trade – aid with dignity – most of it did not survive neoliberal reform in the 1990s and what is left of it (such as the Cotonou Agreement) is winding down and ending in 2020. Globalisation and strong pressures to liberalise national economies meant that the former trade benefactors in Australasia lowered all their tariff barriers, deregulated their own protected industries and sought wider and more liberal trade agreements with all and particularly major trading partners.

Box 3.6 Trade preferences: the Fiji garment industry

Warwick E. Murray

Debatably, and as we argue elsewhere, beyond the official DAC definition of aid, there is a broad range of instruments that can be defined as aid. One such example is that of trade preferences which allow access to markets. These have been very important in the colonial and postcolonial history of the Pacific. The access for Fijian sugar into European markets negotiated under the Lomé convention when the UK joined the EEC is one such example. It successor the Cotonou agreement provided preferential treatment albeit it at slightly inferior terms for the industry. As this has been phased out – and as the sugar sector itself has suffered various crises and inefficiencies – the once corner-stone of the Fijian economy has floundered. There has been a search for diversified earnings – and tourism has provided some promise in that regard. Fiji was also offered preferential trade agreements for its garment sector for access into Australia and New Zealand under the general SPARTECA agreement of 1980 applied to garments later the decade. The USA also offered concessions. These agreements were widely criticised at the time as they were seen to promote low waged, unskilled labour working under poor conditions (Storey 2004). However, at the same time they also offered employment to a section of the society that hitherto had not been widely employed in formal labour markets – urban women. In this sense the preferential agreement and access for garments was seen as something of a double-edged sword, as it also undoubtedly brought in foreign exchange – albeit concentrated in the hands of the few entrepreneurs and owners involved in the garment sector largely in the West of the country. As the author of an Oxfam study undertaken in the early 2000s stated:

> "human development through the garment industry could be substantially enhanced through improved employment conditions for workers which would simultaneously act to address poverty alleviation, gender equality and create more winners through international trade. Trade can work for women workers, their families, and communities . . . but this potential is not currently being realised." (Storey 2004: 50).

The political crises of the 2000s and early 2010s put a complete end to these preferences, although similar arrangements with China were brought into place in the late 2000s and early 2010s. This raises the question as to how we measure aid and how broadly the term might be defined. Further it makes us mindful that when rises in 'formal' and DAC-defined aid take place, if this is accompanied with decline in other preferential agreements – this may actually constitute a reduction in assistance.

Thus, global trade liberalisation put an end to what had emerged as a significant, if indirect, development strategy for many in the Pacific region. Indeed, we argue that the dismantling of this preferential trading regime as a form of aid constituted a significant move by donors against the interests of the Pacific region. It has led to unemployment and in the case of Fiji it has seen a significant contraction of sugar production. Overall the livelihoods of thousands of Pacific households have been negatively impacted. Indeed, when we examine the significant rise in ODA to the Pacific after the turn of the millennium (see above), this could be seen as compensation for the loss of trade-related assistance and its replacement by more direct and financial forms of aid. Although the total net cost of this shift is not knowable, it is feasible that the loss of preferences outweighs gains in direct ODA in the long-run. Nonetheless, the age of preferential trade as a development strategy and a form of aid for the Pacific is likely to remain consigned to the past. Furthermore, as we will see, the replacement of trade-related aid, which gave space to Pacific countries, communities and people to instigate and control their own development within these relatively stable if not durable bounds, was replaced by formal ODA which has involved more direct control by donors. Thus, donors have re-concentrated power in terms of the control of aid flows while cloaking this in rhetoric of ownership and effective endogenous development. This has been combined with a retroliberal trade agenda which ostensibly opens doors, whilst in reality it further concentrating economic opportunities and power in the hands of the larger Pacific Rim economies and their companies.

Other financial flows

To date we have seen aid existing either in the form of ODA (formal, quantifiable and official) or as development-related concessions (migration access, labour schemes or preferential trade) granted by aid donors to Pacific countries. Yet there are other less tangible forms of financial flows that depend on a degree of magnanimous support from donors to help promote development and welfare in the Pacific.

Perhaps the most significant form of such flows in this case can be seen in the case of incorporated territories: those which call on the government reserves of the metropole to pay both for the day-to-day operation of the civil service and public services, and for certain development projects. In the Pacific these are found in territories such as American Samoa, French Polynesia, New Caledonia, Rapa Nui, and Guam for example. Government expenditure is not (or in the case of the two French territories, no longer) counted as ODA. It covers the wages and salaries of teachers, bureaucrats, nurses, doctors, road workers, and a plethora of other government employees or contractors. It covers the cost of new investment in infrastructure as well as depreciation and maintenance of the existing public facilities.

It is expenditure that contributes greatly to the welfare of citizens and it often promotes development through education and infrastructure. It often involves the payment of salaries (especially at higher levels of a bureaucracy) at levels close to, or even in excess, of those at home.[16] In many Pacific countries much of the cost of this expenditure comes from external sources and it counted as 'aid'. Yet in some incorporated or closely integrated territories (though not in Tokelau or Wallis and Futuna) it is not. Instead it is seen as the normal provision of public services for a population equivalent to what might be expected in the metropole. Although it is not possible to count such flows accurately (see Box 3.1 above) we can conclude that such payments represent a very significant net flow of resources from the metropole to the Pacific in ways which maintain high salaries and high levels of welfare relative to other states and territories in the region.[17] Indeed, average per capita incomes in New Caledonia, for example, now exceed those in New Zealand.[18] It can be argued, of course, that these high incomes are not distributed evenly – they may favour expatriate workers or members of the elite and income inequalities are often very high within such territories – yet we should see them as a form of 'aid'.

The portability or otherwise of superannuation and other welfare flows also play an important development role in the Pacific. Again, this is not usually seen as ODA, but it does rely on a concessional agreement granted by a donor country. It is also certainly a measure that has been strongly advocated for by some Pacific governments. We saw above how the extent of migration and remittances by Pacific migrants has a major impact on livelihoods 'at home'. Such migrants often spend much of their working lives overseas where they are often contributors to compulsory and voluntary superannuation schemes. They also pay many years of taxes and gain access to welfare services. Yet, on retirement or in older age, many decide to return to the Pacific Islands to reconnect with their social and cultural roots. In doing so, many have found that their superannuation schemes or pensions, much less access to health facilities, are not as portable as they are as individuals. They are forced to remain in their countries of work in order to access the benefits of their payments in old age or they go back to the Pacific islands virtually empty-handed.

Efforts to liberalise the portability of such superannuation and pensions have had some limited success (New Zealand allows superannuation portability to twenty-two Pacific countries) but this remains a significant issue for many extended families in the Pacific. Increased portability would allow such payments to support living in the islands and, given the generally much lower cost of housing and living generally, the benefits of this income can be spread more widely across families, communities and economies. Although it would incur a cost for the metropole – pensions and superannuation income would not be spent locally and generate multiplier effects – it would also lower some costs (health, public housing etc. for older Pacific citizens) and bring benefits to Pacific economies[19]

Thus, a broad view of development and welfare assistance can illuminate forms of aid that are rarely counted as such, nor analysed systematically in terms of their likely benefits for Pacific people. However, research and practice should take these forms seriously in assessing both the nature and impacts of aid flows to the region as well as the institutions and actors that control them.

Conclusions: aid and the framing of the Pacific Islands

In this chapter we have mapped the way 'aid' has been disbursed in the Pacific Island region over the past fifty years. We focused initially on official flows of development assistance (ODA) and this revealed marked patterns of inequality and concentration. There were marked differences in the volume of aid to different territories and countries and this was revealed in both total and per capita data. There was also a marked concentration of activity by major donors with several distinct 'spheres of influence' being apparent. The maps of these ODA flows show an increase in real aid levels since the turn of the millennium and, until recently, a focus on both poverty-related programmes and the reconstruction of Pacific Island state systems.

Yet when we broadened the analysis of aid to include forms of assistance that are not recognised or enumerated by the OECD – migration and remittance flows, preferential and non-reciprocal trade agreements, other financial flows, and 'South-South cooperation' – some quite different patterns and conclusions arose. Firstly, it could be argued that some of these forms of assistance – especially those resulting from concessions over migration or trade preferences – have been historically more important for the welfare and development of several Pacific Island states than ODA. Secondly, these alternative forms have often delivered a greater degree of autonomy for the region: trade and migration regimes provided an environment in which Pacific Island policy makers, families and individuals could assess different development options and develop their own strategies to exploit perceived. This could vary from the decision of the Government of Fiji to expand sugar cane production substantially in the 1970s and 1980s, or for people in Cook Islands or Wallis to decide to migrate to work overseas and send remittances home to support members of their families who remained.

We conclude that the first decade of the new millennium did represent a considerable expansion in official aid by many donors into the region. As we saw in Chapter Two this was accompanied by a rhetoric – and some practice – that acknowledged the importance of local 'ownership' over development. Pacific Island states developed their plans for poverty alleviation and these were frequently and sometimes generously funded and supported by donors. Following the global financial crisis in 2007–2008 donor policies changed, moving away from poverty alleviation towards more explicit support for economic growth that involved industries and commercial actors in donor countries, which we term 'retroliberalism'. However, aid volumes were largely maintained (at least for the early 2010s). Despite these larger volumes of ODA, there was concomitant reduction in other forms of assistance, particularly non-reciprocal trade concessions, and the deleterious consequences of these cuts in some cases (such as Fiji) were not generally compensated by aid increases. More generally, the shift towards an emphasis on ODA rather than other forms of aid served to close the spaces for autonomous development and local decision-making because recipient counties' options for trade-related growth became more limited in an ostensibly neoliberal global trading environment. It is true that migration options and concessions have remained open for people in some Pacific territories and remittances remain a considerable feature of many Pacific Island livelihoods. Yet we have also seen attempts ostensibly to widen these options (through seasonal labour schemes) which, in effect, simultaneously try to regulate and control migration flows bring migration into the realm of development assistance, and ultimately extract benefits for the donors, in terms of regulated labour supply.

Development aid – ODA at least – is portrayed as a form of altruism. Throughout the early 2000s public campaigns in the West (Make Poverty History, Jubilee 2000, Band Aid +20- etc) asked Western donors to increase aid levels significantly to help end extreme poverty and address issues concerning indebtedness. Politicians listened, and aid budgets increased. However, despite these increases and the rhetoric of local ownership, aid increases in the Pacific were part of a set of processes which redefined development and re-engineered the ability of Pacific Island people and institutions to control their own development. The shift to more ODA and fewer broader aid concessions actually limited the sovereignty and autonomy of Pacific Island states and territories to set their own development priorities and policies.

Furthermore the linking of aid to negative discourses concerning Pacific Island states and people – those focusing on failed/failing states, poor governance, poverty and need – acted to reframe the region (Fry 1997) in a negative light. Seemingly incapable of managing their own limited resources and suffering from corruption, instability and poverty, many Pacific Island states were portrayed as in crisis, in need, and at risk of collapse. Aid – and aid aligned to strategies of state-building or market liberalisation – tied them to donors and often to development strategies that are defined externally. In addition, because, as we have seen above, the map of aid flows was highly concentrated, aid meant that a few key donors became more influential in the development trajectories of recipient states. It is these themes that underpin our theme of inverse sovereignty, to which we will turn in more detail in Chapter Five.

Notes

1 1965 is chosen as the starting date because this is the first year Papua New Guinea (as the largest recipient in the region) was included. These figures exclude data for New Caledonia, French Polynesia and Northern Mariana Islands as these were not included by the OECD after 1999.

2 ODA flows to Palau, Federated States of Micronesia and Marshall Islands were only counted after about 1992. 'Colonial' era expenditure in these territories by the USA prior to this time was not enumerated in this data set.

3 The twelve countries are PNG, Fiji, Solomon Islands, Vanuatu, Samoa, Tonga, Tokelau, Tuvalu, Kiribati, Cook Islands, Nauru and Niue. These countries were consistently included in the DAC aid statistics over the period since 1965 and give a clearer picture of patterns of aggregate flows.

4 Aid to these countries by France and the USA is no longer counted as ODA (see above) but is nonetheless substantial.
5 This group of other donors excludes the major donors (Australia, France, the USA, Japan, New Zealand and all multilateral agencies) but includes Canada, Germany, Finland, Norway, Korea, UK and the United Arab Emirates (all of this latter group each giving between $US 1 million and $US 13 million to the region in 2014). It also includes private donors, the only significant one of which enumerated is the Bill and Melinda Gates Foundation, donating $US 720,000 to the region in 2014.
6 These direct payments to recipient governments did not show up clearly in the DAC categories for general budget support (GBS) probably because they were earmarked for particular ends – programme assistance – rather than unspecified general expenditure.
7 We have used the most recent five-year period (2010–14) for which data are available and averaged the data over this period to even out possible year-to-year anomalies (such as disaster relief).
8 The Treaty of Friendship between New Zealand and Samoa for example recognises not only a (chequered) colonial past but also a continuing commitment to a special relationship. Tonga also has call on some special treatment by New Zealand in terms of migration ties and a large expatriate Tongan population in New Zealand.
9 Niue is technically part of the next group (self-governing in free association or with a compact of association) rather than more closely integrated.
10 Although 'new', we can see clear antecedents for this form of aid. Japan for long had a strong emphasis on infrastructural forms of aid and it and other donors have used tied forms of aid and been keen to promote their own economic interests through aid contracts etc. In fact, a key aspect of the earliest development aid regime of the 1950–1980 period was investment in infrastructure by donor countries in developing nations.
11 For Guam and the Commonwealth of the Northern Mariana Islands (CNMI) we use OIA data on the amount of federal grants in 2013 ($US 373 million and $87.2 million respectively – Office of Insular Affairs 2016: 2). This is a probable underestimate of the way metropolitan funding contributes to the economy. For American Samoa, the territory's statistical yearbook records total federal grants of $162.6 million for the financial year 2013 (American Samoa Government 2014: 115).
12 Quantification of remittance flows is highly problematic. Asian Development Bank (ADB) data relies on estimates built up from various sources such as bank transfers and household surveys. As well as formal money transfers through banks or agencies such as Western Union, however, much remittance income travels through less formal channels: cash is carried by visitors or returning migrants, goods (cars, electrical goods etc) are purchased and sent or carried home, and income from migrants overseas is used to support Pacific household expenditure on things such as overseas travel or the hosting of children attending school overseas. It is likely, then, that many estimates of remittances underestimate the net economic value of migrant income for Pacific economies and households.
13 Such data however may be limited. ADB data, for example, would suggest that countries such as Tajikistan, Kyrgyz Republic, and Nepal, would exceed Samoa and even Tonga (ADB 2014: 214)
14 There were also some negative aspects to this support, such as the expansion of sugar cultivation on to marginal land and associated land degradation (Ward 1985, 1987). The dominance of sugar and relatively good incomes for many may also have provided a disincentive for the growth of other industries – a classic

Dutch Disease effect of resource dependency. Another, albeit much smaller, form of support for the Fiji sugar cane industry came from New Zealand. New Zealand looked to Fiji for much of its sugar needs for many years, especially via a CSR factory (the Chelsea factory) which has operated in Auckland since the late 1880s. In a heavily regulated economy (until the late 1980s), it obtained much of its sugar from Fiji and government policies acted to stall moves within New Zealand to establish a sugar beet industry.

15 The garment industry in Fiji was facilitated by SPARTECA and it also benefited from certain trading concessions granted by the USA for some categories of garments (pyjamas were notable!) from Fiji.

16 French civil servants for example in places such as New Caledonia or Wallis and Futuna are paid a bonus over their normal salaries in France as recognition of a hardship posting to the territories. They are also attractive for local residents: students from Wallis and Futuna who study overseas actively seek well-paying civil service salaries back in Wallis and competition is high. Over time, this is resulting in a gradual replacement of expatriate bureaucrats on lucrative limited-term contracts with locals (Jacobs 2016).

17 Again we note that such payments to New Caledonia and French Polynesia were counted as 'DAC-able' aid until 1999 and their re-categorisation as internal transfers rather than aid was simply by an arbitrary administrative and bureaucratic decision.

18 New Caledonia's per capita income was estimated as $US 38,896 in 2011, whereas New Zealand's figure was $US 32,734 for the same year from the same source (see www.tradingeconomics.com/new-caledonia/gdp-per-capita accessed 31 March 2017).

19 On the other hand it could be argued that if such flows were very high, it could have an inflationary effect on Pacific economies, raising the shadow price of labour and acting as a disincentive to some forms of economic activity. It is also possible that much such income would be spent on consumption or the church rather than be channelled to productive investment. Such arguments, however, deny the rights of people to determine how and where they decide to spend their income and savings.

Royal Palace, Nukuʻalofa, Tonga

4 Sovereignty

Introduction

In Chapter Two we outlined how the successive international declarations on development aid effectiveness between Rome (2003), Paris (2005), Accra (2008) and Busan (2011) embodied the ideas underpinning the phase of neostructuralism in the 2000s. It is actually remarkable that none of these declarations about international collaboration between independent donor countries and independent recipient countries contain the word 'sovereignty'. Instead, the word 'ownership' takes centre stage. The guiding principles from these international declarations define ownership as the recipient countries exercising "effective leadership over their development policies" (OECD 2008: 3). As this definition seems to describe state sovereignty, we are left to wonder why in the world of international development aid, the word 'ownership' apparently has emerged as a substitute for 'sovereignty'.

In this chapter we explore the various meanings of, and the debates concerning the concept of sovereignty. This allows us to see that that the use of the term 'ownership' rather than 'sovereignty' in official declarations about international aid is not arbitrary; rather it reflects the fact that "Sovereignty, as a concept, is in a state of flux" (Jacobsen et al. 2008: xiii). As such, the purpose of the chapter is to provide a perspective on how development, aid and the concepts of ownership and sovereignty mutually interact in the context of the Pacific. This lays the ground for the subsequent chapters which focus on the practice of development aid in the region and the constructions and expressions of sovereignty in recipient countries and territories that arise in aid negotiations.

We will first explore how the contemporary concept of sovereignty can be deconstructed to reveal one comprising three elements that connect the 17th, 19th and 20th centuries. Next, we review five lines of critical analyses of the concept of sovereignty as they emerged over the last decade or so, including debates associated with the concept of indigenous sovereignty. With this understanding as a basis, we then we investigate how social, cultural and economic values and practices of Pacific peoples might form a distinct Pacific or 'Oceanic' concept of sovereignty. Towards the end we will also explore a particular subset of this Oceanic sovereignty; the ways in which non-self-governing islands express self-determination while maintaining constitutional bonds with their former colonial metropoles.

The contemporary concept of sovereignty

The accepted contemporary concept of state sovereignty refers to 'Westphalian sovereignty' alluding to its origins: the peace treaties concluded in the German state of Westphalia in 1648 that ended the wars that had ravaged Europe for many decades. The negotiations towards these peace treaties took several years and in the course of time, hitherto ancient principles of international relations were abandoned, and new ones established. One new principle was the right of a state to 'self-determination' and the corollary, the principle of non-intervention in the domestic affairs of other states. At the time, this led to the wide acceptance of the principle that a state's ruler had the exclusive right to determine his or her state's religion. A Catholic king in one state was no longer allowed to speak for Catholic subjects in a neighbouring state ruled by a Protestant prince, for example. Another new principle with a considerable legacy was that all states were to be treated as equals, irrespective of their size or power. Noblemen representing the Holy Roman Emperor, for example, negotiated for the first time on equal footing with an illegitimate son of a local chief who represented a recently seceded region of a minor Dutch bishopric (Blok and Molhuysen 1924). Protocol rules for negotiations among the 83 diplomatic parties were rewritten – a veritable 17th-century version of UN's General Assembly: "They entered at the same time and same speed, saluting each other in great show and sitting down at the table at the same moment" (Meerts and Beeuwkes 2008: 164). These new principles of international relations constitute the foundational element of today's concept of sovereignty.

The next important element emerged in the early 19th century when nationalist movements began challenging a monarch's right to rule based on hereditary and/or divine rights. Hitherto, even revolutionary or secessionist movements had not challenged the position of a king as the supreme lawmaker. The concept that the king or queen was the sovereign was succinctly expressed by the 18th-century French King Louis XIV's apocryphal: 'I am the state' (*l'état, c'est moi*). This view remained influential, even in ostensible departures from this norm. For example, the Declarations of Independence by the Dutch (1581) and the USA (1776) did not challenge the concept of royal sovereignty; rather, these declarations presented the rebelling populations of the territories as subjects to a king who felt aggrieved by his administrative policies, including taxation. In fact, grievances justifying these rebellions take up more than two-thirds of both declarations (Armitage 2005: 4; Coopmans 1983: 548).

Box 4.1 Westphalia in the Pacific: the Kingdom of Tonga

John Overton

Tonga's King is considered a virtually sacred paramount chief, and, despite recent reforms allowing for an elected majority in parliament, the King and a class of nobles effectively control the legislative and executive institutions of the state. Moreover, Tonga's King retains largely unfettered control over public resources. In many ways, Tonga represents an example of the survival of an early form of Westphalian royal sovereignty: national sovereignty was built upon the authority of a powerful monarch.

After a period of conflict amongst various claimants to chiefly authority, a single throne was established during the mid-19th century. King Taufa'ahau proved adept at combining traditional authority with the adoption of Christianity and careful diplomacy with European powers to put in place a united Tongan state. Political and social reforms enacted by a new law code in 1862 (Campbell 1989: 81) drew both on advice from missionaries and a study of constitutions elsewhere in the region. These formed the basis of the Tongan state and reinforced the supremacy of the monarch, albeit with the institution of a legislature with, at the time quite liberal, provisions for an elected component (under the 1875 Constitution). Taufa'ahau and his advisors were also successful in playing European powers off against each other and, eventually, Tonga was recognised by them as a sovereign state. In 1900 a Treaty of Friendship was concluded with Great Britain which created in effect a British protectorate over Tonga. However, King Taufa'ahau and subsequent rulers, most notably Queen Sālote Tupou III (who ruled from 1918 until her death in 1965), continued to exercise almost complete control, receiving advice but not heavy-handed interference from the British (Thompson 1994: 81). Notably also, King Taufa'ahau – with shades of European rulers two centuries earlier – also established an independent Free Church of Tonga over which the monarch had considerable influence.

Tonga's royalty adopted many of the trappings of European royal families and the associated elite class – the nobles – continued to wield economic and political power. Traditional authority, then, was recrafted and ossified by grafting onto it Westphalian sovereignty in ways which gave external recognition and protection. It kept a model of royal sovereignty that allowed for a high degree of independence throughout the colonial era. Yet Tonga has been slow to move to the later models of sovereignty, notably those which derived legitimacy from either nationalist movements or decolonisation and which vested power in the hands of a popular democracy.

Political pressures for reform grew. Protests and pressure from metropolitan powers – interestingly with implications for development assistance as donors tried to 'encourage' democratic reforms (Osborne 2014) – gradually led to change. Shortly before his coronation in 2008, King George Tupou V signalled some major changes and, in 2010, a constitutional amendment allowed for an elected majority in the Legislative Assembly. However, the King still retains considerable power, including the power to appoint a Prime Minister and additional cabinet ministers so that the Executive branch of the state is still in effect under royal control. Tonga is the country in the Pacific Islands with the arguably the strongest claim to continuous independent sovereignty, yet it remains a country with its social and political order still rooted in a feudal past.

The French revolution of 1789 changed the concept of royal sovereignty drastically. The French revolutionaries not only decapitated King Louis XVI, but also at a conceptual level they removed the head of state as the sovereign power. Their *Declaration of the Rights of Man and of the Citizen* stated: "The principle of all sovereignty resides essentially in the nation" (Legifrance 1789). From then on, nationalist movements from Mediterranean Europe to Latin America justified insurrections against imperial rulers by identifying these rulers as 'foreigners' who had usurped the sovereignty of the people from that land. When, for example, in 1809 the local elites of Bogotá demanded independence from Spain, their claim to sovereignty was simply based on their connection with the land: "We are the descendants of those who shed their blood to acquire these lands" (Pérez Silva 2010).

The third and final central element of the contemporary concept of sovereignty was created in the decolonisation processes of the mid-20th century. Until then, claims to sovereignty in a land were mostly created by a bottom-up process in which local elites challenged the right to rule of distant – mostly European – overlords. The challenges, in turn, often faced a repressive and violent response from these rulers. The 1960 UN *Declaration on the Granting of Independence to Colonial Countries and Peoples* fundamentally changed that dynamic. It declared: "all peoples have an inalienable right to . . . their sovereignty and the integrity of their national territory" (United Nations General Assembly 1960). Although many colonial powers abstained from voting, the fact that no UN member voted against the Declaration underscores how colonial powers felt the 'winds of change', as the British Prime Minister MacMillan noted in the same year (Ovendale 1995: 455). Interestingly, such winds of change had moved fast: a similar United Nations Declaration had been presented in 1952, but was then voted down by Western powers (Reus-Smit 2011: 236).

The first example of decolonisation in the Pacific was in 1962 when Western Samoa was granted full independence under a new and novel constitution.

The new nation signed a Treaty of Friendship with New Zealand, its former colonial power. Yet the Treaty could not disguise a less-than-friendly past. Samoans had resisted rule by both New Zealand after 1914 and Germany before it. The Mau non-violent movement in the 1920s and 1930s involved a struggle for independence from rule by New Zealand but it encountered aggressive and lethal repression by New Zealand's military policy. In 1929 New Zealand police fired on a group of Mau protesters and several were killed. The Mau's motto was *Samoa mo Samoa* (Samoa for the Samoans) and it self-rule on the basis a *fa'a Samoa* (the Samoan way) as handed down by "ancestors . . . who sprang from the native soil of the islands created by Tagaloa-Iagi when he threw down pebbles from the heavens" (Herr 1975: 301). For the Mau movement, sovereignty meant getting rid of an external power. Yet, a little over 30 years after the fatal shootings, its leaders were able to forgive that repression, gain full sovereignty and forge a new special relationship with New Zealand.

In the 1960 United Nations Declaration a universal right to territorial sovereignty was introduced and enshrined. No longer did a people ruled by a distant ruler have to explain in detailed declarations why they wanted sovereignty, but rather these colonial rulers now had to explain to the international community why their overseas territories had not yet aqcuired it. In 2013, for example, the United Nations called on France to explain and "intensify its dialogue with French Polynesia in order to facilitate rapid progress towards a fair and effective self-determination" (UN 2013). In response, France sent a letter to UN member states noting it would not take part in that particular meeting of the General Assembly.

Since the UN remains pre-disposed to support and encourage claims to territorial self-governance and sovereignty, it is important to consider the process involved in becoming a sovereign state. Within the framework of the UN, the path towards recognition of a sovereign state is laid out clearly and simply. According to the UN Charter, any people living in a discrete territorial area that have established organisational forms that represent a state structure can apply to be admitted as a member state of the UN. Five criteria are applied to such applications: a people need to declare themselves to be a state, be peace-loving, accept the UN Charter, be willing to carry out the obligations contained in the Charter, and demonstrate their ability to implement these obligations (UN 1945: Article 4.1). With regard to the first criterion of the declaration of statehood, it should be noted that there is no formal definition of such a thing: "the most widely accepted source as to a definition of statehood is the Montevideo Convention of 1933" (Grant 1998: 403). Article 1 in this convention declares that a state is simply constituted by a permanent population, a defined territory, a government and a capacity to enter into relations with other states. None of the other criteria seems onerous either. As a consequence perhaps, the number of UN member states rose from 98 in 1960, to today's 193 (as of 2018).

The European peace treaties in 17th century, 19th century nationalism in Latin America and the Mediterranean, and the global 20th century's decolonisation processes have left an important legacy to the island states of the Pacific. These elements constitute a basis for international relations that have enabled small and lowly populated Pacific states such as Tuvalu (approximately 10,000 inhabitants) and Nauru (approximately 9,000 inhabitants) to acquire a sovereignty and ostensibly converse on an equal footing in the General Assembly with – and in a real sense are provided with equal seating to – historic and rising powers with populations that are far larger, such as the USA, India, Brazil, Russia or China.

Box 4.2 Sovereignty for Tokelau?

Warwick E. Murray

Tokelau is a New Zealand dependency in Polynesia with a population of just under 1400, though many more people of Tokelauan descent live in New Zealand and elsewhere. Located north of the Samoas and northwest of Cook Islands, Tokelau is comprised of three atolls. It has a scarce infrastructure, with no airport or harbour and communications are limited. Its GDP is approximately $US 1.5 million, and apart from small-scale local agricultural sectors most expenditure is funded by New Zealand aid (Hooper 1993). It is considered a non-self-governing territory by the United Nations which has applied pressure for self-determination through its commission on decolonisation across the years.

Though the islands of Tokelau have a legislative assembly – the general *Fono* – they have been administered by New Zealand since 1926, first as a British colony and subsequently (from 1949) as a New Zealand dominion and then realm territory. Tokelauans are citizens of New Zealand and outmigration has been a constant feature – with the consequence that remittances are very important in the national economy. Tokelau has the British Queen as head of state and New Zealand is represented by the Administrator. The territory has a high degree of effective self-government in terms of day-to-day affairs and there is also a high degree of village-level control (Angelo 1997).

In 1960, New Zealand proposed to merge Tokelau with either Western Samoa or the Cook Islands. Tokelau rejected both options. In 2006 and 2007 referendums on independence were organised, partly under pressure from the UN. These proposed a change to a self-governing state in free association with New Zealand, with loss of neither New Zealand citizenship nor the substantial budgetary support from New Zealand. Both polls failed to get sufficient votes for independence (a two-thirds majority for change was needed), although the margin was very close and less than a score of votes (out of just under 700 cast) would have swung the result in favour of the self-government option.

Tokelau remains on the UN list of non-self-governing territories and some commentators argue that Tokelauans may never vote for anything other than this situation (Hooper 2008). The National Party government of New Zealand (2008–2017) made it clear that it had no intention of undertaking further referenda in Tokelau in the foreseeable future. Tokelau thus represents an example of a society that has chosen not to push for constitutional sovereignty. For some external observers this is a sensible choice due to the questionable viability of such a small political entity. But it seems that for many Tokelauan people, especially the high numbers resident overseas, the (very slight) risk of losing unrestricted entry into, and support from, New Zealand is not worth taking.

Questioning Westphalian sovereignty

While the concept of Westphalian sovereignty is dominant in discourse and practice of international relations, it has been critiqued from various conceptual perspectives. Large-scale human migration, which has grown in proportional and absolute importance over time, has affected the connection to the land and sovereignty of various peoples in diverse and diverging ways. In this regard, many indigenous peoples, in particular, have lost sovereignty over their land largely in favour of European colonisers and have inherited the state system that was imposed following independence. On the other hand, the perspective of sovereignty of the descendants of these settlers who now live in the lands that their forebears occupied is also affected in peculiar ways by their colonial history. Furthermore, the increasing importance of diasporic populations, including both those that involve a diffuse population that spreads out from a territorially demarcated and recognised cultural hearth across international borders, as well those 'nations' of people without distinct territory of any sort, has further complicated the attachment of people to a sovereign territory.

Building in part from considerations such as those above and from a conceptual perspective, we can discern five lines of critique of the concept of Westphalian sovereignty in the literature. First, there is a school of thought arguing that today's appreciation of the principles that underpin the concept of Westphalian sovereignty are merely historical constructs of narrow Eurocentric views. These scholars argue the 17th-century peace treaties of Westphalia were at the time far less innovative than has been depicted: the Treaties of Westphalia can be seen as just one more set of peace treaties in a continuous and evolutionary chain. They claim that the present view of the Westphalian treaties is "really a product of the nineteenth- and twentieth-century fixation on the concept of sovereignty" (Osiander 2001: 251). Others add that the global discourse concerning Westphalian sovereignty reflects a 'Eurocentric metanarrative' predicated on Europe's history of imperial and territorial competition (De Carvalho et al. 2011: 737). This Eurocentric view on sovereignty has little meaning or value to other regions of the world where relations between states are influenced by other principles and different factors, or it ignores the impact European colonialism has had on the "relation between Orient and Occident" (Pourmokhtari 2013: 1767).

A second line of critique contends that the principles of Westphalian sovereignty are only professed when it is convenient, but amount to little in the practice of international relations. These critiques suggest that larger powers continue to violate the principles of non-intervention and 'size-does-not-matter'. The 'war on terrorism' in the past decade has seen larger states intervene in other states not only with diplomatic initiatives, but also militarily without much regard for the principles of Westphalian sovereignty. In addition, the UN – the pre-eminent forum upholding the principles of Westphalian sovereignty – began reconsidering the application of these principles when its Secretary General wrote in 2000: "The principles of sovereignty and non-interference offer vital protection to small and weak states. But . . . if humanitarian intervention is, indeed, an unacceptable assault on sovereignty, how should we respond to a Rwanda, to a Srebrenica – to gross and systematic violations of human rights?" (United Nations General Assembly 2000: 35). From this moment on, the UN has been sanctioning politico-military interventions leading state-building processes driven by external actors as much as by people born and living in a given territory. From this, new principles such as the international community's 'responsibility to protect' (R2P) and the concept of a 'shared sovereignty' were born (Krasner 2005: 74).

A third line of conceptual critique focuses on the impact economic forces since the 1990s are having on the sovereignty of states. Several researchers point out that globalisation is effectively dissolving the borders between states and thereby the boundaries within which states exercised self-determination. Citing a few titles of authors of such an opinion speaks volumes: 'Sovereignty and state . . . in jeopardy' (Jotia 2011), 'Waning sovereignty' (Brown 2010), and 'Sovereignty under siege' (Beeson 2003; Patman and Rudd 2005) and, most famously, 'The end of history' (Fukuyama 1992). However, the opposition to these views is equally vocal. Agnew counters that the story claiming "state sovereignty is in worldwide eclipse in the face of an overwhelming process of globalisation" is "overstated and misleading" (2009: vii). He claims states' sovereignty over their territories has never been has never been as complete as is often portrayed and he suggests states are finding new ways to exert control. Others concur and find the alarm expressed by many is "premature and inconsistent with the continuing power and relevance of nation-states" (Ku and Yoo 2013: 235). Indeed, since neoliberal globalisation has come to dominate economic development, the number of states has increased not decreased.

A fourth line of critique doubting the relevance of Westphalian sovereignty is connected to both globalisation and the increasing frequency of politico-military interventions. Rather than focusing on the loss of sovereignty by the state, scholars in this vein have investigated how the powers of sovereignty have shifted. They present the concept of 'sovereignty without territory' in reference to the increasing authority and capabilities of global bodies whose powers to develop and enforce policies are superseding or displacing the sovereignty of states. These global bodies can be multilateral organisations or agencies overruling state sovereignty on moral grounds – such as UN-sanctioned peace-keeping and state-building missions – but these bodies can also be offices that enforce international trade agreements, ranging from the World Trade Organisation (WTO) to the courts attached to free trade agreements to settle disputes between international investors and nation states (e.g. Otero 2011; Perez et al. 2011). Simpson's review of these new actors in the international arena led him to talk of the many 'guises of sovereignty', before detailing provocatively new adjectives such as: 'metaphysical', 'extraterritorial', 'deferred', 'internationalised', 'incipient' and 'deterritorialized' sovereignties (2008: 69).

If conceiving of 'sovereignty without a territory' is challenging, equally challenging is a final line of critique to the concept of Westphalian sovereignty presented by indigenous peoples: 'sovereignty without a state'. Whereas the decolonisation of the mid-20th century resulted in the formal sovereignty of most peoples who were colonised, indigenous peoples were often excluded from these processes. Interestingly, the 1960 UN Declaration on decolonisation did not seem to apply to indigenous peoples, perhaps because some of them had become small minorities in their lands as a consequence of the immigration of large numbers of European or other settlers. Yet, after decades of struggle and debate, indigenous peoples succeeded in 2007 to get some of their perspectives on indigenous sovereignty recognised with the adoption of UN Declaration on the Rights of Indigenous Peoples. Most of the Declaration predicates itself on the classic elements of Westphalian state sovereignty we presented earlier, but now applies these principles explicitly to indigenous peoples. It refers to the principle of equality irrespective of size – "indigenous peoples are equal to all other peoples" – and it underscores their "right to self-determination." However, the highly contentious character of the Declaration was revealed in the very last article stating that "nothing in this Declaration" can be used to impair "the territorial integrity or political unity of sovereign and independent States" (United Nations General Assembly 2007: 1, 4, 14). In effect, the UN established the concept of indigenous sovereignty as a form of sovereignty without a state.

The last article was added at the very last moment, less than two weeks before its adoption by the UN General Assembly. Indigenous representatives "were well aware that many Indigenous peoples had argued for many years against the inclusion of a provision upholding state territorial integrity in the Declaration," but learnt that without this added article "adoption in the UN General Assembly was proving to be extremely difficult" (Global Indigenous Caucus 2007: 1, 5). They accepted the last-minute article and days later, the General Assembly adopted the Declaration with only four 'no' votes: Australia, Canada, New Zealand, and the USA. At this point in time, the Declaration mostly serves as an inspirational framework to negotiate settlements between indigenous peoples and states. The states that voted 'no' in 2007 have come around and now accept the declaration as a set of "non-binding aspirations" (New Zealand Parliament 2010). Among indigenous peoples, the debates continue about the meaning of the Declaration. Some commentators believe a train has been set in motion that will ultimately lead to indigenous states (Daes 2008: 24). Others believe "for the most parts, Indigenous peoples do not wish to be states" (Young 2004: 187) and find "Indigenous peoples do not see the state 'as the highest and most liberating form of human association'" (Anaya, cited in: Pitty and Smith 2011: 126).

We believe debates and negotiations in the coming years will centre upon what indigenous sovereignty actually means as a result of what the Declaration repeatedly defines as "constructive arrangements between States and Indigenous peoples" (United Nations General Assembly 2007: 2, 3, 13). Furthermore, we acknowledge how much indigenous concepts and practical struggles regarding sovereignty, specifically *tino rangatiratanga* in Aotearoa/New Zealand (Box 1.2) have helped shape our own approach to understanding complex practices of sovereignty in the wider Pacific region. In fact, the peoples of the Pacific are adding their own perspective – an Oceanic perspective – to the concept of sovereignty, as we explore in the next section.

Oceanic sovereignty

For decades now, mainstream views on current or potential sovereignty of Pacific islands have revolved around economic, political and geographic impediments to development of the islands, often adding that social and cultural practices of the peoples in the Pacific also represent barriers. Recently, the problems emanating from climate change – increasing intensity and frequency of storms and droughts, and rising sea levels – have given further weight to the image of Pacific societies as vulnerable and dependent (Barnett and Campbell 2010). In this section we will look briefly at these oft-mentioned impediments to being and acting sovereign in the Pacific, before outlining what we see as the effective and unique Pacific responses to, or interpretation of, what we contest are largely constructed obstacles.

When in 1960 the UN Special Committee on Decolonisation established a list of non-self-governing territories that should be decolonised, most islands in the Pacific and the Caribbean featured. And as with most other territories on the list, many islands became sovereign states in the course of the 1960s and 1970s. However, by the early 1980s the first voices began pointing out that the newly acquired independence did not work out as positively as people on the islands had hoped. Problems resulting from "smallness . . .

and colonially induced dependence" came to "haunt the island states in a way unimagined at the time of independence" (Robertson 1988: 617). For Samoa, for example, it was argued, "access to overseas remittances, and the smallness of the domestic economy serve to perpetuate the deadlock" (Ronnås 1993: 339). Seen with the eyes of analysts in urban or continental environments, the islands in the Pacific are indeed very different. Measured against the yardsticks of peoples who can drive hundreds of kilometres without seeing the sea, most of the islands in the Pacific are isolated and remote, have very small populations, virtually no large-scale commercial agriculture nor abundant mineral resources to finance development. Because Pacific islands lack contiguous neighbours, size and resources, some outside observers have concluded the islands, as with other small island states, must have low levels of domestic activity, no economies of scale and very limited political bargaining power to express their sovereignty on the international stage (Ward 1967; Dommen 1980; Selwyn 1980; Shaw 1982; Lockhart et al. 1993).

A number of largely foreign researchers argued independent island states were left dependent on Migration, Remittances and Aid to sustain their relatively large Bureaucracies. This formed the acronym 'MIRAB' – a putative model for development of small island states (Bertram and Watters 1985 – see Box 2.2 – also Watters 1987; Munro 1990; Connell 1991; Hooper 1993). It was clear development aid for these islands had become a critically important resource to sustain the institutions of the sovereign states they had become. Most aid came from the former colonial metropole that had also become the preferred destination for migration and thereby the origin of the remittances these migrants sent back to the islands. In this light, it is not surprising that many policy makers in the 1990s and 2000s found this MIRAB model of development – largely predicated on islands' deficiencies and dependencies – a useful model to justify external assistance to the Pacific Islands. In today's retroliberal phase, the MIRAB model serves equally well to justify interventions in the Pacific by global powers seeking to extend the reach of continental corporations.

The discourses of modernisation and globalisation consign Pacific Island people to passive, recipient roles where sovereign progress – i.e. progress in forms and directions decided by Pacific people – is apparently not really happening nor possible. In addition, as climate change is becoming an international concern, the international community has singled out small islands as a particularly vulnerable group. Pacific Islands are suffering from environmental change brought about by modern industrialised economies way beyond their shores and yet they must bear some of the worst consequences, some of them threatening the islands' very survival. The image that triggered UN programmes supporting small islands was laid out in a landmark paper:

> Many small island developing states (SIDS) face special disadvantages associated with small size, insularity, remoteness and proneness to natural

> disasters. These factors render the economies of these states very vulnerable to forces outside their control – a condition which sometimes threatens their very economic viability.
>
> (Briguglio 1995: 1615)

On top of the islands' supposed passivity and vulnerability, modernisation and globalisation discourses also argue that the Pacific's diverse cultures are not conducive to sovereign economic and political development. Pacific practices – *kastom* – including for example communal ownership of land and reciprocity in transactions, are deemed hindrances, as World Bank research entitled *Obstacles to Economic Growth in Six Pacific Island Countries* reported in 2006:

> Samoa has constraints on foreign investment in the form of obligations to include local investors in any foreign investment, which must be a significant obstacle to overseas investors (. . .)
>
> Major obstacles to private investment and growth in Vanuatu appear to be the following: difficulties in obtaining secure, long-term rights to land . . .
>
> (Duncan and Nakagawa 2006: 39, 59)

As a consequence, so the orthodox argument goes, Pacific countries and territories struggle to connect with the dominant global political and economic structures. The removal of Pacific cultural practices that are perceived as obstacles by overseas corporations continues to figure in the conversations between the overseas government officials and political representatives of Pacific states:

> We must maintain a[n] . . . economic environment that protects major investment, and grows our international reputation . . . We must prepare to be as competitive as we possibly can be. That is a challenge my Government understands, and increasingly our landowners understand.
>
> (Prime Minister's Office of Papua New Guinea 2014)

> [Papua New Guinea] Minister for Pubic Enterprise and State Investment Ben Micah visited New Zealand to discuss energy and land ownership model issues.
>
> (MFAT 2014b)

However, we would contest these images of the Pacific and the relevance of these arguments. Of course, climate change challenges many Pacific islands, but Pacific people are responding creatively and vigorously. Of course, small countries do not have a lot of leverage in international negotiations, but Pacific representatives are making their voices heard in the UN and gain concessions from superpowers that medium-sized countries could not realistically contemplate. And indeed, the view that communal ownership of land does not facilitate foreign investment by agencies that subscribe to the private ownership of all the factors of production, is not the same view as those who see considerable value and enhanced well-being in maintaining strong communities, historic, cultural and spiritual ties to the land, and the ability to provide for subsistence needs (Crocombe 1972, 1987; O'Meara 1987; Crocombe and Meleisea 1994; Ward and Kingdon 1995; Batibasaqa et al. 1999). It is important to point out that the above images and arguments come from the modernising and globalising discourses, that seek to understand and impose that understanding of islands from the perspective of their arriving 'vessel' with their own, continental, homes at their back. In contrast, fewer people have explained the world from the perspective from the islands next to their canoe looking for arriving vessels and towards the horizon (Voi 2000). What does Pacific sovereignty look like from this perspective?

Epeli Hau'ofa (1939–2009), who lived in Papua New Guinea, Tonga and Fiji, was trained as an anthropologist. In a seminal paper he recounted how he was trained to see the Pacific world – or 'Oceania' as he later insisted – from a European perspective:

> Initially, I agreed wholeheartedly with this perspective . . . It seemed to be based on irrefutable evidence . . . our economies were stagnating or declining; our environments were deteriorating . . . our people were evacuating themselves to greener pastures elsewhere . . . Some of our islands had become, in the words of one social scientist, 'MIRAB Societies'.
>
> (Hau'ofa 1993: 4)

However, as he reflected more on his responsibilities as an educator, he began to realise, "I was actively participating in our own belittlement" and started to explore other, older, perspectives. In precolonial times, Pacific peoples had travelled and traded over great distances for thousands of years, possibly as the greatest seafarers of all times if one compares the level of technology at their disposal. As Hau'ofa argued, with an Oceanic perspective the ocean connected Pacific peoples instead of separating them; the sea is a possibility instead of an obstacle:

> If we look at the myths, legends and oral traditions, and the cosmologies of the peoples of Oceania, it will become evident that they did not conceive of their world in such microscopic proportions. Their universe comprised not only land surfaces, but the surrounding ocean as far as they could traverse and exploit. . . . There is a gulf of difference between viewing the Pacific as 'islands in a far sea' and as 'a sea of islands'.
>
> (Hau'ofa 1993: 6)

Importantly, Hau'ofa revalued Pacific people's personal networks in the past and present. He argued these personal networks – built on the principle of reciprocity – had been resource and a source of wealth in precolonial times. "Theirs was a large world in which peoples and cultures moved and mingled unhindered by boundaries . . . from one island to another they sailed to trade and to marry, thereby expanding social networks for greater flow of wealth" (1993: 8). And, he continued, the end of colonialism "had a liberating effect . . . enabling the people to shake off their confinement and they have since moved, by the tens of thousands, doing what their ancestors had done before them: enlarging their world as they go" (Hau'ofa 1993: 10). They are a mobile and diverse workforce that ranges from peace-keeping in conflict zones, nursing in hospitals across the world, to professional sporting codes in North America, Europe and Australia and New Zealand.

However, it should be noted that the migration of Pacific people does not mean they lose their bonds with 'home' – whether they are recent arrivals or have been 'away' from their homeland for generations, the bonds of kinship, community and culture remain very strong. Modern telecommunications and transport have only assisted Pacific migrants to remain firmly centred on 'home' and in regular contact: remittances continue to play a very important role many generations after the first-out migration of Pacific peoples to the Pacific Rim. Hayes has described societies that "become dispersed in space while still retaining some degree of organic unity at the level of the social system" (1991: 7). This sets Pacific Island people apart in the world of migration. They act quite unlike the people that moved from Europe to, say, the America, Southern Africa, or Australia and New Zealand, who generally lose the bond with the family left behind in Europe. All, it seems, underscores Hau'ofa's point: "the world of Oceania is neither tiny nor deficient in resources" (1993: 11).

Tonga's Futa Helu (1934–2010) was a philosopher, who also contributed to rethinking the way Pacific Island people perceived themselves and their relations with the world. He critically analysed the dynamics of Pacific 'customs' and concluded that many ruling elites throughout the Pacific used a mix of customs and Western institutions to keep themselves permanently in power – a very un-traditional practice. "In many . . . Pacific Islands, power rotated . . . quickly and few regimes held sway, without major modification, for longer than a few decades" (1994: 230). Along with Hau'ofa, he looked at Pacific people's great mobility and questioned whether the Western concept of state sovereignty had relevance for the peoples of the Pacific. As a logical consequence, he argued strongly against what many people consider the bedrock of state sovereignty. "Standing armies as insurance against external aggression are . . . in the Pacific Islands utter waste and folly. What can be gained by an aggressor who invades Tonga, Kiribati, or Tuvalu?" (1993: 326). Helu was equally adamant that Tongans coming from a culture with a "penchant for generosity, consensus, friendship, etc. cannot function successfully with this moral gear in a society where dog eats dog" because this would leave Tongans as "perfect pawns for exploitation by peoples of cultures whose values include acquisitiveness, thrift, drive and so on" (1994: 191). He argued – and taught – that the sovereignty of Pacific peoples can only be upheld if they fully understand the other's and their own cultures when they engage with continental powers. In Futa Helu's 'Atensi Institute students were encouraged to learn both Pacific culture and classic European culture including literature, music and art.

Also at the 'Atenisi Institute Futa Helu ensured that Tongan was a medium of instruction for some of the classes, saying that some concepts are better discussed and understood in Pacific languages. The work of a fellow Tongan, who is part of the same well known political and intellectual family – Konai Helu Thaman – has taken this point further. When discussing how to "decolonise Pacific studies," she contends we need to study how colonisation and globalisation have had

> an impact on people's minds, particularly on their way of knowing, their views of who and what they are . . . western-derived economic and educational developments have destroyed important aspects of Oceanic cultures, including languages, as well as social, political and economic structures.
>
> (Thaman 2003: 1, 7)

Helu Thaman – along with Hau'ofa and Helu and other Pacific scholars such as David Gegeo (1998), Gegeo and Watson-Gegeo (2001) – essentially advocate revisiting the way we look at key concepts that underpin 'our way of knowing' in the Pacific, suggesting we stand on the beach when we do it.[1] There are two key complementary concepts that might be helpful in understanding what sovereignty may mean from a Pacific perspective: *vaka vanua* (the ways of the land) and *vaka moana* (the ways of the sea).[2] We believe that, taken together, the values and practices of *vaka vanua* and *vaka moana* might elucidate a uniquely Pacific concept of sovereignty.

Vaka vanua comprises the intricate and diverse ways in which Pacific communities have structured their local social, economic, cultural and political practices; a complex and inter-connected set of practices in different spheres of life. It can sometimes be seen as a synonym for 'customary' ('the way of the land') and is similar in many ways to the Melanesian expression of *kastom*. It encapsulates the intimate connection people have with their *vanua* – their 'land' in its widest sense (encompassing the soil and water, the flora and fauna, the people and culture, and the past, present and future). It is largely terrestrial and can be strongly territorial, in that different *vanua*, as in Fiji, are associated with different kinship-based groups. Yet *vaka vanua* is not immutable or static, and changes in response to new opportunities or pressures will emerge. Changes in one sphere may, or may not, affect practices in adjacent spheres (see Ward 1965; Ravuvu 1983; Ward and Kingdon 1995). Helu, for example, has discussed convincingly how marketisation eroded most of Tonga's customary collective economic production, but customary socio-political redistribution practices persisted (Helu 1993: 193).

It is important to note that in many cases *vaka vanua* in an island society (in Polynesia in particular) centres on multiple layers of hierarchical governance structures, where most decisions are made at the lowest level of the pyramid where people are held together by intimate kinship and daily interactions. Moving towards the top, layers are increasingly connected through common ancestors. 'Kings' have sometimes emerged as a result of negotiations among groups at lower levels – and were also removed in the same fashion. For example, on the island of Wallis, after extensive deliberations that had lasted months, the customary council of ministers chose Kapeliele Faupala to become king of the island in July 2008. By 2014, discontent with the king's rule was widespread. At the celebration of Maria's Assumption on 15 August, thousands attended Catholic Mass in the Cathedral but only a handful attended the subsequent traditional ceremony at the next-door royal palace. In September, most clan leaders agreed to remove the king from the throne. However, selecting a new king proved contentious. In 2016 the royal succession struggle came to a head and Wallis saw enthronement ceremonies for two, competing, kings. At the time of writing, matters remained unsettled.

This complex array of power and relationships is what Helu meant when he noted that power at the top in many Pacific island polities rotated quite quickly. To illustrate how, generally, the top of the pyramid is less powerful than the base, it suffices to point out that the power base of most Pacific kings or top elites is ceremonial and adjudicatory; they do not command local resources of land or labour. These resources are controlled by kinship leaders at the lower levels. This aspect of *vaka vanua* also explains why some Pacific governments – Samoa, Fiji, Vanuatu – face strong local opposition if they attempt to interfere in the management of customary land, for example. In relation to our discussion of sovereignty, therefore, *vaka vanua* may accord with the Westphalian notions of territory and self-determination and it stresses the pre-eminent right of indigeneity – those who belong to a particular territory and have *mana* (authority) over the *vanua*. Yet it is also complex, layered and flexible: power can be contested, different social levels have different arenas of power, and customary practice, whilst durable, is capable of change and providing a set of principles for dealing with new opportunities or threats.

The idea of *vaka vanua* also explains why many Pacific societies in the past have resisted colonial rule and in some cases fought for independence. In Samoa, German then New Zealand colonial rule was actively opposed over a long period of time and full independence, when gained in 1962, was accompanied not only by a recognition of national self-government but also of the customary local *matai* system which, in effect, limited the power of the new state over local land and village-level governance. Samoan sovereignty draws heavily on the local concept of *pule* (Ulu 2013). In New Caledonia, indigenous *kanak* people have long resisted French rule and control of land and resources. Independence there has been called for and fought for, yet it may well be that a more complex and negotiated form of sovereignty will emerge that falls short of *kanak* independence in a Westphalian sense over the whole territory, yet delivers effective local sovereignty for communities with continued metropolitan subsidies. And in Fiji, although the real reasons may be much deeper and opaque, coups in 1987 and 2002 were in the name at least of indigenous *taukei* sovereignty with regard to a perceived assertion of political power by Fijians of Indian ancestry (Box 4.3).

Box 4.3 Fiji: constitutions and coups

Warwick E. Murray

The case of Fiji shows us clearly that sovereignty is malleable and contested, and that gaining independence by no means guarantees that competing claims are necessarily resolved. After a brief period as an independent monarchy Fiji became a British colony in 1874 – the first in the Pacific region if New Zealand of 1840 is not included. The colony lasted nearly 100 years and it was during this time that indentured labour was imported from another British colony, India. Indo-Fijians came to play the central role in the country's main economic base – the sugar industry, first providing the cane-farming and cutting labour and eventually from 1909 leasing tribal land from ethnic Fijians. This ethnic division of labour would later have significant political implications in the country. In 1970, after a significant period of preparation under which Fiji's first constitution was constructed, Fiji became independent as a member of the Commonwealth of Nations.

The contentious nature of sovereignty was evident in the nature of this constitution which structured the voting system in such a way to give dominance in parliament to ethnic Fijians. The first prime minister of Fiji was a long-respected paramount chief Ratu Sir Kamisese Mara (Lal 1986). In 1987, a coalition of the Fiji Labour Party and the National Federation Party was voted in and Timoci Bavadra became Prime Minister. Although Bavadra was an ethnic Fijian there was a perception – or perhaps a misrepresentation – that the new government was 'Indian dominated'. Lt-Col Sitiveni Rabuka took control of the government and following unrest after the first military intervention staged a second coup which vested control in his hands. Accusations of ethnic dominance were used as a rhetorical device to cover deeper issues of governance and control would be utilised a number of times in the subsequent history of the country (Hagan 1987; Robertson and Tamanisau 1988). This led to the proclamation of the Republic of Fiji and the establishment of the presidency, which in turn saw the country suspended from the Commonwealth. In 1990 a new constitution was put into place which further cemented the dominance of ethnic Fijians in the parliamentary system through the extension of reserved seats based on ethnicity.

In 1992 Rabuka won the elections – this time as a civilian – under the 1990 constitution. With his new ethno-nationalist party, he dominated politics and undertook further reforms to the constitution. In 1997, Fiji re-joined the Commonwealth and a new constitution was established. The 1999 elections saw the victory of the Fijian Labour Party led by Mahendra Chaudary. Although Chaudary was Indo-Fijian and had a labour union background among sugar farmers and workers of the western division of Fiji, the government itself was not, as sometimes claimed, dominated by Indo-Fijians; it was a broad progressive coalition. Nonetheless, the Chaudary government's policies led to the coup in 2000, when a civilian George Speight took control citing ethno-nationalist imperatives. Eventually control was regained by Commodore Frank Bainimarama who took executive control from President Mara and handed it to President Josefa Iloilo (Lal and Pretes 2001; Robertson and Sutherland 2001).

A new election was held in 2001, which led to the Prime Ministership of Laisenia Qarase under the auspices of a new nationalist party called the United Fiji Party (SDL). Qarase was explicitly pro-ethnic Fijian and his Reconciliation Bill of 2005 sought to pardon a number of those involved in the coup and subsequent mutiny in 2000 and 2001 respectively. Having won the election of 2006, his insistence on the Reconciliation Bill saw Commodore Bainimarama intervene with a set of demands to drop the concessions. Qarase did not, and Bainimarama seized power by presidential decree of President Iloilo in 2006, citing racism against the non-ethnic Fijian population and corruption in the Qarase regime. In 2009, the takeover was deemed illegal by the Fijian court of appeal and President Iloilo dissolved parliament and the executive and heralded a 'new legal order' which saw the restoration of the government as it was before the court ruling and reinstated Bainimarama as Prime Minister.

As a consequence of these upheavals, Fiji's membership of the Commonwealth was again suspended in 2006, followed by a suspension by the Pacific Islands Forum in 2009, and there were a range of sanctions from development actors in the region including New Zealand and Australia. It was during this time that the influence of Chinese aid in Fiji, which asked fewer questions regarding domestic politics, became established. The military was criticised internationally for violating human rights, while supporters claimed that it was seeking to restore ethnic balance and root out ethno-nationalist corruption. Bainimarama dissolved the Great Council of Chiefs which he claimed was behind much of the instability that had dogged Fiji since it declaration of independence and which had been established initially during the colony as a vessel for the enactment of colonial power. Following public consultation, a new constitution was proclaimed in 2013 which removed many of the ethnic components of previous ones. Under this constitution, Bainimarama won the 2014 election with his new party Fiji First with approximately 60% of the popular vote. Relations with nearby countries and membership of the Commonwealth were restored.

In its approximately four and half decade history as an independent nation, Fiji has suffered four coups, various constitutional crises and mutinies and has put four constitutions in place. A military officer has been head of state on three occasions and the military has played a role in putting a number of presidents in office. Overall, the instability that has dogged Fiji has often been painted as an ethnic struggle for control, but in reality this has covered up deeper issues concerning the ownership of land, the dominance of the Eastern traditional chiefs over the Western chiefs and a broader struggle concerning for whom sovereignty exists. At times personal agendas and accusations of corruption have played a role as was the case in the intervention of Speight in 2000 and the removal of Qarase in 2006. In a general sense the story of Fiji illustrates the enormous complexity of the terms 'sovereignty' and 'independence', defined often by only those in power at the expense of relatively marginalised groups in society.

Next to the practices and values of *vaka vanua*, we find *vaka moana*, comprising the ways in which the many Pacific communities engage with the ocean and peoples 'over the horizon' – either by receiving them or by travelling to them. Evidently, the concept of sovereignty in *vaka moana* differs in several aspects from the Westphalian concept of sovereignty we discussed at the beginning of this chapter. The most relevant difference here is that Westphalian sovereignty seeks to enforce borders and restrict the free movement of people across these borders in order to protect wealth. As a consequence, states adhering to the concept of a Westphalian sovereignty debate who is included in the state and who is not. In sharp contrast, sovereignty in the framework of *vaka moana* encourages the movement of people, both incoming and outgoing, because the resulting personal networks are perceived to be a source of wealth. Hau'ofa's comment, "For my part, anyone who has lived in our region and is committed to Oceania, is an Oceanian" illustrates this most succinctly (2000: 36). An Oceanic view of sovereignty emphasises the ocean not as a boundary with other states but as a series of pathways to other linked communities and the protocols and relationships which bind communities together.

Vaka moana is associated with the long history of Pacific navigation. Intricate knowledges, sailing technologies, rituals and protocols allowed Pacific peoples to travel knowingly over vast distances of ocean for many centuries, way before Europeans had the technology to navigate such waters. It meant that people were confident voyaging and being an integral part of a rich and accommodating ocean that other people – such as European visitors – saw as huge, empty, strange and dangerous expanse of water. For most Pacific peoples, the ocean is not a barrier but a highway. And *vaka moana* is not just about the means to travel across and live in the ocean; it is also about the relationships that span it. Furthermore, *vaka moana* is not just focussed on the past and the historic achievements of the seafarers of the region, but it also extends into the present. Pacific Island people are found all across the globe, in proportional numbers way in excess of many other societies. The Pacific Islands diaspora has spread far and wide, seeing the spreading of cultures across the Pacific Rim, allowing us to hypothesise the existence of the new Polynesian triangle that includes Santiago, Los Angeles, as well as New Zealand and Australian cities (Barcham et al. 2009).

Within this *vaka moana* concept, the inherent vitality of the Pacific Island peoples shows development in very different ways. People move within and between boundaries and maximise opportunities where they might be found. The isolation and distance that orthodox analyses place at the centre of interpretations of development potential are actually vast resources of opportunity. Furthermore, the cultural traits that conventional analysis might place as barriers to development (e.g. the value put on reciprocity, community and family – ahead of individualism, materialism and commercialism) are actually essential to sustainable development when recast as cultural resources. Globalisation, then, is not a threat to the region; it is something which Pacific people have been dealing with adeptly for centuries.

Central in this analysis is the concrete problem of access to those places where the expanding diaspora can continue to prosper, and this has become an increasing problem given the border controls that have existed periodically and sometimes been accentuated in countries of the Pacific Rim and beyond. Under this approach it is not the territory or the nation-state that lies at the centre of the prospects for well-being of Pacific Islanders, but rather a decentralised networked diaspora. In addition, *vaka moana* involves strategies that seek to identify and expand opportunities for mobility, often involving 'cracks' in the exclusiveness of the sovereignty of metropolitan states. These cracks might include the use of kinship and genealogy to gain migration access to metropoles, and the ways in which Pacific peoples might adopt multiple national identities, yet maintain a strong social and cultural connection with their Pacific hearth. The mobility of Pacific professional sportspeople provides several such examples (Box 4.4). The concept of *vaka moana*, can enrich our understanding of sovereignty in the Pacific. In this sense, *vaka moana* does not stress territory and difference but rather interconnection, reciprocity and shared space. It opens up quite different understandings of sovereignty, as we explore next.

Box 4.4 Pacific athletes and sovereignty: two tales

John Overton

Nathan Mauger was born and raised in Christchurch, New Zealand. He became a professional rugby union player and represented the Canterbury Crusaders. He was selected at the national level and played two games for the All Blacks in 2002. He has Samoan ancestry through his mother's side of the family and one of his maternal uncles, Stephen Bachop, represented both New Zealand and Samoa at rugby. Once it became apparent that his prospects for a long career with the national team were limited, he looked at opportunities to play professional rugby in Europe where salaries were high but there were limitations on the number of overseas players who could be hired by each team. He received an offer to play for the Gloucester club in England encouraged by the realisation that he could apply for a French passport and thus qualify as a local (i.e. European player) and avoid the limit on foreign players. Unfortunately for him, the passport failed to materialise and he was forced to seek other opportunities, playing for a professional team in Japan. The story is of interest, however, for his claim to a French passport was based on the fact that his Samoan grandmother was born in Tahiti, the result of a journey there from Samoa by her family following old kinship links between the two islands. A distant Pacific kinship connection and an established sea journey across the ocean thus became a potential means for this New Zealand citizen with Samoan heritage to circumvent European labour regulations and find lucrative employment in the United Kingdom.

Jimmy Peau was born in Samoa and schooled in New Zealand, following his family's move to Auckland. He became a talented boxer and won a gold medal at the 1986 Commonwealth games in Edinburgh, representing New Zealand. He later turned professional, being based in Australia for much of his career. I remember seeing one his fights televised in Australia, perhaps it was against the American J.B. Williamson in 1989. Before the fight started there were the introductions and national anthems were played. First there was the Star Spangled Banner for the American boxer. Next there was the anthem for Jimmy Peau, who had taken the name 'Jimmy Thunder' when he became professional. The national anthem for Samoa was played, recognising the country of his birth. All seemed ready for the fight to begin, when a second national anthem was played for Thunder: the New Zealand anthem, in recognition of the country he grew up in and first represented in boxing. Again, the crowd readied themselves for the fight. But a third anthem started: this time 'Advance Australia Fair', a nod to the country of his then residence and, no doubt, a good way to get the local crowd on his side. It was an interesting illustration of how this

Samoan athlete could adopt three national identities that represented different and important phases of his life and reflected a life of international mobility to exploit opportunities worldwide.

Sources:

www.gloucesterrugby.co.uk/news/3007.php#.V6FuUbh97cs (accessed 3 August 2016)

www.worldboxingfederation.net/yy-wbf-article0490.htm (accessed 3 August 2016)

When one of the world's wealthier countries offers international aid to a Pacific island territory, it engages with values and practices of *vaka vanua* and *vaka moana*. On the one hand, the outside officials provide the aid to what they believe to be the most powerful representative – king or government – of the island. This approach may not adequately take account of the complexity of representation. *Vaka vanua* is likely to lead to situations where the implementation of aid programmes run into obstacles because important local decision-makers have not been involved or obtained their share. Moreover, international aid provides resources to persons at the top of the pyramid that was previously not available to them and for which *vaka vanua* protocols have not been observed. Quite possibly this generates new dynamics that, again, will probably not be conducive for effective aid programmes. On the other hand, international aid is also likely to be perceived in the framework of *vaka moana*; relationships with the wider overseas world. In this light, aid becomes a gift, a start of a reciprocal relationship. This may work out well if the aid is presented to achieve the donor's geopolitical objectives. However, it is less likely to work out well if the aid is provided as a condition for, or in expectation of, changes in domestic policies of the recipient island; for example legislation concerning land ownership.

Returning to the search for a Pacific concept of sovereignty, we believe *vaka vanua* and *vaka moana* describe a local as well as regional, Pacific or Oceanic, set of identities and aspirations. The concepts are not just connected, but intimately and inseparably intertwined. In the words of Hau'ofa:

> Our diversity is necessary for the struggle against the homogenising forces of the global juggernaut. It is even more necessary for those of us who must focus on strengthening their ancestral cultures against seemingly overwhelming forces, to regain their lost sovereignty. This regional identity is supplementary to the other identities that we already have.
> (Hau'ofa 2000: 33)

This struggle in the Pacific to express and uphold both local identity from *vaka vanua* and regional identity from *vaka moana* in politics, business and culture, often runs into opposition from Western ways of managing public policy, closing a business deal, or presenting an academic argument. Helu Thaman analyses this struggle as "a Tongan woman of the commoner class, schooled in western ways":

> your way
> objective
> analytic
> always doubting
> the truth
> until proof comes
> slowly
> quietly
> and it hurts
>
> my way
> subjective
> gut-feeling like
> always sure
> of the truth
> the proof
> is there
> waiting
> and it hurts
> (Thaman 2003: 4)

We now see three defining features of a Pacific concept of sovereignty. First, a Pacific sovereignty lays the primacy of political agency and power at the lower, kinship-based, levels of governance structures – less with the territory or the nation-state. This holds generally, despite the great diversity of social and political structures across the wider region, from strongly hierarchical kingdoms as in Tonga, to much more 'horizontal' and fluid social forms in many parts of Melanesia. A part of this idea is the very strong attachment of largely kinship-based groups to their *vanua* and the protection of it from outside interference. Second, the purpose of Pacific acts of sovereignty would be to make borders more porous for the mobility of people of the lower tiers of the pyramid easier – not enforce border controls for people. In this sense the relatively brief history of nation states (and their borders) in the region in the past 150 years has cut across and interrupted the mobility, interaction and interdependence that were hallmarks of pre-national forms of sovereignty. Third, success for Pacific sovereignty would be defined by the breadth of global personal networks and the wealth it generates for all who participate in it – not by the penetration of global private capital and the profit it generates for those who own it. Interestingly, Oceanic people were practising what we recognise as globalisation long before the term entered the popular conversation. The difference is that the form of globalisation in action here is one that is more holistic, comprising social, cultural and economic objectives. For decades, researchers have described these global personal networks as "transnational corporations of kin" (Bertram and Watters 1985; Munro 1990; Connell 2014; Marcus 1993; Voigt-Graf 2008), and although we do not think the term 'corporation' quite captures the character of these personal networks, it does emphasise the communal nature of Pacific development and juxtapose them effectively with 'transnational corporations of capital'.

An emerging 'Islandian' sovereignty?

Reminding ourselves of the critiques of the classic concept of Westphalian sovereignty and having now looked at the opportunities afforded by unique Pacific concepts, it becomes clear that the orthodox way of thinking about sovereignty can be substantially challenged. In fact, we believe Pacific peoples and states have been exerting their sovereignty and shaping relations across borders – including development aid – in unique ways that may be embodied in the term 'Oceanic sovereignty'. In addition, at the closing of this chapter, we would like to pause for a look at another unique form of sovereignty that we see emerging in islands across the globe, including the Pacific; the oxymoron of the sovereignty of non-self-governing islands.

The 1960 UN declaration on decolonisation not only declared that all colonised peoples have the right to sovereignty over their lands, but it also presumed that "the peoples of the world ardently desire the end of colonialism" (United Nations General Assembly 1960). And indeed, throughout the 1960s and 1970s, many colonised peoples successfully completed their struggles to become sovereign states – including many islands. In the Pacific, Samoa was the first to secure sovereignty and independence from New Zealand in 1962. However, the decolonisation process seems to have stalled since the mid-1980s. Since then, none of the about forty remaining non-self-governing islands associated with colonial metropoles in Europe, the USA or New Zealand have obtained independence. In fact, since the mid-1980s, non-self-governing islands have rejected independence when presented with the choice in referendums and in diverse forms, they have all maintained constitutional or treaty-based bonds with the colonial metropoles (Prinsen and Blaise 2017). People on these islands undoubtedly wanted an end to colonialism, but they did not 'ardently desire' an end to the bonds with their colonial metropoles (McElroy and Pearce 2006; McElroy and Parry 2012).

Some of the refusal or reluctance of non-self-governing islands to seek full, Westphalian, sovereignty may be understood when in the late 2000s, researchers compared socio-economic indicators of sovereign and non-self-governing islands and it became clear that the latter were doing much better in a material sense than the islands which had seen sovereignty as a road to prosperity and better lives. McElroy and Parry, for example, compared twenty-five non-self-governing islands and thirty sovereign islands over the 1985–2010 period and found per capita income on non-self-governing islands was approximately double, and infant mortality half (2012: 415) of those of sovereign island states. These patterns were confirmed by others (Baldacchino and Milne 2006; Dunn 2011; Oostindie 2006). This triggered a review of the MIRAB model we discussed earlier.

A new perspective emerged, suggesting peoples and governments on islands were not so much dependent or passive recipients of outside resources such as aid and remittances, but active global actors, "making their own often transnational futures through international networks that mobilised opportunities, resources, identities in multiple sites" (Marsters et al. 2006: 35). Baldacchino spoke of islands "tapping the hinterland" (2010: 65 – also Baldacchino 2006a, 2006b). This new perspective argued that islands carefully manage their powers with regard to People, Resources, Overseas paradiplomacy, Finance/taxation and Transportation to tap into their hinterland: a model represented by the acronym 'PROFIT' (Oberst and McElroy 2007). In this model, both sovereign and non-self-governing islands engage actively and creatively with the world, but islands that are not fully self-governing have proved to be much more successful at negotiating their interests than sovereign islands.[3] This seems contradictory, if one had thought sovereignty provided a higher degree of self-determination, but this finding parallels our 'inverse sovereignty' premise at the start of this book.

When reviewing the connections between many of these non-self-governing islands and their colonial metropoles, Baldacchino and Hepburn concluded these islands expressed "a different appetite for sovereignty," in which they are negotiating 'innovative autonomy arrangements' rather than seeking Westphalian state sovereignty (2012: 555). Studying the arrangements within France's colonial heritage, Mrgudovic drew the same conclusion when she wrote that these islands explore "the fine line between autonomy and sovereignty" (2012a: 456). In reply to Hepburn's question, "what makes [these] islands so special?" (2012: 123), we believe there are five distinct patterns in how non-self-governing islands engage so successfully with their metropoles expressing what we would define as an 'Islandian', sovereignty (Prinsen and Blaise 2017; Prinsen et al. 2017).

1 They vote 'no' in independence referenda. After the people of the Federated States of Micronesia voted 'no' to Westphalian independence in their referendum in 1983, people on all non-self-governing islands have consistently voted against independence and in favour of retaining close bonds with the old colonial metropole, sometimes repeatedly: Curaçao (1993), St Maarten (1994, 2000), Puerto Rico (1993, 1998, 2012), US Virgin Islands (1993), Bonaire (2004), Saba (2004), Bermuda (1995), Curaçao (2005), St Eustatius (2005), Tokelau (2006, 2007), Mayotte (2009).
2 They continuously negotiate variations on already exceptional constitutional statuses. When a colonial metropole possesses far-flung islands, each of these islands tends to have a constitutional relationship that is

different from the one the other islands have. Moreover, these statuses are often renegotiated in the course of time. Mrgudovic reviewed the French territories and concluded that the islands' constitutional statuses, "have evolved progressively at different speeds and degrees" and "there are as many statuses as there are overseas territories" (Mrgudovic 2012b: 85, 95). For the United Kingdom, more or less the same applies (Hintjens and Hodge 2012). The six Caribbean islands of the Netherlands have four different statuses – ranging from 'countries within the Kingdom' to 'special municipalities' – and these statuses have been renegotiated in 1954, in 1986, and again in 2010 (Kochenov 2012). The USA has different constitutional arrangements with seven Pacific territories. Three have 'compacts of free association', but each of these compacts differs in essential aspects and each has been subject to renegotiations (Van Beverhoudt 2003). New Zealand has different connections to the Cook Islands, Niue and Tokelau (Angelo 2002).

3 They 'get away' with bending the metropole's rules. Aside from the constitutional exceptions non-self-governing islands secure, they are otherwise expected to abide by the laws of their metropole. However, there are numerous examples of islands effectively bending, ignoring, or postponing the application of these laws to an extent that is inconceivable for other sub-national entities. For example, Dutch law recognised same-sex marriages since 2001, but the Aruba island council argued this was against Aruban morality and refused to comply. Years of legal wrangling ensued, and in 2013 Aruba was forced to 'register' same-sex marriages. Since then, it has been arguing it does not need to give these marriages the same legal effect in the provision of public services (Rasmijn 2013). New court cases have been started. In another example, France has a strict constitutional separation of church and state: "The Republic neither recognises, nor salaries, nor subsidises any religion" (Legifrance 1905). Yet, on Wallis and Futuna the entire primary education system is managed by the Catholic Mission and the French state has paid the Mission for all related expenses since 1969 (Lotti 2011).

4 They manage their own public budgets, but shortfalls in domestic revenue are covered by financial transfers from the metropole. Governing bodies of non-self-governing islands often have a large degree of autonomy and responsibility in local tax collection. Most have a recurring budget deficit, but secure arrangements with their metropoles cover or underwrite the shortfalls in domestic revenue collection. For example, the Federated States of Micronesia receives budgetary support from the USA, averaging an annual $US 113.7 million between 2007–2011, or $US 1,100 per capita (US-GAO 2013: 74). The budget of the Cook Islands government for 2012–2015 will be supported by New Zealand's aid transfer of $US 69 million, amounting to $US 1,325 per capita (Government of the Cook Islands 2013; MFAT 2014a). France's state's

expenditure in New Caledonia reached $US 1.72 billion in 2013, about $US 6,800 per capita (IEOM 2014: 67).

5 As a fifth and final pattern, we note that many non-self-governing islands can sign international treaties beneficial to them but uncomfortable to their metropoles. In itself it seems odd that non-self-governing islands sign up to international treaties, which is conventionally the prerogative of sovereign states. Nonetheless, islands which are not sovereign states can and do sign international treaties, even if these diverge from metropolitan interests. As an example in 1980 the Cook Islands government was working towards a trade agreement with the government of the Republic of China in Taiwan. New Zealand felt this deal might reflect badly on its business with the People's Republic of China and put pressure on the Cook Islands. Elegantly, the Cook Islands saved New Zealand's face without losing its trade deals by signing the agreement with Taiwanese companies, instead of the Taiwanese government (Smith 2010).

The values and practices of *vaka vanua* and *vaka moana* go a long way to explain these five patterns of an Islandian sovereignty of non-self-governing islands. The refusal to vote for independence is consistent with *vaka moana* in maintaining and expanding networks across large distances; severing such carefully developed relationships would be contrary to *vaka moana*. Signing as many international treaties as possible is similarly consistent with the *vaka moana* way of expanding personal networks. The values of *vaka vanua* may explain why Pacific Islands are continuously renegotiating the relationship and 'getting away' with bending their metropole's rules; it emanates from the principle that sovereign power ultimately rests with the base of the pyramid and not with the top. Finally, expecting the metropole to help out when an island's budget falls short seems logical when reciprocity is a key principle in both *vaka vanua* and *vaka moana*. Haven't Pacific societies not always helped out their metropoles when those metropoles needed labourers, nurses, rugby players, musicians and soldiers? As such, this Islandian sovereignty can be seen as a subset of what was earlier described as Oceanic sovereignty. Furthermore, close association with the metropole is not a sign of weakness, poverty or passivity; rather it is the reverse – astute and far-sighted strategies to maintain connections and tap external resources yet retain a high degree of autonomy and effective self-determination.

Overall, we find that since the beginning of the 1990s both sovereign islands and non-self-governing islands have developed innovative ways of asserting forms of self-determination with the metropole. In the process, it seems that non-self-governing islands have been most successful because they have been able to leverage to their advantage the constitutional bonds with their colonial metropoles. We should not necessarily see this as a forsaking of independence, for by using these bonds within the metropolitan states, we have seen the evolution of unique forms sovereignty without the establishment of a sovereign state. This islandian sovereignty is based on flexible and constructive arrangements with a metropolitan state.

However, these expressions of 'islandian sovereignty' apply to only some Pacific territories (Wallis and Futuna, Niue, Tokelau, Cook Islands, Palau, FSM and Marshall Islands and, arguably, French Polynesia and New Caledonia), though we might also see similar processes at work in the more fully integrated islands such as Rapa Nui and Hawai'i. For most Pacific Island states and peoples, the reality is of fully independent sovereign statehood. The 'sweet spot of sovereignty' in the Pacific that we explore in this book, comes about because of the search for outlets for inevitable and inherent expansion and interaction that has characterised the Oceanic peoples since their arrival in the region many millennia ago that frames their sovereignty. Oceanic sovereignty is expansive, flexible, open, interactive, kin-based, communal and sustainable. What we need to do is explore how Pacific institutions and representatives seek to exert their own forms of sovereignty within such diverse constitutional relationships. The concept of islandian sovereignty gives us an insight into the way sovereignty is not simply a matter of formal political independence – a state of being – but rather a complex and continual process of relationship building and negotiation that asserts local decision-making and secures external resources. Sovereignty goes much deeper than constitutional definitions.

Box 4.5 Flags of the Pacific

Warwick E. Murray

Flags have come to be central expressions of national identity. Evolving from standards used on battlefields in the middle ages as the European system of kingdoms and eventually nation-states became established from the 1600s onwards, they have become a central signifier of nationhood. Many nation states have their flags described in national constitutions and complex international protocols have evolved concerning their use in diplomacy, war and events of national significance. European colonialism, and in particular by the British Empire, was instrumental in the diffusion of the use of flags by the end of the 19th century. During the period of independence following the culmination of the two World Wars and the establishment of the United Nations the number of national flags has proliferated, and they were used as salient expressions of independence and sovereignty, despite their colonial roots.

Flags do not just represent nation states, both dependencies and associated territories have flags, and these can be powerful symbols of independence or a distinct identity. Rapa Nui, for example, is a fully integrated part of the state of Chile, thus the Chilean flag is the official flag. An alternative exists featuring a red pendant worn traditionally by women of the island and was adopted by the independence movement in 2006. In Wallis and Futuna, the French flag is the official flag of the territory, but the territory often flies its own unofficial flag alongside. This unofficial flag of Wallis and Futuna, as well as the three distinct flags for each of the three kingdoms, all have the French Tricolour in their top left corner. In New Caledonia, due to its particular and recently turbulent history there are two official flags – the French and the Kanak varieties.

In the Pacific seven countries or territories have the British Union Jack in the top left corner of the flag. Six of these are found in Polynesia, most probably given the longer association with ex-British colonies of New Zealand and Australia. In the case of New Zealand an unsuccessful referendum was held in 2016 to replace the national flag in part to remove the Union Jack in a country that still has significant but increasingly distant formal ties with the United Kingdom. Curiously, the US state of Hawai'i also contains the Union Jack, given its former status as a British protectorate prior to the American takeover in 1898. This has been maintained until the present following full integration into the United States of America in 1959.

In other cases, especially in Melanesia there has was a clear attempt to break with the colonial past and design flags that have unique national symbols. The flag of Vanuatu has the colours of the party that led the independence movement in 1980 (Vanuatu Pati) and a boar's tusk and ferns to signify peace. The flag of Papua New Guinea, for example, has red and black which is associated with various tribes and the raggiana bird of paradise. The PNG flag also illustrates continuity; the Southern Cross featured in various colonial versions of the flag, and the bird of paradise also features in the national coat of arms and earlier flags. As such, flags are important rhetorical devices for instilling and at times imposing a sense of collective identity that allows the exercise of power and the communication of signals concerning a perception of sovereignty and both links and breaks with past.

Conclusions

Virtually all international declarations on development aid over the last decade avoid using the word 'sovereignty'. At first sight, this may seem rather puzzling because development aid generally involves a transfer of resources from one state to another in exchange for which the recipient country agrees to undertake specified project or policies, which suggests aid affects the recipient's state sovereignty. However, a more careful consideration discovers that the concept of 'Westphalian sovereignty' – as an accumulation of ideas of 17th-century peace treaties in German Westphalia,

19th-century nationalism, and 20th-century decolonisation – has been challenged from diverse angles at about the time as development aid was established. As the number of sovereign states doubled to almost 200 since decolonisation, several rather disconnected processes increasingly seemed to undermine state sovereignty 'on-the-ground'. These processes ranged from international bodies that could overrule a sovereign government (varying from UN-sanctioned humanitarian interventions to multilateral free trade bodies) to global corporations. Furthermore the broader process of neo-liberal globalisation with the freer movement of capital and products also reduced a state's sovereignty. In addition, indigenous peoples secured rights to 'self-determination' in the 2007 UN Declaration on the Rights of Indigenous Peoples, which effectively implied an indigenous sovereignty without the trappings of a state. In all, it is perhaps not so surprising that international declarations on development aid avoided the word 'sovereignty' – the concept has become increasingly ambiguous and contentious.

Nonetheless, for the peoples of the Pacific, the struggle for sovereignty and self-determination has been an enduring feature of the region for the past century and more. Colonialism was sometimes embraced but often resisted. There were violent clashes in New Caledonia, Samoa, Fiji and elsewhere as Pacific people fought to resist foreign control. Political independence was strongly argued for, and gained, by Pacific leaders in Vanuatu, Papua New Guinea and Fiji. Yet it was also carefully negotiated in ways which often aimed not only to enshrine local rights but also maintain a degree of responsibility by the former colonial power for the new country. Political independence has been a key means for Pacific people to gain greater control over their own resources and destinies and the appeal for self-determination has been consistent and compelling (Wesley-Smith 2007).

Yet it is important to note that while non-sovereign territories in Europe, Africa and Asia continue to seek full Westphalian sovereignty, island territories across the world seemingly have stopped this search. Even more remarkably, while continental territories such as Montenegro and South Sudan recently obtained sovereignty by secession after bloodshed, island territories, for example the American Virgin Islands, the Dutch Antilles, and New Zealand-administered Tokelau, refused to vote 'yes' to sovereignty from their colonial metropoles when presented by their metropoles with referendums on independence. This is probably explained by the findings of several comparative analyses that on average non-self-governing islands have better indices for health and wealth than sovereign island states. Debate rages whether this is mostly due to a dependence by islands on their (former) metropoles, or whether this is because the people on non-self-governing island are more adept at 'tapping into their colonial hinterland'. Perhaps voting 'no' is not rejecting sovereignty but rather asserting it. Saying 'no' to an imposed notion of a state is a sign that more effective self-determination and control over well-being may lie in other more complex and connected political relationships. We identified a handful of patterns in the relationship between non-self-governing islands and their metropoles that leave us to conclude that these islands may not possess a Westphalian sovereignty, they have developed an 'Islandian' sovereignty that serves them well and in which they act with a high degree of self-determination. Other Pacific countries, and by far the bulk of the population of the region, however, do have a high degree of political sovereignty in the form of full independence. Here debates about sovereignty exist in other ways: there are regional conflicts (as with the attempts by Bougainville to gain autonomy from Papua New Guinea), ethnic divides (the continuing differences between indigenous peoples and later arrivals in New Caledonia and Fiji for example), and accusations that political control is in the hands of powerful elites who, in effect, exclude the bulk of the population from both political and economic power (as demonstrated in student protests against the Prime Minister of Papua New Guinea in 2016).

Thus, in the Pacific, debates about sovereignty rage passionately and profoundly. However, we argue that these debates should not be framed in narrow terms regarding the definition of sovereignty nor in ways which fail to recognise the active agency of Pacific people. Are the islands small, isolated, resource-poor, and vulnerable, or quite the contrary, are islands in fact a huge, inter-connected, resource-abundant, and resilient community of Oceania? We side with those Pacific scholars who argue their communities do not live on 'islands in a sea' but in a 'sea of islands' (Hau'ofa 1993) and in a large ocean world, not small island states. Moreover, we concur with the same scholars, arguing the need to decolonise our minds – involving the symbolic shift in gaze to look at the Pacific from a beach next to our canoe looking outward instead of looking at an island from an arriving European vessel – when we think about what sovereignty might be in that sea of islands.

We suggest the Pacific *vaka vanua* (ways of the land and its local identity) and the *vaka moana* (ways of the sea and its shared regional identity) are two pillars on which a Pacific perspective on sovereignty is built and can be expanded; an Oceanic sovereignty. In particular, this means forms of sovereignty that are predicated both on the political primacy of local communities in accordance with *vaka vanua*, and a sovereignty that seeks to expand personal networks of Pacific peoples across a region without borders; the *vaka moana* concept. These two principles allow for a diverse range of sovereignties to be developed: sovereignties which may be layered; sovereignties which will vary according to local environments, cultures and world views; and sovereignties which may overlap. The strength of these Oceanic forms of sovereignty would be measured in the number of people participating in these networks, the wealth it generates for them, as well as their durability and permanence.

Notes

1 Karin Ingersoll (2016) has also recently added to this body of work, drawing on Hawai'ian identities and ways of knowing and arguing persuasively for a 'seascape epistemology'.

2 We propose these terms based on the fact that both include words that are quite common in Austronesian languages and have comparable meaning across the Pacific. Here we use the variants as they may appear in Fiji but they will be reflected in other languages (*waka whenua* and *waka moana* in *te reo* Maori for example). Mali Voi joins the terms '*vaka moana*'. He notes *vaka* is not only a common word for 'canoe' but also refers to a group of people linked by common descent and migration and *moana* a common Austronesian word for 'ocean' (2000: 208). In using the Fijian 'vaka' however we do not mean a word for canoe but rather the form '*vaka*', '*va'a*', '*whaka*', '*fa'a*' etc denoting 'in the manner of'. Many have also employed the term *vaka vanua*, particularly in relation to customary land tenure practices in Fiji (Bryant-Tokalau 2010: 8; Ravuvu 1983, 1988; Batibasaqa et al. 1999; Eaton 1988; Overton 1987, 1989, 1999) *Vanua* is a common Austronesian word for 'land', though this English noun fails to encapsulate the complex and holistic concepts inherent in *vanua* (or its variants *whenua, 'enua, fenua* etc.).

3 A further model has been added with the recognition of tourism as a major source of income for some small island states – see McElroy (2006) and McElroy and Hamma (2010).

RAMSI Mural in Honiara, Solomon Islands

5 The inverse sovereignty effect

Introduction

The notion that the aid regime that emerged in the early 2000s resulted in an 'inverse sovereignty' effect provides a central theme for this book and is explored in depth in this chapter. This notion posits the idea that, despite the adopted principles of the Paris Declaration, and indeed, rhetoric regarding ownership of the development policies and programmes by the recipient country, the actual practice of development assistance has resulted in diminished room for the countries and territories in the Pacific to determine their own policies. A new set of conditions has been imposed on Pacific governments, which significantly undermines recipients' ability to exert self-determination, thus reducing their sovereignty. This is particularly the case for the Pacific's small island states where implementing the global template for aid effectiveness has been especially fraught given problems of scale and capacity.

In order to explore the themes above, we examine the case of several Pacific Island states and territories. In doing so, we analyse how the application of development assistance works in practice. We assess this not so much in terms of high level agreements, frameworks and policies but at the level of common and everyday performances, acts, routines, interactions and agendas. Of particular concern is the position of Pacific Island officials, politicians and civil society representatives who have to negotiate and implement development programmes. They are in the critical position of linking global and donor frameworks, resources and objectives with the capacity of local institutions and the aspirations, capabilities and values of local people. In order to gain some understanding of the nature and influence of their local roles, we draw upon a number of case studies resulting from our research project, for there is no aggregate quantitative data on how such people and institutions manage their complex roles and responsibilities. Such studies give us rich and specific material which illustrates the daily realities and intricacies of aid in practice in the Pacific Islands.

This chapter begins by reviewing the issue of the 'new aid conditionalities' that have accompanied the implementation of the neostructural aid regime.

It suggests that there are *process conditionalities*, to do with the way recipient governments are required to put in place certain policy documents and management systems that conform to donor requirements. It also considers *political conditionalities* that focus on issues of human rights, democracy and public consultation that donors also require of recipients. We then turn to the issue of the *burden of consultation* and how the putative need for such interaction, grounded in principles of both efficiency and justice, has rendered the political landscape more complex, leading to pressures and conflicts. This complexity and multiplicity of aid institutions is then examined, noting how Pacific Island institutions are forced to deal with not only a large number of donor agencies but also a shifting donor world replete with personal and political objectives. Finally, we relate the issues of donor-required conditionalities, consultation and complexity to the realities of Pacific Island capacity. As noted above, this is particularly pressing in, but not confined to, the smaller-scale countries and institutions in the region that can experience severe limitations in terms of the ability to cope. We conclude by suggesting, on the basis of these cases, that the effective development sovereignty of many states and territories has been undermined, not improved, by the nature of the implementation of numerous aspects of the aid effectiveness agenda in the Pacific Islands region.

Compliance: the new conditionalities

Aid conditionalities are not new. It could be argued that any aid relationship (especially government-to-government) comes with strings attached between donor and recipient; that donors give aid with the expectation of some sort of return as a consequence of giving financial resources to another jurisdiction. The return may come in the form of political support, economic benefits or structural change that aligns recipient economies and polities more closely with donors. Through the fundamentally unequal interaction, both in the past and present, sovereignty of the recipient is likely to be compromised in some manner. However, development aid from richer to poorer countries has also been provided on humanitarian and altruistic grounds, rooted in Judeo-Christian and Islamic concepts of charity, in socialist calls for solidarity, or in liberal notions of fairness. Moreover, richer countries can also provide aid to poorer countries on the basis of enlightened self-interest: aid may increase political stability across the globe, increase the worldwide web of producers and consumers, or slow down the pace of global warming.

Whatever the motives for giving aid, since development aid became a matter of public concern and policy in the 1950s, the explicit imposition of conditionalities has waxed and waned. Their use was especially notable, for example, during the early neoliberal era when the 'Washington Consensus' led to the imposition of Structural Adjustment Programmes (SAPs) that required recipient states to restructure fundamentally their political-economies in order to liberalise markets and trade and to diminish the scope and size of the state. In Chapter 2, we assessed some of the ways neoliberal aid conditionalities were implemented in the Pacific in the 1990s. Such explicit, and arguably harsh, conditions of aid dispelled the comfortable notion that aid was driven mostly on humanitarian grounds and mediated through relationships of trust between donor and recipient.

The neostructural turn at the advent of the new millennium seemed to mark a rejection of these heavy-handed conditions. It was recognised that they caused much resentment and alienated many developing world policy makers from the new mainstream of development. It was also recognised that severe neoliberal reform had adversely affected the viability of recipient states to provide basic public services and, critically, ensure that market mechanisms could function effectively. Yet neostructuralism, in turn, has required that recipient states are reconstructed in particular ways with an associated set of objectives, policies and institutions. This has led to the imposition of 'new conditionalities' (Gould 2005) that have been less explicit and, thanks to the discourse of 'ownership' within the Paris Declaration, passed off as implemented and maintained in a partnership agreement with recipient states themselves (Buiter 2007). Such an approach seemed 'kinder and gentler' superficially. However, we contend that they were as powerful and pressing as the neoliberal conditionalities they replaced (also Larmour 2002). Furthermore, in recent years as the retroliberal aid environment has unfolded, these conditions have been reshaped, often in ways which have seen a return to a more obvious heavy hand among donors. In what follows we consider process conditionalities before turning to political conditionalities.

Process conditionalities

Whereas neoliberal structural adjustment programmes imposed conditions on recipients that were explicit and aimed at high level policy objectives and fundamental institutional restructuring – freeing of markets, deregulation, privatisation, downsizing of government agencies and services, balanced budgets – the conditionalities of the neostructural era were less overt and aimed instead at the detail of government systems and practices. Here we outline the key elements of required changes at strategic, policy, institutional and systems levels during this latter era (see Table 5.1 for summary). Below we consider process conditionalities at various levels – strategic, policy, institutional and systems.

At the high **strategic level**, new conditions were largely encapsulated in the requirement that recipient states develop Poverty Reduction Strategy Papers (PRSPs).[1] PRSPs were instituted in the mid to late 1990s by institutions such as the World Bank to replace the unpopular SAPs. The IMF and the World Bank, together with many bilateral donors, required that PRSPs be in place before substantial aid packages were agreed to. Unlike SAPs they were required to address specifically the issue of poverty: what strategies would governments put in place to identify, address and alleviate poverty (largely defined by the MDGs) in their countries? They also had a longer-term view of reform, focusing on strategic medium term goals rather than on immediate crisis-related changes. Yet they also retained many of the elements of the old SAPs: plans were required to facilitate the free operation of markets and restrict the ability of states to operate outside fiscal and monetary parameters articulated by the IMF and World Bank (Craig and Porter 2006). PRSPs required recipient countries to undertake a comprehensive analysis of poverty in the country and also ensure wide community consultation. Furthermore, there was much emphasis on 'results-oriented' strategies and a demand for data collection (such as through participatory poverty assessments and household surveys), monitoring and reporting. In the Pacific Island region, PRSPs were instituted rather later than elsewhere in the aid world but they were rolled-out over time, even if they were not officially called PRSPs – Pacific governments seemed to prefer talking about 'development plans' rather than the value-laden 'poverty reduction strategies'. Samoa, for example had its 'Strategy for the Development of Samoa (SDS)', Fiji had a 'Strategic Development Plan (SDP)', Tonga a 'Tonga Strategic Development Framework' (TSDF), and Solomon Islands a 'National Development Strategy (NDS)'. Significantly though as the largest recipient in the region, Papua New Guinea produced a 'National Poverty Reduction Strategy' in 2002. Behind many of these documents was the Asian Development Bank, offering advice and assistance with their drafting.

In many ways, PRSPs were not unfamiliar. In earlier times, many Pacific states were encouraged to develop their own five-year plans. These five-year plans were written by respective central planning agencies within countries, often with the assistance of overseas staff and consultants. Countries such as Fiji, Tonga and Samoa had a succession of such plans (nine in Fiji) spanning the late colonial era (late 1960s, early 1970s) until the late 1980s. They identified key development needs, listed and prioritised development projects, sought to integrate and rationalise such projects and, importantly, provided a 'shopping list' of projects that donors could examine and decide to fund or not. In many ways they were a precursor to PRSPs. In some cases – again Fiji was an exemplar – the national development plans were accompanied by a number of regional development plans, aiming to promote the dynamism of the regions by integrating infrastructural projects with the provision of government services such as education as well as linking remote areas with expanding national and global markets. These strategic plans were based on the developmentalist-era predilection for state-led development and put much emphasis on infrastructural provision and the hope that rural communities would 'modernise' through a dual strategy of better welfare provision (education and health) and increased surplus production for markets. Neoliberal reforms in the late 1980s and 1990s led to the end of these national planning approaches and laid the ground for the introduction of PRSPs which were less state-directed, more focused on poverty alleviation, and had a greater emphasis on market liberalisation and export-led growth.

Yet PRSPs were not the only requirement placed upon states at the strategic level from the 1990s onwards. The United Nations Environmental Programme (UNEP) and its regional equivalent, the South Pacific Regional Environmental Programme (SPREP) oversaw the production of National Environmental Management Strategies (NEMS) in the late 1990s. These plans were largely funded externally and conformed to the UNEP's templates, but they required Pacific Island states to address issues relating to environmental management and sustainability. Although critical to the well-being and future of people in the region and useful in identifying policy needs and priorities, these NEMS represented an early example of the way Pacific Island policy strategies were shaped and pushed by external agencies. Though the strategy plans did aim to address local concerns and resources, the broad agenda was largely set externally, and a degree of conformity was established across the region.

Complementing and following strategic level requirements, donors also required that recipient government develop **policy level** operational plans and documents. These flowed from the priorities set by the PRSPs and required particular sectors of government activity to identify key objectives and plans. Thus, for example, departments of education would be asked to set out base line data (many informed by the MDGs – primary school enrolment rates, participation of girls in education etc.), identify key needs (such as teacher training) and set their own strategic plans in place. Box 5.1 presents the example of Vanuatu's Employment Law Reform: a process instigated by the country's need to comply with international labour standards, following its joining of the International Labour Organisation (ILO). These sector plans and policy documents, again with a high degree of outside advice from consultants and officials, became the basis for sector wide development and financing, through basket funds or, eventually, Sector Wide Approaches (SWAps). Such plans and policy documents could severely tax the human resource capabilities of Pacific government departments of education and health (and at times others such as police, customs, public works etc.) and there was a danger, if left to external consultants to write, that they could be seen as not well grounded or 'owned' by local institutions and communities. Notwithstanding, they often performed a valuable role in requiring states to address longer-term goals and priorities. They also gave donors a clearer idea of what could and should be funded. In addition, as these policies and plans were formulated (with the use of external advisors and templates), donors benefitted from a level of familiarity in style and format. Policy level conditionalities, albeit in a rather subtle way, helped ensure that recipient policies aligned with the objectives, priorities and systems of donors.

Box 5.1 Vanuatu's Employment Law Reform and policy conditionalities

Potoae Roberts Aiafi

The Employment Law Reform was instigated by the Government of Vanuatu in 2008 to enact domestic law changes in compliance with international labour laws. Vanuatu became an ILO member in 2003 and is a signatory to eight ILO conventions. Implementing these conventions requires domestic policies and laws. The ILO Decent Work Country Programme (DWCP) has been the key policy document utilised to specify domestic labour reforms and Vanuatu's DWCP helped guide its Employment Law Reform (ELR).

As part of the ELR, Vanuatu's Employment Act of 1989 was amended in 2008 to enact increased employees' benefits. These included a 300% increase in severance allowances and a 100% increase in maternity leave. The necessary Bill was passed through Parliament but, following hostile reactions from the business community, it was suspended. Other attempts were made to amend the 2008 provisions, lowering the level of allowances, but the dispute has dragged on and it remained unsettled when research was conducted in 2012. This research asked a number of officials about the process of labour reform and it became clear that local 'ownership' was limited. Participants were asked why local actors accepted these policy impositions:

> *There's always this conditionality . . . If you don't meet them then we still don't access funding. If there's no policy, there will be no funding. If you have a policy, yes. It's a performance-based thing for them – these UN or semi-UN agencies . . . They bring aid and super programs but also ask us to abide by their rules.*
>
> (Respondent A)

The ELR's ideological foundation has been the ILO standards and principles upholding the rights of individual workers in formal employment. Promoting human rights in employment relations involved adopting contractual and bargaining arrangements, tripartite structures and union representation. These arrangements, based on a well-developed private sector, were problematic for Vanuatu's economy with an undeveloped private sector and where unions are small.

Participants saw the ELR as not adjusted to Vanuatu's situation in terms of its development status. In particular, aspects such as the large redundancy provisions were seen as unsuited:

> *This redundancy payment is inappropriate . . . Or economy is quite superficial. We depend a lot on aid. We don't produce money*

for our budget. So we need to be careful about the way we legislate our employment because it affects the private sector, the government and its finance. It's very delicate.

(Respondent B)

This case demonstrates the risk of adopting a policy when a country is not ready for it. While the intentions may have been good, contextual factors question the suitability of advancing such reform at this stage.

There was an order to say rather than a rational consideration of what should be done, a consensus was reached, for everyone to save face and feel happy. This is somewhat problematic. This policy, this DWCP, has been driven by the ILO and not contextualised. If you look across the Pacific you will find they are all similar. If you look at the problems internationally . . . it's a mess.

(Respondent C)

(modified from Aiafi 2016: 77–79, 83)

At another level, new conditionalities led to a response at an **institutional level**. In this case, there was a need – or a demand – that new or reformed institutions be put in place. This was part of the project to reconstruct states in ways which conformed to ideals articulated in certain 'good governance' principles (see also Chapter 3). These related to aspects such as transparency and accountability. Thus, donors were keen to support the establishment or rebuilding of offices such as that of an Ombudsman to provide an independent monitor – and an avenue for complaint investigation – of government services and agencies (Larmour and Barcham 2005). Similarly, anti-corruption offices and policies were supported by donors and there was widespread training for reform encouraged within police, customs and judicial agencies. Within this broad category could also be included many of the gender and development projects supported by donors; projects which aimed to increase women's participation in Parliaments, local decision-making bodies and economic enterprises as well address key concerns concerning gender-based violence, and gender inequalities in education and maternal health for example. Similarly, later neostructural aid policies, despite a strong emphasis on state-building, favoured the use of civil society organisations as channels for citizen participation and consultation (Mawdsley et al. 2015). Institutional strengthening within and outside the state thus eventually became a prominent aspect of aid in the first decade of the 2000s. In all, these institutional-level conditionalities aimed to construct more robust political and economic systems that strengthened the ability of citizens to participate in decision-making, see more clearly what government institutions were doing and make their governments more accountable. However, they also served the interests of donors keen to ensure that newly reformed government agencies were limited in their ability to act without open scrutiny. As a consequence, institutions – such as NGOs – that owed their existence and continued financial support to benevolent donors would be more likely to align with, and be less critical of, the sorts of policies that donors promoted.

Finally, we can identify **systems level** conditionalities. These were directed at the particular management systems that governments put in place to ensure that their higher level strategies and policies were followed efficiently and openly. They aimed to institute consistent and transparent systems for accounting for financial flows in particular. Such systems involved the minutiae of government administration and the myriad of paperwork that they operated by: terms of references, role descriptions, regular reports, memoranda of understanding, inter-departmental communication, etc. Furthermore, such systems had to be subject to open and independent audits and, ideally, public scrutiny. In many cases, such systems were prescribed by the norms of international practice for public sector administration and financial management. The World Bank, for example, provided guidelines for administrative 'wiring diagrams' and reporting systems (Koeberle and Stavreski 2006). Central to such systems conditionalities were strict donor requirements for financial management systems (FMS), so that flows of funds within and through government departments could be tightly accounted for and audited along accounting trails. Indeed the auditing requirements opened up a high level of external dependence in terms of inspection – there was often an insistence that large international accounting firms (rather than local ones) conduct the required audits. Again this promoted the use of international systems and standards, and this was also a major expense.

The aim of such systems, at one level, was to address issues of corruption within recipient governments and political systems. This was part of a broader political and rhetorical device to address what was identified as a significant problem in many non-Western political systems, portrayed as being run by politicians and bureaucrats who plundered government coffers and aid funds (as well as receiving bribes). The discourse of corruption created an environment in which it was deemed acceptable (in donor countries) to impose strict conditions on recipient countries and their leaders and officials. In the Pacific Island context, the discourse was fed by media reports of bribe-taking by politicians in Papua New Guinea or Solomon Islands and analyses which argued that communal systems based on kinship and *wontok* led inevitably to nepotism and corruption (Larmour 1997; Hughes 2003). However a more benign explanation for the imposition of strict conditions regarding FMS concerned a wider issue of fiduciary risk: a general concern that aid funds should not only reach their intended targets, but they should also achieve their objectives.

Not only did the new aid agenda of the early 2000s require the wide range of strategies, policies, institutions and systems to be put in place as identified above, the agenda itself generated its own set of specific conditionalities. Whether concerning the MDGs or the Paris Declaration principles, donors asked that recipients report back to them on appropriate milestones and the degree of progress that had been achieved in meeting them. Indeed, in an ironic example (Box 5.2) recipient agencies were asked to report back to donors in detail on how well their degree of ownership had been achieved! Such conditionalities ensured that although the principle of ownership was stressed by donors, this example highlights the way accountability for that principle did not trickle downwards to citizens but rather it was directed upwards to donor agencies.

Box 5.2 The Pacific Islands Forum and indicators from Paris

Gerard Prinsen

The Pacific Islands Forum – often referred in the region as 'The Forum' – is a regional organisation of the sixteen sovereign Pacific nations. The Forum is driven by regional goals in spheres of economic growth and political governance and security. From its inception in 1971 onwards, the Forum has been marked by a continuous tug-of-war over the sovereignty of the Pacific Island nations vis-à-vis the former colonial powers and today's regional powers. Sometimes the tug-of-war takes place overtly, sometimes it manifests itself in seemingly technical discussions. Before illustrating how the delicate exertion of – and resistance to – power extends into something as innocuous as collecting data on development targets, a brief review of the Forum's history offers a context.

The Forum was established in 1971 as the 'South Pacific Forum' by the leaders of five sovereign Pacific countries: Cook Islands, Fiji, Nauru, Tonga and Samoa. New Zealand and Australia were, initially, observers to the Forum but were admitted as full members in 1972 (Lal and Fortune 2000: 332). Arguably, this South Pacific Forum was the successor to the South Pacific Conference, an annual gathering since 1947 where representatives from the Pacific region's six colonial powers of the time – organised in the South Pacific Commission – invited representatives of the peoples of the Pacific to share their views on the colonial policies.

Shibuya's review of the South Pacific Commission's practices concluded not much consultation took place on the policies of the colonial powers. The views of Pacific peoples were mostly solicited 'after the fact' (Shibuya 2004: 103). With political independence for several Pacific Islands came a drive to liberate themselves from the influence of Australia and New Zealand. Fiji, Tonga and Samoa (later joined by others) established the Pacific Islands Producers Association in 1965 in a conscious attempt to improve "their bargaining power with New

Zealand" and "promote Island interests in ways which, especially in the light of French reluctance, they could not do, or felt they could not do, within the South Pacific Commission" (Ball 1973: 241). There were increasing voices in "protest against the structures . . . which ensured the dominance by the colonial powers" (Fry 1991: 173).

Throughout the decades, one recurring dividing line within the Forum has seen Australia and New Zealand on one side and the Pacific Islands on the other. The regional security affairs of the 2010s saw Australia and New Zealand concerned about an alleged growing influence of China among the Pacific Islands whilst many of the latter seeing benefits in making deals with China. All this sounds remarkably similar to Australia and New Zealand explicit policies of the 1970s to use the Forum to counter the threat of the Soviet Union after it signed agreements with Tonga in 1976 (Herr 1986). Similarly, the Australian and New Zealand responses to military coups in Fiji have been condemnation and boycott, whereas other Forum members generally sought understanding and dialogue (Kelly 2015).

Against this backdrop of pushing and pulling, the negotiations in the Forum over the development indicators associated with the 2005 Paris Declaration on Aid Effectiveness acquired a new meaning. From the early 2000s, the OECD ran a series of high level international meetings to agree on ways to improve the results from international development aid programmes. The 2005 Paris Declaration established a set of important principles for aid practice and targets to which donor countries and recipient countries agreed to adhere. The Forum was one of the original signatories to the Paris Declaration in 2005.

To measure progress in achieving these targets, specific indicators were also agreed. For example, one of the principles was that international aid needed to be harmonised: different donor countries agreed to harmonise among each other their respective aid policies to a particular country and speak with one voice when negotiating with the recipient government. One indicator for this harmonisation principle was the number of 'joint donor missions' to the recipient country. A target was set: 40% of the donor visits to a country needed to be 'joint missions' by the year 2010 (OECD 2008).

The Forum's Secretariat was charged with the task of collecting information to track progress with regard to the twenty-one targets set in the Paris Declaration. However, it soon became apparent that the Secretariat was unable to collect all the necessary data. It only started collecting data in 2008, and by 2012 it was still struggling to produce annual data or even a base line. The challenge, of course, was formidable. The Secretariat's requests to each member country to report, for example, the number of 'joint missions' was adding to the workload of the many small national bureaucracies.

By 2013, the Forum Secretariat was under increasing pressure to produce progress reports. However, it lacked the staff to collect data. A few consultants were approached to assist. Nicki Wrighton and I were among these consultants. However, by the end of the year it became clear the Secretariat had been under other pressure too, being questioned in whose interest the data was about to be collected. The upshot of the diverging pressures upon the Secretariat was that the Secretariat never produced the report on progress towards achieving the targets of the Paris Declaration.

From informal conversations we learned that several Pacific Island members did not see the need to invest resources in producing a report that in no discernible way was going to enhance the policy sovereignty of the smaller Pacific Islands. Some Pacific policy makers argued that 'joint donor missions' were actually a bad idea for national sovereignty because it took away a country's ability to play off one donor country against another. The parties most interested in collecting the data were, it seemed, Australia and New Zealand who could have used the data to show their compliance to the OECD. In the wider context of the tug-of-war in the Forum, it is no wonder we will never know where Australia and New Zealand ranked in the OECD in their 'joint missions' rating.

Political conditionalities

Alongside the above process conditions, donors have also engaged in what might be termed political conditionalities, focused on the political mandate of recipient governments to devise and implement policies whilst respecting the political rights of their citizens. To a large extent, these conditionalities have resulted from pressures from the domestic political constituencies of donors rather than issues concerning transparency, efficiency and reform that underpinned the formation of process conditionalities. At this level, donors have often been sensitive to issues based on human rights and social justice foundations. Thus criticisms of certain recipient states for not holding – or rigging – elections, detaining opponents, excluding some groups from the vote or political participation, instituting discriminatory policies, or overlooking the rights of women or minority groups, have provoked action. Donors have threatened, and sometimes undertook, to withdraw or reduce aid, or they have redirected aid away from governments they deemed to be lacking legitimacy or instituting unpalatable policies.

In the Pacific region, several examples of relatively strict political conditions on aid have been instituted. When donors have had concerns over the political legitimacy of Pacific states, they have often found themselves in an uncomfortable position. On one hand, they have been reticent to cut ties completely, not only believing that aid can be used as an incentive to encourage states to undertake political reforms, but also because diplomatic principles dictate that relationships between states need to be maintained if possible. On the other hand, donor governments have had to maintain an eye on their domestic constituencies, and these constituents have often been critical of aid budgets being spent on supporting what some regard as undemocratic or unsavoury institutions and processes. The case of Tonga is illustrative of the way donors – in this case New Zealand – have applied pressure through their aid allocations in order to bring about political change (Osborne 2014).

Although concerns over issues such as corruption, discrimination or exclusion have arisen in various instances over past decades, in the Pacific it has been the case of Fiji that has provoked perhaps the most direct action by aid donors. Following an initial military coup led by then Lt-Col Sitiveni Rabuka in May 1987 which overthrew a democratically elected government, donors acted quickly to criticise the new regime and cut their support for it. Over the subsequent years, donors (Australia, New Zealand and the European Union in particular) continued to exert pressure, easing off when reforms, new constitutions and fresh elections were held but applying renewed pressure when other uprisings and/or coups were staged as in 2000 and 2006. Figure 3.12(b) shows the flow of ODA to Fiji from different OECD donors over time. The imprint of the country's coups and subsequent responses can be seen clearly in the aid profile.

Prior to 1987 Australia had replaced the United Kingdom (the former colonial power) as the main donor, and along with New Zealand, Japan and multilateral agencies, it provided the bulk of the aid inflow to Fiji. The two coups in 1987 led to a sharp reduction in aid, the major absolute cuts enacted by Australia and multilateral agencies. Aid recovered the following year and particularly in the early 1990s (coinciding with the development of a new constitution) with donors keen to encourage Fiji to seek a return to democracy. Following this, aid began to decline in real terms. Interestingly, the sharpest fall in aid, particularly influenced by Australia, occurred between 1992 and 1996. This was not so much a response to political changes in Fiji but rather the result of more widespread neoliberal aid reductions. The decline continued thereafter (despite the launch of new constitution in 1997 based on widespread consultation) and reached a nadir following the uprising led by George Speight in 2000 and the subsequent imposition of martial law. A return to democracy after that unrest and the general neostructural increases in the early 2000s led to a sharp rise in aid which again plummeted following Bainimarama's military coup of 2006. Only with moves towards democratic elections (held in 2014) might aid flows – from traditional donors and in traditional forms – begin to recover.

The nature of political conditionalities is further illustrated in the case of Fiji by the manner in which aid donors redirected flows away from the state. A study of aid sanctions in Fiji and the increased involvement of NGOs (Hanks 2011) showed how donors attempted to maintain some links with Fiji by supporting elements of civil society but did so by avoiding direct financial support for what they regarded as an illegitimate state. The example of Fiji and the funding of NGOs – instead of the state – raises another issue regarding political conditionalities. When donor governments are not comfortable with a transfer of their development aid to a particular government, local civil society organisations can be given – or actively acquire – a key role in maintaining some relationships between countries. International donor support for local civil society organisations helped foster the growth of NGOs in the Pacific over the past twenty years, sometimes against the wishes or interests of the state. Some NGOs have become heavily dependent on aid funding. They perform an important function in the aid world in the Pacific, able to portray themselves as alternatives or complements to the state, and occasionally as a deeper conscience of society. They also help donors demonstrate that processes of public consultation have been undertaken, supposedly through the intermediary of well-grounded civil society groups.

However, such funding flows and relationships between international donors and local NGOs (sometimes transacted through sub-contracts given to Northern NGOs to link with local partners) can create tensions and problems for the local NGOs. When donors do not recognise a national government as entirely legitimate but support local NGOs, the latter may be at risk of suppression or worse from national government and have to negotiate a tight line, sticking to core welfare functions and shying away from advocacy roles, for example, which may be regarded as too political (Llewellyn-Fowler and Overton 2010). NGOs also face their own concerns of capacity: dealing with donor systems of contracts, financial management systems, audits, reporting and the like can be extremely burdensome for small organisations without the staff to perform such specialist tasks. Furthermore in volatile political environments, there are issues of sustainability: the funds might flow relatively freely to NGOs when a local government is regarded by donors as a pariah but with a return to democratic elections and renewed positive state-state relations, NGOs can fall quickly from favour, their job being seen by donors as completed and thus their income no longer assured. Such a phenomenon was recognised by some NGOs in Fiji during the era of sanctions and they were aware of the fragility of the situation (Hanks 2011). We suggest, therefore, that civil society has been drawn into the web of aid relationships (Wallace et al. 2007). Though largely marginalised by the strong state-to-state neostructural model, they have faced similar issues to those of recipient states (conditions, systems and priorities defined by donors), but their position has been vulnerable. In short, the involvement of civil society in aid relationships has done little to strengthen a sense of local ownership and sovereignty.

The much vaunted principle of ownership by the recipient of aid thus appears to be fungible: 'ownership' can be recognised and respected when it is vested in the hands of a governments that are seen in the eyes of donors governments and their domestic audiences as democratic, open and responsive, yet when there are doubts about the democratic credentials of that government, 'ownership' instead can also be seen to be exercised, however imperfectly, through a wide variety of both civil society and government institutions. The conclusion is that aid does not simply go through the channels and processes prescribed by agreed international declarations; it can – and it does where it finds motive – seek to restructure and redirect those channels and processes. Furthermore, aid is given and received in both overt and covert ways: sometimes the political pressure exerted through aid is public and explicit (as with aid sanctions to Fiji); at other times is much more subtle and diplomatic (seen in the reticence of donors to increase funding or support particular projects favoured by local leaders, or the move to quietly support certain civil society organisations ahead of the state).

Political conditionalities can also be seen in the way aid negotiations are carried out. The art of diplomacy and the principles of inter-state relationships in the Pacific region – the 'Pacific Way' (Mara 1997) – recognise the equality of states: each is respected as a partner. Yet, in practice, power is exercised in ways which assert the dominance of donors. The example of the Cairns Compact (Box 5.3) demonstrates how accepted protocols and principles have sometimes been pushed aside so that the wider political objectives of, in this case, Australia have been 'foisted' on Pacific regional agreements concerning development assistance.

Box 5.3 A personal perspective on the Cairns Compact

Peter C.L. Eafeare

Let me at the outset declare the views I express here not attributable to the Papua New Guinea Government past or present, as they are my personal reflections and recollections and I alone am responsible for them. I was present at the Pacific Islands Forum Leaders Meeting in Cairns in 2009 as part of the Papua New Guinea delegation.

To begin with I want to quote from the late Professor Ron Crocombe:

> *Aid meets the needs of donors as well as recipients, so international relations in the Pacific is played with aid cards on the table, or under it. There is genuine goodwill in aid relations, but also self-righteous, self – serving hypocrisy on all sides.*
>
> (Crocombe 2007: 213)

In other words, there is a common thread that links all development partners: they are promoting their interests by contributing to sustainable economic growth and poverty reduction, whether it be in infrastructure, agriculture, effective governance policy, institutions, education, health, building resilience in humanitarian assistance (natural disasters), gender equality and empowering women and girls.

On reflection, the way in which the Cairns Compact was foisted on the Forum Leaders began with the discussions at the Forum Officials Committee (FOC) meeting, and the manner in which they were conducted, prior to the Cairns Forum in 2009. At this FOC meeting, the Australian leader of delegation announced under 'Any Other Business' that their Prime Minister would be presenting a paper during the Leaders Retreat that would be dealing with the streamlining of aid, which would include, amongst other things, a review of its effectiveness. He even stated that, "My Prime Minister would like to deal directly with the Leaders on this issue at the Retreat."

At that time there were no reactions or discussions from FOC members, although informally I did ask the Australian delegation why we were being 'ambushed' with this issue and the fact that normally all issues are dealt with at the FOC and that recommendations were provided to the Chair of the Forum by way of a letter from the Chair of FOC. I was told that they were instructed by the incoming Chair to announce what they did and nothing more. I presumed that what they meant was that our Leaders would have their say on the matter when it was directly raised by the Chair.

Immediately following that I thought of what this meant in terms of FOC proceedings and the larger implications it would have on the 2005 Paris Declaration which was significant in that it was by the initiative of donor countries, including Australia, which felt the need to set new principles and parameters for effective and efficient aid delivery through the following principles of ownership, alignment, harmonisation, managing for results and mutual accountability.

More significant in the impact this would have on our Development Cooperation Treaty (DCT) with Australia was the unilateral decision, in my view, by Australia as host of the Forum in 2009 to adopt the Cairns Compact on Aid Effectiveness. This had significant impact on, amongst others, aid modalities whether it be project aid (direct donor intervention) or programme-based (mutual consultation).

At the Cairns Leaders Retreat, normal practice was not followed. Normally, every decision that was taken would be provided as a draft to the senior officials who were camped outside of the Retreat room to be put into the draft communiqué. I was with my Head of Department, whom I advised to inquire with the Australian officials' delegation as to why we were not being given at least a paragraph on the Cairns Compact. The same response was provided, namely that it was the Chair's call. We reiterated that for Papua New Guinea, the Cairns Compact had a serious impact on the Development Cooperation Treaty (DCT) we had with Australia.

To my disappointment when I pointed these implications to my Prime Minister following the Retreat, his response was quite stern, in that he had spoken to the Chair and that our concerns would be dealt with bilaterally. I also learnt that there was not much reflection on the matter and that there was unanimity amongst the Leaders to adopt the document.

When it comes to matters like this, is it right and proper for formal processes at the FOC to be bypassed, because a member of the Forum, which no doubt provides the bulk of financial support to the Forum, decides unilaterally to discuss issues at the Leaders retreat, rather than discussing it at the Officials Committee Meeting, which is normally the case? Furthermore, at the Retreat, should Leaders have questioned the manner in which issues were introduced and agreed to, in this case the Cairns Compact?

I reflect on part of Professor Crocombe's comment, maybe 'the self-righteous and self-serving interest' is not on all sides, but on the part of the donor and not the recipient.

Indeed, I should know, and it is a fact that development assistance, aid effectiveness, and so on, are not charitable endeavours, they are really serving the interests of those who provide it by contributing to sustainable development and poverty alleviation.

To this day, I will not know the exact discussions or debate that may have taken place in the Leaders Retreat, if any, on the Cairns Compact. However, the reality is that it was adopted and is now part of the literature that guides the discussion on aid effectiveness in the region, in concert with the 2005 Paris Declaration, the 2008 Accra Agenda for Action and the 2011 Busan Declaration.

In summary, it can be argued that the requirement from donor governments that recipient governments comply with broad political and economic objectives outlined in international declarations – initiated and drafted by donors – as well as make sure their financial and administrative systems are transparent and conform to international – not national – standards has amounted to a new and highly complex set of conditionalities. Such conditionalities have unravelled in multiple ways in the Pacific Islands. At one level, there have been examples of aid being used by donors to openly 'encourage' political change and reform. At another, more common yet also more opaque level, requirements of donors forced or guided Pacific governments to put in place plans, policies, offices, documents, systems, audits, personnel and reports that amounted to a radical change in the way Pacific Island states operate. Aid has required institutional reform not only because aid systems require such changes in order to flow more efficiently and transparently but also because aid is used as an instrument to incentivise and enforce more widespread shifts towards governance systems that reflect ideologies that the donors desire.

The burden of consultation

An important element of the neostructural aid regime was an apparent commitment to democracy and public consultation. Whereas the early neoliberal regime had been very much a top-down enforced programme of radical change, neostructuralism rested on the argument that change could not be sustainable unless there was public acceptance of an involvement in that change. The strong advocacy for good governance of public institutions by the World Bank in the late 1990s also included a vigorous support for public participation in policy making and public participation was required for PRSPs. Perhaps to underscore its embrace of public participation, the World Bank engaged some of its fiercest critics to advise it on why public consultation was important and how it could be structured. The publication of its 2000 report titled "Voices of the Poor: Crying Out for Change" (Narayan et al. 2000) not only illustrated the shift from a neoliberal to a neostructural aid regime, but also highlighted the importance that was now attached to hearing and listening to the public when making policies.

To some extent of course, public acceptance and involvement regarding policy making could be illusory or superficial – the deeper programme of change may well have been predetermined and 'public consultation' and 'participation' used as mere devices to co-opt political agendas. Cooke and Kothari (2001) describe such strategies as a 'new tyranny of participation', designed to create the semblance of acquiescence. In some ways, this rhetorical strategy also lies at the heart of our concept of 'inverse sovereignty'. Nonetheless, neostructuralism-inspired aid has involved a high degree of consultation on policy making at both rhetorical and operational levels.

Consultation and participation are also expensive and time-consuming. Local officials and planners need to demonstrate that they have interacted with a wide variety of 'stakeholders': recipient government officials and politicians but also church groups, women's groups, local communities, special interest or advocacy organisations, youth, media and many, many others. This imposes large costs on the business of aid and it can significantly lengthen the time it takes to plan and implement development projects and programmes. Yet, superficial or not, consultation has become part of the expectations and practices of development aid, and this is particularly the case in the Pacific Islands.

In some ways, the Paris Declaration should have provided a recipe for dealing with this issue of consultation between donor and recipient governments. As we saw above, 'ownership' seems to be clear when governments are legitimate, have a clear mandate, are responsive to the public and are meet specific requirements regarding transparency and accountability. Here, the second and third principles of the Paris Declaration become important. Once a recipient country shows a degree of ownership of a national development strategy, donors should then 'align' their aid budgets with government strategies, policies and systems. In parallel, donors should also 'harmonise' their approaches and processes in a particular recipient country so that they act in concert, not each requiring a separate processes of consultation. For example, the Paris Declaration expects donors to coordinate their in-country visits and have joint missions, thus reducing the burden on the recipient government of having to engage with many consultative missions. Unfortunately, in practice the objectives and targets set in the Paris Declaration are rarely realised in this way. The Declaration stipulated that 'joint missions' was an indicator for the third principle of 'harmonisation'. A target was set that by 2010 at least 40% of the donor missions needed to be undertaken jointly. In practice, the number never exceeded 19% (OECD 2012) and Wrighton (2010a) noted that in Tuvalu, harmonisation was rare, though sorely needed. Harmonisation, though, may be difficult to achieve in practice because it may require coordinating several donors when dealing with just one donor would be more efficient. Boxes 5.4 and 5.5 present two examples from Tonga in which working with many donors on a 'harmonised' programme seemed to slow and complicate it due to the persistence of separate requirements and processes.

Box 5.4 Harmonising in practice: Tonga and energy policy

Helen Mountfort

It is well documented throughout the aid effectiveness agenda that the divergent interests of donors can serve to undermine the priorities and objectives of recipient countries (Armon 2007; Eyben 2007). Without a harmonised and aligned system, which embodies a common set of objectives, rules and standards, the Government of Tonga has less space to lead the process. The OECD Paris Declaration Draft Survey (OECD 2010) found that only 16% of reported donor missions in Tonga were jointly coordinated in 2010, which does not reflect the call under the Paris Declaration for harmonised efforts of development practices, routines and/or standards.

The Tonga Energy Road Map – a sector wide plan to reduce Tonga's reliance on imported petroleum – illustrates an example of where the donors' rhetoric concerning harmonisation and alignment only goes so far. The Meridian Energy project to develop a 1 MW Solar PV Plant was initially planned to be partly funded by the New Zealand Government and Meridian Energy and partly through a third donor party. The European Investment Bank (EIB) expressed an interest and offered funding to this project. However, what transpired was a very delayed and cumbersome process.

After a year and a half of going back and forth with the EIB, processes of the EU and various other constraints meant that Tonga was unable to receive this funding and the EIB withdrew from the project. The World Bank then stepped in and agreed to grant fund the money that was originally promised by the EIB. However, in order to provide funding to this project it had to follow World Bank procurement guidelines and therefore a review of the process had to be undertaken.

Interviews with a Tongan official noted that there was "little coordination from donors on specific projects, but at a macro level, an increased coordination among donors to adopt development areas that they would focus on." This same respondent stated that "the more donors you have on a project the more administration burden there is" as each donor has "different processes and protocols. It is much simpler to have one per project" (Mountfort 2013).

In the final arrangement, it was agreed that the New Zealand Government was to entirely grant fund the 1 MW Solar PV Plant. A development partner official stated that this project "was an example where harmonisation takes you so far until you stump up against these individual requirements [of the donor organisation]." Lord Sevele also felt that a lesson learnt from this project was that "although the TERM is a multi-donor funded sector approach to development, when it comes down to individual projects within the TERM, these are best developed and funded by individual donors" (Sevele 2011).

Ensuring that the government, development partners and regional organisations are working together on one sector wide plan is a positive change to development practice and highlights a movement away from ad hoc project or programmes approach which has occurred in the past. While harmonisation of development actors can be successful at the initial stages of putting together a plan, Solomon (an official in Tonga) argues that at the project level, coordination between the government and their development partners needs to be done through a bilateral agreement (Solomon, personal communication, May 2012).

Box 5.5 The Green Climate Fund and the 'harmonisation paradox'

Faka'iloatonga Taumoefolau

In the sixteenth session of the Conference of the Parties (COP) to the United Nations Framework Convention on Climate Change (UNFCC) in 2010, it was agreed 'to establish a Green Climate Fund (GFC) to provide financing to projects, programmes, policies and other activities in developing countries via thematic funding windows'(UNFCC 2010). The mechanics of the Green Climate Fund are extremely rigid, and a number of stages were needed to conceive its function, find funding, and define its processes. Nonetheless, the GFC manifests the collective will of 194 countries and is relevant to Oceania as an opportunity to carry weight within global action to address climate change. In one light, it is possible to reach the conclusion that the Green Climate Fund is an example of the neoliberal (or neostructural) triumph in aid design, a triumph in that GFC adheres to principles of transparency, accountability, efficiency, ownership and effectiveness which exists in the governing instrument (GI) and subsequent GFC documents.

In practice, however, for countries to tap into this important resource of funding an arduous journey is needed. For example there are basically six stages to the final GFC proposal approval, starting with nominating a National Designated Authority or NDA. It is likely for a variety of reasons that a developing country may want to change the NDA due to a lack of capacity or strengthen the NDA through the GFC 'readiness programme'. This GFC readiness programme is another layer of funding involving another layer of processes. Added to this mix is an accredited entity (AE), the status of which requires the GFC to look at fiduciary, environmental and social standards of an application. Compounding the complexity and need to align, there may be other priorities at different stages of the process requiring certain auxiliary steps to be added, such as a feasibility study which then devolves into a contracting arrangement. All of these processes can take months, both consecutively and concurrently.

Indeed, no justice can be done within this brief account to explain adequately the processes involved for a GFC proposal approval, suffice to say it is lengthy and robust. Rather, the point is that the GFC has the potential to reveal what can be termed the 'harmonisation paradox'. As mentioned earlier the OECD's Paris Declaration Principles are evident in the GFC official documents. The paradox is that in practice this funding more or less contradicts those very principles and (lack of) harmonisation stands out the most. The complexity of the application process, the long delays and the need to involve many different institutions and sectors makes harmonisation difficult to achieve in practice.

With this in mind, however, another perspective then presents itself. Is the 'triumph' mentioned earlier a triumph of neoliberalism or is it a triumph of Oceania? Metaphorically, we in Oceania are inherently navigators, 'we sweat and cry salt water, so we know that the ocean is really in our blood' (Teresia Teaiwa, quoted in Hau'ofa 2008: 41). We have the ability to endure the impacts of climate change whilst steering the arduous journey of tapping into this and other important sources of funding. Global agreements and resources, such as the Green Climate Fund, inadvertently impose further challenges and constraints. Yet the ability of us in Oceania to see these and carefully work their way through them is evidence of skills in a new form of navigation: reading the signs, preparing the resources, setting the course and patiently steering through global winds and currents of bureaucratic and political oceans.

To be clear, this account is not a critique of the GFC nor its processes, for given both the language of the GFC documents and the spirit of its objectives, it represents an admirable collective will. Within Oceania projects in Vanuatu, Samoa, Fiji, and (through the 'accredited entity' the Asian Development Bank) the Cook Islands, Federated States of Micronesia, Marshall Islands, Papua New Guinea, and Tonga have been approved by the GFC board. These illustrate a small but meaningful victory for Oceania.

It can be argued that, in the neostructural aid regime, donors and recipient governments have agreed to seek public participation in policy making and agreed to have detailed consultation the about the flows and processes of aid is laudable, strengthening democracy and a sense of ownership in the long-term. It is certainly a major improvement on the way SAPs were imposed. However, in the Pacific Islands it could be suggested that in practice the processes of participation and consultation have not worked very well and are instead creating pressures and demands that cannot be effectively dealt with. The starting point for the research project that this book has sprung from was Nicki Wrighton's uncovering of the extent to which Tuvalu had to cope with a proportionally massive influx of visiting officials, missions, researchers, journalists and professionals all seeking appointments to 'consult' with local officials and others (see Box 1.1). This is partly to do with the complexity of aid institutions (see below) but it is largely a matter of the sheer volume of visitors and meetings relative to the size of the local agencies and their complement of staff.

The weight of expectation for consultation sometimes give rise to local resentment that officials or local leaders have to say the same things to different visitors many times over or that visitors do not prepare themselves adequately by reading documents that answer the questions that they seek verbal responses to (Wrighton 2010b). Yet often the local cultures of hospitality and the associated desire not to give offence mean that meetings on similar subjects are repeated, all so that visitors can report that they have consulted, conducted research or ensured the participation of local stakeholders according to the requirements of their missions.

The pressures and inconveniences are felt particularly in smaller countries and territories of the Pacific – as we saw in Tuvalu. In these cases, the whole edifice of government is in place with separate ministers, departments, policy statements, reporting systems and audits. Yet rather than staffing levels of several hundred (as may be found in Papua New Guinea, Fiji or the major donor countries), states such as Cook Islands, Niue, Tuvalu or Wallis and Futuna may only have a handful of staff in a given department or section. Typically also in small and tightly connected societies, people in senior positions are well known locally and expected to be accessible. Government ministers or permanent heads may literally have an open door to their offices, visitors are expected to be made welcome and a relatively high level of informality can often be present. Furthermore, these same senior or mid-level officials and political leaders are the ones who are frequently invited – sometimes pressured – to attend meetings overseas, again as part of the consultation process required in representing their countries. In the Tuvalu case, being absent overseas to attend meetings and conferences, combined with the vagaries of infrequent air services, meant that such individuals were away from their desks and daily duties for substantial proportions of their time: a two day meeting in Europe could mean they were away from home for up to two weeks, for example (Wrighton and Overton 2012).

This situation, argued Wrighton (2010b), led to key people being overwhelmed by visiting missions, consultants, researchers and journalists. Such consultation displaced time and energy for what might be considered their 'normal' work, which in a diminutively staffed office had to span a wide range of functions and subjects. The ability of local officials to resist or limit such burdens of consultation is constrained both politically and, indeed, culturally. Donor officials have to be met and talked to or continued aid allocations may be threatened and there is awareness that reporting is a critical concern of donors. As such, local compliance is forthcoming for what are often cumbersome and time-consuming processes of 'evidence' gathering and 'participation'. To some extent these demands on officials can be managed when separate aid coordination offices are established within recipient bureaucracies (see chapter six below) yet donor agencies, consultants and researchers will still wish to talk directly to those who directly manage aid-funded development programmes in the respective ministries.

Similar pressures are felt in circles outside government agencies. Civil society organisations and local communities are often the targets of evaluation teams and donor missions seeking to gain alternative or additional views to those they get from local officials. In reality, and especially in smaller countries and territories, this falls mostly on a small number of organisations and individuals who are deemed to be reliable and worthwhile as alternative voices. As with officials, such individuals outside government may be reluctant to refuse to talk with visitors because their continued funding may rely on keeping good relationships with external agencies.

A recent addition to this range of consultation is local business people. Given the evolving shift towards a retroliberal approach to development aid, the desire of donors to consult with and involve the private sector is likely to put a growing pressure on business enterprises locally. Thus people such as the chairperson of a small local chamber of commerce, a prominent local entrepreneur or the representative of a women in business group are often now on the calling list of visiting missions and consultants. Such people are sought to gain an understanding of local economic conditions and opportunities or to link with overseas business interests. They, like the pool of civil society leaders and officials, perhaps even more so, may be comprised of a very small number of people and the calls on their time may be frequent.

Another recent consultative requirement has resulted from the global concern for climate change. Considerable donor interest in this issue and the attention of a wide range of international environmental NGOs, academic researchers and journalists has seen the Pacific Islands region become a focus of attention. This is due to the threat of sea level rise leading to erosion, vulnerability to extreme events, saline intrusion, swamping and eventual potential disappearance of low lying atolls and sand cays. In many ways, low Pacific Islands have become a global symbol of climate change and sea level rise leading to claims that whole nations (such as Tuvalu and Kiribati) may not exist in the future and that a flow of environmental refugees is imminent and inevitable. Such communities are portrayed as victims of global climate change, and indeed some may even actively cultivate this (Barnett and Campbell 2010; Connell 2003).

Given this, aid donors have sought ways to channel ODA funding – sometimes substantial – towards efforts to mitigate the effects of climate change and to 'build resilience'. As with other forms of aid, this has required a high degree of consultation to identify the effects of climate change locally, to select local agencies to work with, and to seek possible projects to be supported by aid. In addition to aid-related consultation, climate change has led to a new stream of visitors: journalists and researchers keen to find evidence of the effects of climate change and sea level rise. The Cartaret Islands in Papua New Guinea, for example, have been covered extensively in the global media and portrayed (falsely) as the first place on earth to be evacuated due to sea level rise (Connell 2016). It could be suggested that these rather remote islands have become some of the most visited and written about on a per capita basis. Although the climate change concern raises important broader questions of how Pacific Island peoples and countries are framed by the rest of the world (Barnett and Campbell 2010), for the purposes of analysing aid, we suggest that it has added significantly to the volume and complexity of external agencies active in the region and magnified the pressures on over-burdened people and institutions locally.

The encumbrance of consultation has a geographical as well as social dimension. Socially, the people who are business leaders, politicians, senior officials or civil society heads may be drawn from a small local elite. In a Pacific island setting, those in such a group are likely to be well known to each other and in many cases tied by kinship links. They are also generally located within the main urban areas on the islands that host them, rather than on outlying islands or remote rural areas. In this sense remoteness may prove a hidden blessing for leaders in peripheral areas as the disadvantage of neglect may be balanced by a lesser need to spend time with visiting consultation teams! The importance given to climate change, though, has created a partial exception here: although such attention still requires meetings with officials, politicians and civil society leaders, it does seek out apparently vulnerable islands and communities.

Therefore the burden of consultation is keenly felt in many parts of the Pacific Islands. It involves a plethora of missions, meetings, documents, agreements, plans and reports providing evidence of in-country participation in development activities. It meets the neostructural demand for public participation and consultation between donors and recipients. 'Consultation fatigue' is felt by many officials, politicians, NGO and community leaders and business people, whether in the form of crowded appointment diaries, demands for periodic reports, yet another participatory workshop, or even frequent social events at the High Commissioner's or ambassador's residence. The burdens are also highly uneven, falling on key personnel, at key times of the year (when budget allocations need to be planned or spent) and in particular places. Furthermore the considerable activity in terms of meetings and discussions, though giving the appearance of broad-based participation and broad-based local ownership, may be rather narrow in practice, given the small size of Pacific Island polities and elites. A small pool of people – senior officials, politicians, women in leadership roles, NGO heads and, in the current retroliberal period, business 'representatives' – are repeatedly the targets for visitors. Participation and consultation in the Pacific, then, are definitely a burden; in many cases the Pacific people involved may feel it is a treadmill.

The proliferation and complexity of aid institutions

Whilst consultation focuses attention on what is often a small group of people and agencies locally, aid more generally involves an ever-increasing range of external agencies. In Chapter Three we mapped ODA flows in the region and this revealed both a dominance of a small number of bilateral donors and a concentration of links between key 'patron' state donors and their recipient partners. The bulk of ODA flows in terms of volume, then, are transacted through a relatively small number of bilateral linkages. Yet the reality of aid in the Pacific involves a multiplicity of agencies on both donor and recipient sides, even if the associated flow of resources is relatively minimal in many cases.

As discussed in Chapter 3, the aid and diplomatic agencies of Australia, France, the USA, New Zealand, China and Japan are the main bilateral players in the region providing the majority of funding. However, other bilateral and multilateral institutions are also active. These include the European Union, the United Kingdom, the Republic of China (Taiwan), Canada, and smaller donors such as the United Arab Emirates and India. Multilateral institutions are also prolific: the United Nations and its component agencies (UNDP, UNEP, UNICEF, FAO etc), the Asian Development Bank, and the World Bank, provide significant levels of funding. Table 5.1 provides a list of all official donor agencies in the Pacific region recorded in OECD ODA data. It shows that, beyond the dominance of a few bilateral donors, there is a very long 'tail' of donor institutions in many parts of the region. There are bilateral and multilateral agencies and

Table 5.1 Donor agencies in Oceania 2005–2014
ODA as recorded by the OECD
(total disbursements $US million current)

	2005–2009	*2010–2014*	*% change*
Australia	3110.36	5224.11	68.0
New Zealand	642.49	1093.00	70.1
United States	921.85	953.91	3.5
Japan	424.96	695.37	63.6
France	646.41	672.05	4.0
EU institutions	410.80	509.04	23.9
Asian Development Bank	24.45	303.87	1142.8
World Bank (International Development Association)	28.25	244.52	765.6
Global Fund	91.90	156.10	69.9
Global Environment Facility	64.07	64.51	0.7
Germany	8.59	49.75	479.2
United Kingdom	28.89	40.69	40.8
World Health Organisation		35.55	
UNICEF	32.21	35.53	10.3
UNDP	31.97	28.45	–11.0
IMF	8.94	27.51	207.7
Korea	9.16	23.57	157.3
Global Alliance for Vaccines and Immunisation	5.18	17.75	242.7
Canada	40.64	16.45	–59.5
Adaptation Fund		15.49	
United Arab Emirates		14.90	
UNFPA	13.63	10.12	–25.8
Norway	5.13	9.22	79.7
International Labour Organisation		9.19	
OPEC Fund for International Development	0.89	8.39	842.7
Russia		8.23	
UNAIDS	5.46	7.63	39.7
IFAD		7.12	
Bill & Melinda Gates Foundation	2.76	6.06	119.6
Spain	3.40	5.42	59.4
Finland	0.65	3.96	509.2
Austria	5.53	3.17	–42.7
Italy	26.80	2.79	–89.6
UNHCR		2.62	
Turkey	6.70	2.30	–65.7
Sweden	1.07	2.19	104.7
FAO		1.99	
Climate Investment Funds		1.49	
Switzerland	1.47	0.66	–55.1
Israel	0.21	0.94	347.6
Luxembourg	0.07	0.61	771.4
Ireland	0.22	0.55	150.0
Thailand	1.03	0.52	–49.5
Greece	3.90	0.38	–90.3
UN Peacebuilding Fund		0.28	

(*Continued*)

Table 5.1 (continued)

	2005–2009	*2010–2014*	*% change*
Global Green Growth Institute		0.18	
International Atomic Energy Agency		0.13	
Estonia		0.13	
Poland	0.04	0.07	75.0
Slovenia		0.04	
Czech Republic		0.03	
Netherlands	6.91	0.02	–99.7
Belgium	0.17	0.01	–94.1
Romania		0.01	
UNTA (technical assistance)	150.40		–100.0
UNHCR	1.39		–100.0
Denmark	0.24		–100.0
Czech Republic	0.11		–100.0
Portugal	0.10		–100.0
Iceland	0.08		–100.0
Cyprus	0.03		–100.0

Source: OECD: Stat 3 August 2016

private institutions. And, being OECD data, it omits what we know are other significant donors, primarily the People's Republic of China (and the Republic of China is also not enumerated). In some cases the budgets that these agencies bring may not warrant the attention they require and burden they create locally.

Another feature of the above table is some marked changes between the two periods. The years 2005–2009 covered the time when the neostructural area of poverty-focused development assistance was at its height, whereas the later 2010–2014 saw the more retroliberal approach become in evidence. Between the two periods there were some donors who withdrew from the region or cut their aid substantially (Canada, the smaller European bilateral donors [Switzerland, Italy, Netherlands, Belgium, Austria and Greece as well as Turkey] and some UN agencies [technical assistance, UNFPA]); the USA and France as two of the major donors had relatively static volumes of aid (amounting a small decline in real terms); three of the other main bilateral donors increased aid to region significantly (Australia, New Zealand and Japan); Germany and the UK also oversaw increases; whilst the largest significant increases came from multilateral agencies promoting economic growth (Asian Development Bank, World Bank, IMF).[2] As well as this long list of state or multilateral donor agencies there are the non-government agencies, some of which can be significant and/or long-standing partners. There are large global NGOs, such as Oxfam, World Vision and Greenpeace; there are NGOs based in particular donor countries, such as New Zealand's VSA or Australian Volunteers for International Development (AVID); there may be an influx of humanitarian agencies such as Red Cross following a natural disaster; there are small agencies (Surf Aid, Engineers Without Borders, Child Fund); faith-based organisations (Caritas, Tear Fund, Christian World Service) and a very large number of local NGOs who may partner with these international donors. All these require some degree of local contact. Government officials will usually want to monitor and approve the activities of such agencies, whilst the NGOs will usually want to consult with government. Often, local NGOs have to deal many of the international organisations: they can be an important source of funding, yet they also may impose heavy demands on stretched local staff. They operate in a complex network of relationships between donors (both state and NGOs) and local government institutions and policies (see Box 5.6).

Box 5.6 Civil society and aid effectiveness in Tonga

Pedram Pirnia

Close examination of the civil society in the South Pacific reveals that the sector has, indeed, a unique role in bringing together stakeholders, building and sustaining partnerships and playing a complementary role in assisting donors and partner governments to deliver development aid. Civil society in Tonga, for example, is effective in connecting the global to the local. The sector has managed to inform the communities, and introduce a range of new ways of cohabitation and living to the people, such as goal setting, and inclusion of women in decision-making, which have been highly positive. Women in Tonga as elsewhere in the Pacific have actively used civil society as leverage to mobilise, advance their rights, influence governments and challenge traditional norms (Pirnia 2016).

Aid and development funds in Tonga are channelled to the government of Tonga and to the Ministry of Finance primarily, which is the focal point for donors within the Kingdom. The World Bank, the ADB and the UNDP all maintain offices within the Ministry of Finance in Nuku'alofa, and the Tongan Government has been successful in coordinating donors. Reforming of the aid structure (by implementing the Paris Principles) in Tonga has resulted in receiving all of the allocated budget support from major donors since 2012, which is a rarity in the region.

Tonga is an excellent example of how the 2005 Paris Declaration has changed the delivery of development aid. It is virtually alone in the region in that it has succeeded in coordinating all allocated aid from only one government department, the Ministry of Finance.

China, Australia, New Zealand, the EU, the World Bank, and ADB each work very closely with the Government of Tonga through the Ministry of Finance. Research in Tonga and interviews at the Ministry of Finance revealed that the staff in general are frustrated over the processes and the 'soft power' of the donors and especially the lack of appropriate consultations with the stakeholders, especially civil society and the project beneficiaries, before identification of development plans. In addition, the staff at the Ministry highlighted the fact that their main challenge is dealing with donors such as China who tend to ignore national development plans and priorities. Instead they bypass normal procedures and tend to reach out and establish relationships with individual Ministers.

The main problem for civil society in Tonga as elsewhere in the South Pacific, aside from inadequate resources and absence of capacity, seems to be lack of appropriate consultations with the potential project beneficiaries in formation of development policy. Government consultations especially with the civil society are rare and normally made at the last minute, and aid is normally allocated and distributed based on political interests of the donors – which is why development results are often ineffective and lack ownership by stakeholders especially the project beneficiaries.

Observing the civil society of Tonga thus reveals that CSOs are in a challenging position: their hands are tied because of their dependence on donor funding and their focus is spread between meeting the standards of accountability of their donors and meeting the needs of the communities they serve. Unless development CSOs counter their resource dependence, they will fail to win the cause of the people they aim to serve and will be abandoned by the people over time. It is by cultivating local ownership of development assistance that civil society will be better able to ensure effective and sustainable development aid.

To government and NGOs we must also add a range of individual players in the aid landscape; people who wish to gain information on aid and development or investigate special interests, such as environmental degradation or youth rights, or overseas journalists who are frequent visitors to the region, especially when there are natural disasters, political conflicts or high level meetings: all times when local systems are already stretched. In addition, academic researchers, including postgraduate students, comprise a small but sometimes very needy group, anxious to gain detailed data that is not readily available or to seek the views of local people and communities. Not only do such researchers request much in terms of time and information from local people but often they also need to be hosted and mentored in, for them, unfamiliar environments. Again, Pacific cultures of hospitality result in costs in time and resources being borne whilst accommodating these visitors – costs which we would argue are rarely compensated for in terms of the beneficial results of such research for the communities that covered these costs.[3]

Not only is the very high number of external agencies and individuals vis-à-vis the relatively low number of local resource people an issue for the Pacific region, but also at question is the diverse range of operating methods that they bring. Donor agencies will often adopt different templates and procedures for similar feasibility studies, environmental and social impacts, financial reporting, consultation and monitoring and evaluation. All these have to be negotiated by local counterparts, often anxious to keep donors satisfied but weighed down by varied processes and formats being used time and again for similar information. Such issues regarding the large number of agencies involved in aid can take on particular expressions in certain cases; special arrangements between donors and recipients can bring benefits to both but they can complicate relationships.

Compliance demands, in terms of planning, reporting and financial management systems, may be high and onerous for recipient agencies but the return has been increased funding, longer-term commitment, consistent management systems and, overall considerably fewer overhead costs (in terms of meeting times, reporting and external micro-management) per dollar received. However, although these programmes have involved large portions of bilateral aid budgets, they have not included the wide range of other donors. Wrighton (2010b) noted that some multilateral agencies continued to have a bad reputation for imposing their demands for consultation and reporting – each with their particular systems and templates – even though the size of their aid budgets was small compared to the major bilateral donors.

The other aspect of the harmonisation principle is that it may prove to be a double-edged sword in terms of power relations. The gains from efficiency improvements in terms of a more harmonised consultation and reporting process and higher net aid receipts may be balanced by a loss of local negotiation power. When the many donors are not communicating among each other, or even in disharmony, there is an opportunity for recipients to play off one donor with another and encourage a degree of donor competition for the 'best projects' or status of 'most favoured partner'. With harmonisation, that negotiation space for recipient countries has been much more restricted when it comes to their engagement with the twenty-nine DAC member states who signed up to the Paris Declaration. However, room for negotiation and playing off diverse donors still exists with regard to other players, especially those outside the DAC such as China, Taiwan and more recently Middle Eastern donors. On balance, however, it seems the gains for Pacific Islands of harmonisation outweigh the losses: greater certainty, fewer transaction and overhead costs, more long-term programmes and higher aid inflows – even if having to answer to a single powerful donor advocate – are better than a costly and imprecise process of bargaining with many different agencies and systems.

A final issue to discuss with regard to the complexity of aid institutions has to do with the volatility of donors' aid policies. Achieving development outcomes usually require concerted and consistent policies and flows of resources over a relatively long period of time. For example the goals of reducing maternal mortality or improving literacy rates require sustained investment in the training of health workers and teachers, the construction of health centres and schools, and the provision of books and equipment. Further up the development planning scale, significant gains in achieving high level goals such as a healthier and more professionally skilled population may only be seen over a generation or more. The neostructural aid environment seemed to involve recognition of this complexity and longer-term perspectives through the promotion of multi-annual aid modalities – SWAps and GBS – and the guiding hand of national strategic development plans such as PRSPs.

Yet, over the past fifty years donors have proved to be fickle in in terms of general direction and this has caused disrupted and repeated redirection of development strategies in the Pacific as much as elsewhere. The policy changes by donors from one aid regime to another have been precipitated by broader ideological shifts in public policy making as we outlined in Chapter Two. Thus the abrupt switches from the neoliberalism of the 1980s and 1990s to the adoption of the poverty agenda in the year 2000 to a retroliberal approach to aid that explicitly serves donors' geopolitical and commercial interests following the GFC of 2008, have been played out in both general and specific aid relationships and policies in the Pacific. Officials on both sides of the aid relationship have had to adapt to the changing instructions and policies of their political masters. Local aid officials have had to deal with sudden shifts in the strategic plans of aid donors and in the personnel they deal with. Even if the faces of donor officials stayed the same, those officials may have had to alter direction as the result of sudden changes in instructions from their governments. Such was the case, for example, with New Zealand aid officials following the election of the John Key National Government in 2008 and the appointment of Murray McCully as Foreign Affairs Minister. Very quickly, the poverty focus of the New Zealand aid programme was changed and the aid agency – NZAID – was disestablished (Banks et al. 2012; McGregor et al. 2013). In addition, there was a high turnover in donor staff, as some either left the agency or were re-assigned, and new staff took their place. Government personnel in the Pacific dealing with New Zealand aid counterparts had to very quickly adjust to the new political landscape. Thus aid relationships involve local officials not only dealing directly face-to-face with donors in the present; they also have to read and react to political shifts in the metropole over a longer time frame.

Such shifts in policies in the metropole can have major impacts in the Pacific. Despite efforts such as the Paris Declaration to improve the predictability and effectiveness of aid, donors have demonstrated that they are prepared to override these principles when wider geopolitical considerations are present or, indeed, when domestic political priorities intervene. One striking example of this is offered by the agreements negotiated between Australia and Nauru in 2001 with regard to the accommodation in the latter of refugees and asylum seekers (Box 5.7). Here it appeared that development assistance from Australia (aid which rose spectacularly after 2001 – see Figure 3.12f) was part of a direct political deal, aid as payment for a service provided to Australia by Nauru, rather than development aid guided by the principles of aid effectiveness.

Box 5.7 Nauru and the 'Pacific Solution'

Warwick E. Murray

The case of refugees and asylum seekers arriving by unauthorised sea-vessels has long been a fraught political issue in Australia. Since the influx of Vietnamese refugees – the so-called 'boat people' – driven out by conflict in Southeast Asia in the mid-1970s, policy has come under scrutiny both within and outside Australia. In August 2001 the MV Tampa, a Norwegian cargo ship arrived in Australian waters with close to 450 asylum seekers from Afghanistan rescued when the stricken fishing vessel they were travelling on sank. After a stand-off, the intervention of the Australian special forces and a diplomatic incident between Norway and Australia, the asylum seekers were redirected to Nauru and Papua New Guinea where they were to be held until their cases could be heard.

This represented the beginning of what came to be known as the 'Pacific Solution' (Pérez 2003). Under this policy, first brought in under the Howard Liberal government (1996–2007), the maritime resettlement borders were redefined and those arriving in unauthorised vessels were either turned back or sent for processing in a number of detention centres in the Pacific including notably Manus Island in PNG and Nauru. It was made clear that without passing through these offshore detention centres, refugee settlement in Australia would be impossible. The 2001 terrorist attacks of 9/11 made this increasingly palatable to the electorate and parliament and the policy was approved just eighteen days after the attacks.

In the case of Nauru – with just over 10,000 inhabitants – negotiations with the then President René Harris led to an agreement that aid would be allocated in exchange for the establishment of two centres – Topside and State House – which together would have a total capacity of close to 1,200 refugees. Initially, aid flows of A$20 million per annum were established. The centres were operated by a private 'infrastructure maintenance' company called Transfield Ltd, which later became Broadspectrum Ltd with assets of over A$2.2 billion. Asylum seekers arrived in the camps from countries such as Afghanistan, Iraq, Iran, China, Vietnam and later Sri Lanka and Myanmar. The Nauru government was keen on the flows of Australian development aid that would accompany this agreement. The country's economy had declined from one which was dependent on phosphate mining and which enjoyed a relatively high income per capita in the mid-1980s, to one where mismanagement of sovereign funds, the decline of mining and an inherent difficulties with diversification had seen rank amongst the poorest economies in the world. Nauru had become especially dependent on aid from Australia, Taiwan and New Zealand.

Between 2001 and 2007 the Pacific Solution acted as a significant repellent to unauthorised arrivals to Australia, but from its very beginning it aroused deep criticism from domestic opposition in Australia as well as international organisations such as Amnesty. Given this, Labour Prime Minister Kevin Rudd closed the centre in 2008. This provoked severe concern in Nauru where the prospect of unemployment and aid decline worried local politicians in an economy where over 90% of the labour force was employed in the government sector – many connected to the refugee centres. Arrivals of refugees to Australia increased again following the softening of the policy and it became a significant component in the dispute within the Labour Party. When Julia Gillard took up power she re-opened the centre. During this period it was roundly condemned and riots in the centre took place in 2013. Later that year the newly elected Tony Abbott government put into place Operation Sovereign Borders which sought to bring the arrivals down from the peak they had reached in 2013 (of approximately 20,000 overall in Australian waters) through a much more aggressive policy of policing the waters. In Nauru, the population of detained refugees reached a peak in 2014 at over 1200. However, the conditions for detention were reconsidered and by 2015 the approximately 620 detained persons were free to move around the island of Nauru.

Overall, the arrangement between Nauru and Australia to establish refugee centres on Nauru led to some significant returns for Nauru in terms of aid inflows but there was also a need to deal with a new range of overseas personnel including immigration officials and private security companies. Moreover, the massively increased aid also resulted in increased scrutiny – and arguably interference – in Nauru's internal affairs, such as through New Zealand's questioning of Nauru's handling of political rights and the subsequent suspension of New Zealand aid. Ultimately it created a new dependency, which continues to have domestic political ramifications in Nauru given that a considerable proportion of its national income is associated with this policy. Increases in development assistance through special relationships may result in much more complicated aid relationships and a diminishing of local sovereignty when the donor is able to dominate.

The issue of capacity

It is one thing for donors to use process or political conditionalities to seek to bring about change in recipient states and for a multitude of donor agencies to be involved; it may be quite another thing for Pacific Island states to be able to actually implement the changes agreed upon. The question of capacity of local institutions and the capabilities of their staff to deal with the whole machinery of international development aid is a pressing one for most Pacific Island states. It has to do with the absolute size of government departments, the abilities and experience of staff, the remoteness from centres of power and the particular way capacity deficits are dealt with.

In this regard, the size of Pacific Island territories and their administrations is an important consideration and the huge difference in size between Pacific administrations and donor administrations is critically important. Countries and territories such as Tuvalu, Niue or Nauru have a level of political independence, albeit defined differently, that requires them to establish and maintain the comprehensive machinery of government. This includes a legislature, an executive (with a Prime Minster and cabinet ministers), a foreign service (with representatives overseas), core ministries such as health or education (employing many nurses and teachers), a customs service, a judiciary, law and order agencies and a range of services flowing from these (public works, telecommunications etc., even if some of these are contracted to the private sector). In this political sense, there is a minimally required scale of bureaucracy and government, and a base line set of agencies and institutions that is an essential feature of being a country of territory – other things being equal the smaller the state or territory the larger the relative size of government.

However, in an absolute sense in countries with populations less than 50,000 or 100,000 people, it is no wonder that the various component government offices have only a handful of people staffing them who by necessity are expected to multi-task to cover multiple responsibilities. As already noted, for small Pacific Island states and territories, specialist civil servants are a luxury as a small number of staff has to cover a wide range of portfolios, yet they also have to have sufficient skills to understand expert reports and advice. An official in the Ministry of the Environment in Niue, for example, may have to switch in the duration of one day from considering projects for water quality, to developing a plan for the removal of asbestos building materials or rusting used vehicles, to responding to a request for information on the effects of climate change.[4] The sheer lack of staff numbers creates a significant mismatch vis-à-vis donor countries and agencies, and Pacific bureaucracies may find it difficult to deal on equal terms with donor entities (Box 5.8).

Box 5.8 Sovereignty and size: observing a regional meeting

John Overton

In 2001 I attended a meeting of the Forum Economic Ministers in Rarotonga, Cook Islands, being asked to present a paper on land tenure. Over the two days of the meeting, I was able to observe some of the sessions and see how the Pacific Forum worked in practice. One meeting in particular was notable as a way of illustrating of how the principles of sovereignty and equality were compromised by matters of scale.

The session, a plenary of the Forum economic ministers, was chaired by the then Prime Minister of the Cook Islands, Dr Terepai Maoate. In keeping with the protocols of the Forum, the meeting was arranged with an inner ring of tables with one representative of each of the countries (Australia, the Cook Islands, Federated States of Micronesia, Fiji, Kiribati, New Zealand, Niue, Palau, Papua New Guinea, Republic of the Marshall Islands, Samoa, Solomon Islands, Tonga, Tuvalu and Vanuatu) (PIFS 2001), though from memory not all these countries had a representative at this particular meeting. Most of the Pacific Island representatives were the respective ministers of finance and the two main donors, Australia and New Zealand, had senior diplomatic staff. This 'one seat, one voice' arrangement of the inner ring clearly respected the equality of status of the Forum members. These were the only people who were permitted to speak at the meeting. Yet outside this inner ring were arranged seats for the advisors of the country representatives and this presented a very different picture. Some states had no-one sitting behind the minister and some of the smaller countries had only one advisor. Larger countries such as Fiji may have had three or four, but New Zealand had around six and Australia perhaps ten.

The meeting worked through an agenda drawn up by the Forum Secretariat and it addressed issues that seemed to have been put forward by Australia in particular as potential areas for reform: there was the vexed question of land and its possible opening up to commercialisation; intellectual property rights; governance and accountability; improved statistics; financial sector issues; and commercial law (PIFS 2001). The neoliberal imprint was clear, and it was not obvious that any of the agenda items had been proposed or prepared by one of the smaller member states.

As it progressed the dynamics of the official representatives on the inner ring versus the specialist advisors seated in the outer ring became apparent. The meeting moved at a quick pace through the agenda items and draft decisions and there were large piles of documents at hand to consult. A clause would be raised for discussion and often the representatives of donor countries had an amendment to propose or a question to raise. At these points the donor official could call on an advisor to pass a paper or a suggestion – the team of advisors behind the Australian official was particularly well briefed and organised. Amendments had been drafted and the relevant section of supporting documents was referred to. For the representatives of the smaller Pacific countries, there was clearly a problem in matching this. Ministers or their single advisor had to quickly work through the documents to find the relevant piece and often by the time they had caught up, let alone considered the material, the meeting would have made a decision and moved on.

It was a meeting that involved consensus: there was no open debate or disagreement and the meeting proceeded to adopt a summary document. All countries had a single vote (though votes were not needed in practice). There was no speaking down from one party to another. However, it was clear to me as an observer that the agenda and the detail of the meeting was dominated by the donor agenda – Australia and to a lesser extent New Zealand – simply by virtue of the sheer weight of the number of personnel acting as advisors. They briefed their official representative speaking from the inner ring, they had relevant material at their fingertips, and they were in control of the material to such an extent that the other official representatives at the inner ring simply could not keep up. In fact, the formal recognition of sovereignty of the countries represented at the inner ring could not disguise the reality of asymmetries of power and fundamental inequalities in institutional capacity.

Staff competencies reflect the problem of capacity. This is not to say that Pacific officials are underqualified or lack ability; rather they face the challenge of having to cover a much broader range of tasks and responsibilities than their counterparts on the donors' side, which may leave them relatively less well versed with a specific topic at hand. For example, even junior staff may be required to represent their countries and present reports at overseas meetings simply because no one else more senior is available (Box 5.9). Pay rates offered by Pacific governments are well below what a skilled and trained person may be able to get working overseas, and citizens of many smaller Pacific states (Wallis and Futuna, Niue, Tokelau, American Samoa) have the option of migrating relatively freely and working for higher wages in the metropole. Here, and elsewhere in the Global South, also emerges the phenomenon that state sector wage and conditions often cannot match those offered by external agencies working in-country. Thus, a bright younger member of a government staff may well be offered a good position as a locally employed member of staff of an overseas aid agency. However, even these locally employed staff (valuable assets for donor agencies seeking access to local knowledge and networks) face constraints in the way their careers can develop (Box 5.10). There may also be openings in the political sphere for ambitious staff in the public administration. These conditions are likely to lead to a high degree of mobility for able Pacific officials and it can be difficult to retain a skilled and experienced complement of staff who can deal with the demands of external donor agencies.

Box 5.9 Pacific regionalism and issues of capacity

Potoae Roberts Aiafi

The Pacific Plan was adopted by Pacific leaders in 2005 to promote and implement a Pacific regionalism agenda. My doctoral research (Aiafi 2016) included a study of the implementation of this Plan and it involved interviews with several officials across the region. Official reviews of the Plan pointed to its limited success in achieving regionalism. Amongst the reasons for this, officials pointed to the issue of limited local capacity and the way this was tackled by using overseas consultants.

Most Pacific Island countries did not have the qualified people or technical skills to implement this ambitious regional, let alone national, development agenda. Existing policies were largely seen as incompatible with the capacities of small bureaucracies in the region. Some national services are affected because officials can be absent for long periods of time trying to accommodate international and regional commitments.

> *The issue with the Pacific is small administration . . . The rate of turnover of our people is very high. You can see the rate at which our people go for regional and international meetings. Most of the country . . . 60–80% of the time. They put in the plan, but nobody is there to implement it. That's why a lot of policies fall down.* (SIDA 6)
>
> *The country's capacity is understood in terms of its internal resources and the financial ability of development partners to follow through. The term 'capacity' is often overused because it's not just about training people. It's the number of people. That's a real issue for small island states.* (RPS 17)

Implementation is not just about having the resources but also how they are utilised within the systems of programmes, projects and activities to bring about change. A key implementation modality in donor-funded programmes has been the use of consulting services. As well, the capacity needed to progress effective change has often been underestimated. This modality means that effective donor-local actors' relationships are vital for a policy to survive and sustain an implementation process in the local context. These were often not sufficiently attuned culturally or logistically:

> *Countries have wonderful policies . . . but they aren't implemented well. It's also the modality of implementation. Partners operate through overseas consulting services to implement. They go into countries and the locals are turned off. There's no ownership and*

commitment. But then there are issues with donors not listening. Some projects failed because of personality clashes between consultants and locals. Often donors undermine the amount of resources that are needed for capacity building – they think that by putting someone in there for six months would resolve the issue, no. A lot of projects are about changing legislation. That's not easy because it's political, you need to do consultation and work with countries in changing mind-sets. That's not an overnight thing. But projects are finite. (RPS 8)

There is a need then to address the issue of capacity in the Pacific Island region. One possibility is to improve the use of communication technologies, for example through the use of online meetings and exchanges in order to limit the amount of time that local officials spend out of country. In addition, the use of short-term placements of consultants should not be seen as an effective substitute for the long-term development and use of local expertise.

(modified from Aiafi 2016: 177, 204–205, 258)

Box 5.10 An invitation for dialogue: locally engaged staff career pathways in Oceania

Faka'iloatonga Taumoefolau

Foreign embassies and missions in Oceania commonly employ Locally Engaged Staff (LES) to augment the work of expatriate staff. Although their jobs vary in purpose, role and duties, local staff are vital to the work of donors. They can provide a very good source of information regarding local conditions and personalities and their access to local networks can be invaluable, especially as many expatriate staff are on limited-term postings and do not have time to develop long-term local relationships or language skills.

On the other hand, finding a job for a foreign agency can be attractive for the local staff. It can be a window of opportunity that establishes a strong foundation for a professional career. For many who have walked down this path the reward is measured in being exposed to a competitive work environment that encourages the drive to succeed. The draw therefore is the opportunity to acquire skills and experience that are internationally recognised, something particularly attractive for local people who have recently completed higher education studies abroad (many are themselves recipients of donor scholarships). The roles that are offered range from administration, policy, research and management. Linked to this, the duties and performance of a LES, just like other categorised staff, are reviewed under performance appraisals and incentives. From a wider organisational perspective, LES employment acts as a milestone to a longer career path in administration or policy for citizens of a particular country.

Be that as it may, more appreciation is needed not only of the contribution of LES but also of the openness or not of career pathways within the donor organisation. From personal observation, discrepancies exist in contracting, policies and opportunities. For example, often the method employed to determine salary is market-oriented (measured against local pay rates) rather than merit-based (based on expatriate scales of performance). Curiously the argument for this reason varies itself with some quoting or unconsciously adopting elements of the Noblemaire principle (that all staff should be remunerated at the rates of the highest paying country) whilst others have formed their own policies and regulations.

Salary wise, the situations that can arise from this is that LES are employed in developing countries at a disadvantage compared to those employed in developed countries. Because of the market-oriented principle for salaries, this disadvantage to LES can involve them being paid under the minimum wage acceptable in the country of origin of the organisation. Of particular concern are circumstances where what are considered norms and fundamental basic entitlements in the organisation's home work environments, and even practiced in the local market, are absent in the LES practices of an organisation. For example, paid maternity leave for three months may be denied to LES due to it not being stipulated in any specific local legislation.

The human resources policies adopted also lack clarity for many LES (or it can be argued creates an unfair application of policy principles). For example, the language of the human resources manuals may use words such as 'all staff' but in practice exclude LES, often falling back on distinctions drawn with regard to security classification. This limited or selective application of a policy also restricts career pathways and opportunities for LES. Despite being paid below the donor's minimum wage, LES find it difficult to progress upwards, even when delivering very high and quality work. In other circumstances too, LES frequently have more experience and a wealth of institutional knowledge (compared to their expatriate counterparts), that is underutilised or underappreciated. An interesting aspect also involves the categorisation itself, namely the 'local' in LES. LES may be nationals of the country in which they work but sometimes LES can be expatriate staff hired locally (as a result of residing in country). Whilst the latter may be able to progress up the career pathway and find jobs elsewhere within the organisation, the progression of the former is blocked because they are not citizens of the country of the donor organisation (or covered by reciprocal agreements with other donors).

Local staff are hired by donors because they add value given their understanding of local political, social and cultural nuances. This added value was acknowledged in a recent evaluation of New Zealand's aid programme in Tonga (New Zealand Foreign Affairs and Trade Aid Programme 2016). One finding was that there are several structural constraints, including security clearance, that inhibit LES in understanding the strategic intent. Furthermore, in this report it is acknowledged by some senior level officials that they are not only aware of this situation but feel more LES could and should be used more strategically (New Zealand Foreign Affairs and Trade Aid Programme 2016: 38). This acknowledgement is welcome and perhaps an invitation for dialogue that could lead to a more open career pathway, within the organisation, for LES in Oceania.

Geography is a further constraint with regard to capacity. Although less of an issue in recent times with improved communications and the advent of the Internet, it is still the case that Pacific Island bureaucracies are remote from the regional and global centres of power and the direct or informal access to information sources and personal political relationships that underpin global policy making. Officials may travel from the Pacific to international meetings and some Pacific people are employed by global institutions[5] but Pacific officials are rarely able to engage closely and regularly in the debates and discussions that shape aid policies internationally. Pacific institutions are thus largely 'policy takers' in the aid world and even when they are consulted and given a chance to participate, scale and remoteness may conspire to limit the effectiveness of their contributions.

These capacity issues have led governments on both sides to push for capacity building efforts to build the size and quality of Pacific institutions. Whilst admirable in terms of building domestic capacity to manage and direct the development agenda, these capacity building programmes do raise some questions with regard to sovereignty. Firstly, short-term schemes involving the placing of expatriate staff to train and mentor local staff has proved problematic in some instances, as with Australian officials working in Papua New Guinea. The loyalty and accountability of such staff is sometimes questioned and the short-term placements and generous remuneration (mostly paid at expatriate salary rates with allowances) can create local resentment and poor personal relationships, which limit the effectiveness of the training of local officials. At the end of 2015, the government of Papua New Guinea removed from office some fifteen Australian officials who were working as senior advisors in a number of ministries, stating that such officials were making local officials 'lazy' and "They're not able to takeover civil decisions, they are over-dependent on consultants and advisers and sometimes many of those decisions are not . . . in the best interests of our nation" (*The Guardian* 2015). That said, there are also instances where the use of expatriate staff can prove very useful for Pacific Island governments. They can bring skills and networks. Many prove to have a strong professional and personal commitment to the Pacific Islands and their people in ways that may even lead them to openly criticise their home country development agencies.

In-country training programmes can also be an effective means of capacity building. When Pacific staff stay at home they remain linked to their social networks and affiliations, yet they can have access to good quality training offered by overseas institutions and teachers. These courses are often short-term and intensive specialist training programmes aimed at introducing particular skillsets. Disruption and removal from one's desk is minimised and people tend to stay grounded and committed to their local careers. Not all such training programmes are successful though with regard to enhancing local ownership of development. Box 5.11 presents a case study of media training in Solomon Islands where Australian-backed schemes to train local journalists in the RAMSI environment revealed one aspect of a heavy donor-driven agenda to reshape local media so that they would join the drive to attack corruption and promote good governance, rather than develop more locally appropriate media and subjects of interest.

Box 5.11 Media development to influence behavioural change: a case study in Solomon Islands

Adele Broadbent

> *The Solomon Islands Media Assistance (SOLMAS) program ran from 2008 to December 2013. The program was strong on outputs but sustained outcomes are harder to see. The program was ended due to poor performance . . . The program ended up creating a degree of dependency by media organisations looking to SOLMAS to fill critical gaps (DFAT 2014).*

SOLMAS was funded by AusAID and formed as part of the civilian programme instituted by Regional Assistance Mission to the Solomon Islands (RAMSI). RAMSI was the Australian-led multinational police-centred force organised in 2013 by the Pacific Islands Forum following civil conflict in the Solomon Islands. RAMSI began a phase-out process in 2011 with the withdrawal of its military component in July 2014. SOLMAS had already been closed but during its time based out of Honiara it had a mixed reaction from local journalists.

My research examined the use of media development by development donors working under a state-building mandate in the contemporary aid modality (Broadbent 2012). It questioned how world views in developing countries can be changed and shaped by the direct and indirect influence of development donors on local media. By concentrating on media development as a development tool, the research looked at the motivation and understandings of the Western donor-funded trainers, the trainees and the consumers. It asked: can this donor initiative successfully be a 'one size fits all' model across different cultures and contexts, and is this agenda as clear cut as it professes?

The research uncovered growing unease concerning the priorities of programmes used in this area. There were charges that its religious adherence to the commercialised neoliberal model of media threatened to marginalise the poor. There were concerns that media development was seen by donors as a way of training a local 'watchdog' to oversee the funds that they no longer had control over in the new aid modality.

The SOLMAS programme's mandate was detailed on the RAMSI website after the launch in 2008 -

> *[SOLMAS] provides support and coaching to Solomon Islands journalists, to build their skills in reporting and editing to ensure they undertake their work efficiently, ethically and with integrity, and to give them the confidence to ask the tough questions.*

SOLMAS had been operating for just over a year as the lead up to the crucial 2010 elections began. This seemed a bit of a flashpoint for local journalists who questioned the motivation of the Australian trainers under SOLMAS.

One senior journalist I spoke to was circumspect about how SOLMAS approached journalism training leading up to the elections. This contributor felt SOLMAS 'fed the media' an agenda during the elections, although he admitted their trainers asked journalists to come up with their own questions as well.

> *They will organise this forum, and then hide under MASI, the Media Association of Solomon Islands – and get these people [the candidates] yeah – then they would go and hide in the glass room and feed us with ah questions and what we are supposed to ask these people.* AB: They fed you questions? *J 1: Yeah they prepared the questions and then, but then it is not closed I mean, 'these were some of the questions you might want to ask but we would like you to ask these questions'. [Or] 'these are very good one'. Its open to journalists to ah reporters to ask whatever questions they like. Yeah so actually they are running the show behind the curtain.*

This journalist articulated an uncomfortable feeling that the news media was being coached to ask the questions the donors wanted answering, or the donors wanted the country to ask. It is a theme that

Allen and Stremlau (2005) raise about experts from the donor countries setting standards for 'truth' and 'justice' while ignoring local realities. The contributor goes on to recommend Solomon Islands media may need to adapt a model to suit their own worldview and culture.

> *This is our country – this is our people. What is effective for us is what we should apply here. Not only in media. It is not because we are saying it is our country – it is because we want to put effective measures in place where the country can develop. So in terms of media definitely I think we should look at ways that we can disseminate information effectively to people. It's good to adapt the Western style but then is it going to work is the big question. So we should really look at ways that we can reorganise how we can do it and really should get the message down with people.*

As can be seen by the negative analysis of the now closed SOLMAS programme at the beginning of this piece, evaluation of such media development programmes seems to be based more around the efficiency of media development in a neoliberal sense of 'growing' economies than in terms of culturally appropriate or evolving media, or 'growing' resilient communities.

If donors accept that a strong local media is important in overall development, and if they also buy into the 'ownership' rhetoric, then they must apply it to the media sector of developing countries within which they are working. This means opening spaces for heterogeneous, indigenised media model that can be explored and developed to engage local realities in an informed and critical way.

Another strategy for capacity building is to provide scholarships to train staff overseas either through specialist short-term courses or with university-affiliated qualifications. Whilst resolving some capacity problems of the Pacific's public administrations, this strategy may also generate its own set of problems. Key staff may be absent overseas for two to three years and some return with heightened expectations that cannot be satisfied by returning to their old jobs. Moreover, with good qualifications, comes the ability to earn higher salaries and after experiencing life overseas some choose not to return home. If they have to serve out a bonding period, they may seek emigration after that requirement is fulfilled. More fundamentally though, overseas training may serve to orient staff towards those overseas values, systems and ways of life rather than reinforce their commitment to local cultures and societies or adapt their newly acquired skills to Pacific realities. For donors of course, it is very useful to have counterpart staff in the Pacific who are imbued with Australian or New Zealand (and more recently

Chinese) culture, language and ways of working. In this sense, capacity is created but strongly in the image of donors.

The capacities of donors

The issue of capacity in aid relationships is usually framed in terms of recipient capacity. Recipient officials and institutions in the Pacific are seen as often too small or lacking the skills to implement appropriate policies or manage complex financial systems. Such a view has led to numerous training programmes for local officials or to the placing of overseas technical advisors within Pacific bureaucracies.

However, capacity may be a more complex and two-sided issue. 'Expertise' may involve specialist skills and experience in fields such as financial management, public policy or human resource management. Yet for any development work, an appreciation of context is critical: public policy has to work through local institutions, cultures, languages, knowledges and social relations. When donors or foreign consultants lack such understandings and knowledge, and lack the ability to work with local agents who do, then no amount of technical capacity will ensure project success. The example of researchers working on food production systems in a post-storm situation on Chuuk (Box 5.12) demonstrates the need to recognise the worth of local ecological knowledge.

Box 5.12 Chuuk and post-storm aid

Alexander Mawyer

Heightened concerns about insular ecological and cultural resilience, agricultural and food security in the face of ongoing and intensifying climate change is a current feature of contemporary Pacific Islands conversations about the present and the imagined future.

Climate scientists in the marine, atmospheric and meteorological sciences have established striking evidence that for some Pacific regions, ongoing climate change will see a significant increase in the number, intensity and impact of tropical storms and cyclones. As ocean waters and the atmosphere above them heat due to global climate change, the amount of energy available to produce and maintain dangerous storms will increase proportionately as will their disruptions and damages.

In each of the last three years alone, significant storms have caused enormous damages in different parts of the region. In February 2016, Category 5 tropical cyclone *Winston* passed through densely inhabited areas of Vanuatu, Tonga, Niue and Fiji causing an estimated $US 1.4 billion dollars in damages. In 2015, *Pam*, another Category 5 storm, passed through Fiji, Kiribati, Solomon Islands, Tuvalu, Vanuatu, New

Caledonia causing an estimated $US 360 million dollars in damages. In 2014, *Ian*, a similarly powerful storm, passed through Fiji and Tonga with $US 48 million dollars in damage. In each of these cases, the Australian, New Zealand, and US governments, as well as other regional aid donors, played key roles in relief and support for local communities. However, the periodicity and depth of need is not always easy to meet.

I saw this for myself, in 2015, in the aftermath of another storm, so-called 'Super-Typhoon Maysak' in the North Pacific, which passed through the Federated States of Micronesia and the Philippines and directly through FSM's Chuuk State once again causing massive damage. I arrived in Chuuk lagoon in July as part of an interdisciplinary team piloting a project for the Recovering Voices Programme of the National Museum of Natural History in Washington, DC. Our project "Explorations of Ethnobotanical Knowledge in the FSM (Chuuk Lagoon)" examined contemporary *atake* 'homegardens' as dynamic ethnobotanical spaces of local ecological knowledge and drew on a suite of methods from anthropology, linguistics, ethnobotany and museum studies to investigate the status and transmission of language-encoded biological and ecological knowledge. In an age of significant loss of linguistic competence and cultural expertise due to outmigration and the ongoing legacies of previous US colonisation, this was important, primary research on ecological knowledge 'on the ground'. However, we had not planned for our work to be contextualised by heightened local concerns about small island ecological and cultural resilience and food security in the wake of Maysak. The post-superstorm context in summer 2015 emerged in our interviews not only because Maysak caused a tragic loss of the productive agroforest and devastated home gardens in many villages but because cutting-edge aid regimes included active forestry and garden transfers seen as a response to historically more common high-calorie/high-fat and unsustainable food transfers.

Post-storm aid had a number of visible dimensions operating at the community level including support for the repair and improvement of the electrical grid, roadways and water systems. Post-storm aid also included direct aid to individuals and families in the form of food transfers notably rice and other packaged processed foods. However, direct aid also included the transfer of breadfruit and other productive tree saplings facilitated by US Federal or other regional agencies and managed, in part, through the Federated States of Micronesia Forestry Service. The presence and evident breadth of distribution of these saplings suggested an enormous and exciting effort to promote 'sustainability-through-aid' by transfer of trans-Pacific and inter-island cultivar transfers. However, this very welcome aid demonstrated some of the limits of inter-agency cooperation and coordination.

Here are three examples in which the limits of this ecological aid emerged. Early in this field research period, I was invited to address the Chuuk Women's Council to share our research plans. At the end of this meeting, I was thoroughly welcomed and a number of those present noted that they would be happy to participate in our research. However, many present asked whether the researchers would be in a position to help them identify and offer advice and instruction on growing some of the saplings recently received through aid initiatives. It appears that no instructions were included on how or where to plant the received saplings. Were they intended for near-shore or upland gardens? Were they salt-tolerant or not? Friendly to loamy soils or sandy, high-ph or low, wet or very dry? The saplings were wonderful, several of those present suggested, but the absence of instructions made them difficult or impossible to transplant.

Later, in the garden of one of the most respected Chuukese cultivators in Tonowas village on the island of Weno, we found a wonderfully enclosed, fenced off and highly valued, breadfruit sapling thriving. This happy tree was another aid transfer whose generous donor was not identifiable – our consultant noted that it was difficult to know which agency was responsible – US Ag, US Forestry, or the University of Hawai'i were all offered as possible donors. In this case, the landowner was very happy to have this tree and had the expertise needed to see the sapling thrive. However, unlike all of the other breadfruit in his garden, he noted that he had not received information about it and could not guess which of many *Artocarpus* species he had added to his land with a many-year commitment. Species information or guidelines about the seasonality, fruit qualities, disease susceptibility, or other highly useful details would have allowed him to make even more nuanced choices about where to plant this tree or how to enact his long-term plans around it.

Finally, on Fefen, we encountered an entirely different response to sapling-focused agricultural aid. Sometimes called the 'island of gardeners', Fefen sits across Chuuk lagoon from the main island Weno and is locally famous for its traditional myth about the expertise of Fefen men and women as agricultural connoisseurs and experts whose alacrity for gardening led them to erect a ladder so high that they could harvest the moon. At a large meeting of gardeners, Fefen men scoffed at the idea of sapling aid. Trees are a many-year commitment and in a post-storm context, locally surviving 'hero' trees are totally sufficient for re-propagation of the agroforest. What is needed they said is a transfer of seeds for commercially marketable short-term produce which can be sold in the local markets on the main island Weno in order to purchase school supplies, clothes and other basic needs. This was something, they thought, the state or national government (or international aid regime) was unwilling to provide for reasons unknown.

All three of these examples suggest that even exceptional, thoughtful, productive and sustainability-minded aid requires close multi-agency consultation to 'close the gap' between highly informed ecological agricultural aid projects and community need and knowledge.

The issue of donor capacity may go further than a lack of knowledge of local context. The assumption that international technical advisors 'know best' has been challenged by Pacific officials (Box 5.13). Whilst there may be many examples of technical assistance that is effective and appropriate, particularly when experienced advisors and consultants develop understandings and relationships over a long period of time to complement their particular skills, there are also instances where even the technical skills may be lacking.

Box 5.13 Technical Advisors and capacity building in Tonga

Sisikula Sisifa

Capacity building has long been the focus of public sector development in order to cultivate more resilient and productive civil servants and improve their ability to perform projects activities. Technical Advisors (TA) are a fundamental feature of capacity building. Donor agencies utilise foreign consultants as TAs and solicit these consultants to develop and train civil servants. Capacity building should build not only local skills but also enhance a degree of local ownership and long-term control.

However, a study of three capacity building projects in Tonga (Sisifa 2015) revealed that they were assessed as unsustainable and the training provided was insufficient in developing local capacity to manage projects. In the case of the Urban Planning and Management System project there was a notable lack of skills transfer. Local participants pointed to some of the problems in the design and delivery of the project:

> *We were not provided with training manuals or guides and it was really difficult to follow the advisor because the advisor knows the technical side of urban planning inside out, but he didn't know how to train people. So most of the training sessions he was just doing all the work himself and not really talking about the steps or why he was doing what he was doing. So then when we are asked to do it again we didn't know how to because the advisor just did it all and didn't really train us.* (Participant 10)

This brings to light questions regarding the quality of TAs. There has been a long-standing discussion in the literature with regard to issues such as the ability of TAs to replace themselves with locals and the importance of adequate time in the field. Little attention has been given though to variation in the quality of TAs, the training they deliver, or the sustainability of the skills transfer. Another civil servant stated:

> *When the project ended all of the foreign guys left and it left a big gap in the PUMD [Planning and Urban Management Division] because we were so used to them dealing with all the work. I mean they are very uptight about standards so at times it was easier for them to do it rather than for our guys to do it because then they will have to come back and fix our mistakes.* (Participant 9)

Donors also expressed reservations regarding TAs and one pointed to the issue of staff 'churn':

> *I think TAs are not effective because it is not as sustainable as building the capacity of ministries from scratch. They are then able to manage and coordinate their own development. You know that staff turnover is quite high, so you lose a lot of institutional knowledge [and] you have a lot of new people.* (Participant 33)

Yet there was also a lack of a framework to measure the quality of TAs and their selection was left in the hands of donors. Local officials commented:

> *The advisors' capacity in the project was very poor but you know it's always different with the advisors. We have a pool of commonly used ones that we like to use but at the end of it [it] has to be signed off by the donors, so it's really not up to us.* (Participant 15)
>
> *I have to admit the quality of TA is slowly deteriorating and I have to put it down to donors for having these haphazard ways of selecting consultants. I remember eight years ago when I was involved in providing financial advisory services for this project and I remember clearly the project manager was a very young consultant who had all of these degrees but no experience in financial processing. It's very bad.* (Participant 28)

(modified from Sisifa 2015: 217–220)

Looking forward: inverse sovereignty and retroliberalism

We have investigated the implications of the neostructural aid regime on the Pacific Islands during the first decade of the 2000s. As outlined in Chapter Three, this regime has been displaced and we are now forced to ask what has happened with the rise of what we term elsewhere a 'retroliberal' approach to aid (Murray and Overton 2016a). At one level, it could be that the space for local control has widened: adherence to the Paris Declaration principles by donors has weakened and some of the process conditionalities, such as developing poverty reduction strategies, may have become less burdensome. With less emphasis on state-building, there is perhaps less of a spotlight on the ways states should comply – Australia for example has strongly questioned the value (its return on 'investment' – see Hayward-Jones 2014a) of its considerable expenditure under RAMSI and in public sector reform in PNG. Mirroring politics in the metropole, there is less emphasis on public consultation and perhaps donors are now less interested in pressuring recipient governments to ensure they have wide public participation. Aid is now more blatantly supply-led: the desire to 'do business' opens opportunities to conduct affairs in ways which parties can agree on particular projects and collaborations rather than be tied in bureaucratic knots. It could be argued that aid negotiations are now more explicit in their goals (donor self-interest is now more overt rather than denied or disguised), even if their transparency is less apparent.

On the other hand, we contend that, whilst the objectives of the retroliberal aid regime are markedly different from the neostructrual period and aid modalities may have changed, the overall phenomenon of donor power has not diminished and may even have increased as donors seem to have retreated from the Paris principles. The concern for participation may have weakened – apart from the need to 'consult' local businesses perhaps – but there is still a very strong concern that financial management systems are tight and there seems to be an even stronger focus on Pacific public sector inefficiency and corruption (for example Hayward-Jones 2014b). Given greater prominence now is the 'results-management' principle. Political leaders in Australia and New Zealand have demanded that they be given 'evidence' of 'value for money' and a good 'return on investment'. This has generated much activity to develop indicators that can generate such data and arguably it has also affected the type of aid that is funded: long-term programmes involving attitudinal shifts (gender-based violence, the use of trained birth attendants, girls participation in secondary education) are harder to measure in the short-term, whilst 'kilometres of highway constructed', 'proportion of households with access to electricity' are seen in more favourable light. A cynical view of this might point to the term 'policy-based evidence making' rather than the intended reverse.

Finally, there is evidence that the new aid approach has been accompanied by an expressed willingness to use aid as a tool to pressure policy change in the Pacific. Australian aid cuts in 2014–2015 have communicated displeasure about progress towards reform in PNG and Solomon Islands and New Zealand suspended its aid to Nauru in 2015 over a relatively minor incident. Political conditionalities then have certainly remained in the retroliberal era and the burden of consultation, together with capacity issues, remain very important.

Conclusion

This chapter has mainly addressed the issue of how local sovereignty has been usurped in practice within the broad neostructural aid regime of the first decade of the 2000s. In particular this has centred on the new conditionalities that accompanied developments such as the MDGs and the Paris Declaration. This era of aid put stress on public participation and democracy, poverty alleviation and the building of state capacity to deliver basic services and welfare. It acted with strong rhetoric regarding local ownership – in effect it claimed to build local development sovereignty: the ability of Pacific institutions to articulate, manage and promote their own long-term well-being. Yet the practice of the new conditionalities, coupled with the burdens of consultation, the proliferation and complexity of multiple aid institutions and the sometimes limited capacity of Pacific institutions, meant that sovereignty was often undermined in practice. Neostructuralism in subtle ways, we argue, acted to increase donor power and lessen the ability of recipients to manage their own development strategies and activities effectively. We contend that the current regime, retroliberalism, will further this process of concentrating control in the hands of the donors. In many ways it could be argued that 'inverse sovereignty' is a long-term implicit objective that transcends aid regimes pursued wittingly by, and in the interests of, donors.

Notes

1 PRSPs came in other forms such as Poverty Reduction Strategies (PRS) and were often written as national development plans. Yet they had to conform to a certain template and address certain specified requirements and issues. And they had to receive the approval of donor agencies, such as the World Bank. We use the term PRSPs in a general sense to cover this range of plans.
2 Interestingly, this increase from the three major international financial institutions – and their use of soft loans – paralleled the rise of the People's Republic of China as a donor at this time which we note below.
3 This ethical issue of research imposing burdens on local people and institutions without returning benefit was a concern for this particular research project. For Wrighton and Ulu's work in particular there were efforts made to minimise demands on time but also to ensure that local participants were able to 'opt-in' to the research, rather than having to respond to a direct request to participate. Nonetheless we are aware of the contradictions and conundrum inherent in this research regarding the imposition of an external research agenda!
4 This hypothetical schedule of activities is drawn from talking informally to an official about the nature of his daily work.
5 For example, Dame Meg Taylor, the present Secretary General of the Pacific Islands Forum, held a senior position, that of Vice President, Compliance Adviser Ombudsman for the International Finance Corporation (IFC) and the Multilateral Investment Guarantee Agency (MIGA), of the World Bank.

French and French Polynesia Flags, Papeete, French Polynesia

6 Asserting Pacific sovereignty[1]

Introduction

In Chapter Five, we explored the hypothesis that aid in practice in the Pacific has acted to undermine the effective development sovereignty of island states, despite the overall principles of recipient ownership embedded in the Paris Declaration. The 'inverse sovereignty' effect was revealed in the way various conditionalities were required of states that, in turn, often lacked the capacity to handle the multitude and complexity of donor agencies and obligatory interactions. Yet we are also mindful that this is far from a simple picture: during the course of research it became apparent that the agency of Pacific Island officials and institutions is strong. We have witnessed a range of ways in which officials are able to resist, subvert or mould the way aid is designed, implemented and evaluated in the region. Furthermore, there is no monolithic donor world: donors react and interact in different ways. Some are not only responsive and empathetic in the Pacific context, they may also be complicit in subtle ways in the way Pacific Island agents are able to assert sovereignty.

In this chapter we explore this countervailing set of processes and practices and suggest that sovereignty is, in many ways, being asserted through practices and relationships in the aid sector. It is important here to recap the broad concept of sovereignty we adopt in this study. As noted earlier, we see sovereignty not as a simple one-dimensional legal definition circumscribed by a Westphalian concept of an independent state. Rather, we suggest, there are 'layers' of sovereignty from that broad constitutional formulation of a state, through sets of government policies and institutions, to daily practices. It is these performances of sovereignty that give meaning and substance to the way Pacific agents in the aid sector exercise control. In this sense, we are particularly interested in 'development sovereignty' – the way Pacific people and agencies exercise effective control over the development and well-being of their citizens and communities.

In the discussion that follows, we begin with the macro-scale practices and negotiations that define sovereignty in terms of the relationships between Pacific Island states and their donors or metropolitan partners. At this level, we then consider the ways in which Pacific officials engage with the global aid environment and adapt these concepts and principles to regional and national contexts. We then move to what might be termed meso-scale sovereignty: the way state policies and institutions are developed to enhance local control and the way civil society is involved or not in 'owning' development. Finally at the micro-scale, we explore the attitudes, practices and daily interactions that illustrate the ways in which Pacific Island agency is exercised and performed with regard to development sovereignty.

Box 6.1 Rapa Nui and Chile

Warwick E. Murray and Karly Christ

The dominance of the metropole over the affairs of the periphery is truth by definition. However there are cases where influence may be exercised back in order to obtain conditions which somewhat disrupt this hierarchy. The case of Rapa Nui further illustrates some of the contradictions and tensions in these two-way relationships as evidenced elsewhere in this book.

Rapa Nui (known in English as Easter Island, or Spanish as Isla de Pascua) has a long and distressing history of colonialism. Following its discovery by the Polynesians at least 800 years ago, and quite possibly many more centuries prior to that, this eastern-most point of the traditional Polynesian triangle became home to a fascinating and complex society. At its high point the construction of the world-famous Moai and their enormously complicated relocation across the island represented a very significant cultural and baffling technological achievement. Subsequently, economic and environmental pressures which were both caused by and led to conflict precipitated population decline and social distress (Bahn and Flenley 1992). This crisis-ridden society was first visited by Europeans in 1722 with the arrival of a Dutch ship. Subsequent contact with Europeans, including Spanish and British led to significantly worsening conditions. A combination of Peruvian slave raids and introduced diseases almost destroyed the population in the mid-1800s. By 1870 the population had declined to just over 100 people, which at its peak had been over 15000. It was annexed by Chile – who purchased it from a trader in a deal that today some argue should not be binding – in 1888 for geopolitical and maritime reasons. Through much of the Chilean period the indigenous Rapa Nui people were confined to the town of Hanga Roa and the land was rented to a sheep-raising farm company in 1953. This practice desecrated many sacred sites and Hanga Roa remained fenced and policed until 1966. Abuse and mistreatment was amplified significantly under the Pinochet dictatorship of the 1970s and 1980s which declared martial law. It was during this period, for example, when an emergency runway for the US Space Shuttle was built under a deal between the US Reagan administration and the military junta which resulted in a further loss of land and sacred sites.

Not surprisingly, given the above, a significant independence and resistance movement has evolved in Rapa Nui (Christ 2012). This was brutally repressed in the dictatorial period (1973–1990) but following the return to democracy it has re-surfaced and is better organized. Symbols, such as an independence flag, are widely visible. In 2010 the movement occupied the Hanga Roa hotel until they were forcibly removed by armed Chilean police. Yet there is no consensus regarding independence among the approximate 5500 that live on the island. The island has been given 'extraordinary status' since 2007, whereby it is both a *comuna* (district) and *provincia* (province) in the Valparaiso Region of Chile in the Chilean local government hierarchy. This means that the political representation of the island, ostensibly at least, is significantly greater proportionally than is the case in other similarly ranked local government units. This 'concession' is seen by some as a necessary compromise given the fraught history and the strengthening economic links between Chile (known in Rapa Nui as '*el continente*') and the island. Although this compromise situation satisfies some, there is still considerable resistance on the island itself and the issue remains one which is pending and is likely to resurface in the future.

National sovereignty: engaging with the global aid environment

Being an independent state or being partially independent in a constitutional sense is very important in the Global South but to give substance to these forms of sovereignty requires engagement with the wider world which recognises and asserts that sovereignty. This has two main aspects: the knowledge of, and engagement with, global aid agencies and agreements; and adaptation of the global development agenda to local priorities and conditions.

As for the first aspect, we noted in Chapter Five that requests for Pacific Island officials to attend meetings overseas and meet with visiting missions imposes burdens on overstretched offices and officials, particularly in smaller territories. Yet, attendance at an overseas conference or a meeting with a delegation may yield long-term benefits. Also the small size of Pacific bureaucracies means that those staff working in the aid field are those who travel to meetings and attend conferences – knowledge and experience is concentrated in the hands of a few staff who develop a good broad understanding of the aid environment. This is often in contrast to staff in large donor aid agencies where staff tend to specialise more, both in a thematic and geographic sense. As such, a single Pacific Island official over the space of a couple of years may well have attended the Busan High Level Forum on Aid Effectiveness in 2011, participated in regional meetings of the Pacific Forum, been engaged in early meetings concerning the draft SDGs, been invited to an international climate change conference, visited Paris for consultation with the DAC and visited another Pacific Island country as part of a Forum peer review team. By contrast, aid officials in Wellington or Canberra may have been to only one or two of such meetings and their job description as part of a specialist team would have confined them to, say, environmental or trade policy aspects of aid.

When this broad and close engagement is combined with continuity of service in an aid office (the case for some Pacific Island countries such as Cook Islands or Samoa – but not others), Pacific Island officials may possess a breadth of knowledge and experience about aid policies and negotiations that surpasses that of their donor counterparts. They become broad-based experts, at relative ease in international meetings and well-versed in the detail of global agreements. By contrast, larger donor aid agencies may have staff who move through the aid division as part of their career development in a bigger foreign affairs ministry: they spend less time in the aid world, they move from one regional 'desk' to another and they have more layers of seniority above them with regard to high level decision-making and participation at important meetings. Although this situation is far from universal in the region – some Pacific bureaucracies may not contain a specialist aid division nor have experienced staff and some donors do have experienced staff with considerable Pacific expertise – it can mean that specialist skills and astute deals are possible where very experienced Pacific officials can take the lead.

The example of the Cook Islands and the tripartite agreement over the water supply project on Rarotonga, involving Cook Islands, the People's Republic of China and New Zealand (Box 6.2), is perhaps illustrative of how innovative aid activities can be developed where officials and politicians are aware of the global political environment, confident in their abilities to initiate and manage the process, and able to build on existing aid modalities to institute a novel approach to development cooperation. A further innovative example of asserting sovereignty is provided by the way Samoa developed a new policy regarding electric vehicles (Box 6.3). These two examples show how there are opportunities for Pacific Island politicians and officials to be closely connected to international trends and ways of working, 'read the signals' effectively and then institute responses that attract the interest of donors and their resources in ways which meet locally-articulated needs and preserve local control.

Box 6.2 *Te Mato Vai*: Cook Islands and the Tripartate project

John Overton

Cook Islands receives aid from a number of countries, but New Zealand has traditionally been its largest donor. Recently, however, the People's Republic of China has emerged as a significant donor. Although data on aid from China is extremely hard to gather and analyse, Cook Islands Government budget data is helpful in this instance. The country's Minister of Finance, Hon. Mark Brown, has provided data showing that whilst New Zealand accounts for 28%of Cook Islands ODA receipts and Australia 7%, China provides the equivalent of 17.95%(Brown 2015). Aid from China has come in the form of grants (for infrastructure projects such as the construction of a new courthouse and police station in Rarotonga and various items of equipment) and concessional loans through the Export Import Bank of China. Loans have been used to help construct a sports arena and a government building. Chinese aid is also likely to be used to help reconstruct the Apii Nikao school, though at the time of writing progress on this has been slow.

However the most significant example of Chinese aid in Cook Islands is the *Te Mato Vai* water reticulation project. China is often portrayed as a secretive and self-serving aid donor in competition with traditional OECD donors, yet this project will see a strong 'tripartite' approach, linking China with not only the Cook Islands Government but also New Zealand (Brown 2016). *Te Mato Vai* is a substantial project for the country, involving the reconstruction and extension of Rarotonga's water supply. The project was launched in 2014 and addressed the inadequacies of the old and neglected water supply system on the island. Cook Islands and New Zealand will each provide around $NZ 15.3 million and China $NZ 23 million and this will fund reservoirs, water treatment, and major water pipe 'rings' around the island (see Zwart 2016). Although not an explicit part of the project itself, the government has signalled that it will introduce water metering and charging at its conclusion to help with cost recovery.

The tripartite agreement has involved a great deal of negotiation in order to agree on different roles, modalities and standards. As noted, China is providing a loan of approximately $NZ 23 million (at 2% interest) over twenty years to finance the main 'ring' of pipes and Chinese contractors and workers will be employed for this work. This is acknowledged as being tied aid "linked to the fact that China is itself a developing country with more poverty than the Cook Islands" (*Te Mato Vai* n.d. p. 4). New Zealand's contribution is a grant and along with the Cook Islands Government's own funds will be used to procure further materials and work in open contracts. The project is one that the Government of Cook Islands took a strong leadership role in and it has been successful in coordinating the two main donors and overseeing the project. It has been a novel exercise in combining Chinese and Western aid practices and has been seen as model for possible future Chinese involvement and investment in the region. As

Mark Brown noted about the project with regard to discussions with donors:

> *The Cook Islands would be like a bicycle. We've got the frame and the driver, but we need a back wheel and a front wheel. One of you guys provide the back wheel, one of you provide the front wheel, and then we've got a bike we can go around the island on. But the wheels have got to be the same. You can't provide a different wheel for this one and a different wheel for that one. And I'm going to be the driver, not you guys.*
>
> (Brown 2016: 205)

Box 6.3 Electric vehicles in Samoa?

Klaus Thoma

In developing island nations such as Samoa, the integration into the global economy has led to severe fossil fuel dependence. One important fuel use is the country's growing transport fleet. The Government of Samoa is well aware of the importance of the transport sector for development aspirations and the nation's limited options to finance ever growing fuel imports. The Pacific region's already severe exposure to climate change, however, has provided first-hand experience of the downside effects of fossil fuel combustion. This all too tangible antagonism between Samoa's economic aspirations and its stark environmental realities has provided focus for government policy to consider non-fossil fuel options for the transport sector.

The successful introduction of electric vehicles (EVs) would allow the Samoan transport sector to reduce fuel imports if experiences from other island nations are applicable to Samoa. Such a paradigmatic change in mobility, certainly from the point of the socio-cultural realities of an underdeveloped island nation, would require several more layers of imagining electro-mobility in the Samoan context. Reliance on reductionist technology transfer and a market-based approach may, however, not achieve the intended adoption of electro-mobility. Similarly, extra-paradigmatic step changes such as a potential switch to electro-mobility may be difficult for established aid institutions to appreciate and accept. If there was a possibility to deconstruct electro-mobility into its technical but also socio-cultural components and then re-assemble it into Samoa's everyday realities, the island may well be able to swiftly achieve a transition away from a fossil fuel dominant and dependent economy.

My study provided a platform for a wide group of Samoan stakeholders to imagine such steps towards alternatives to a fossil fuel dominant transport sector (Thoma 2014). I was tasked with identifying alternative transport options and Samoan stakeholders engaged in a series of workshops, learning from each other and 'outside' information.

Insights from the workshops highlighted that electrical vehicles charged with diesel generated electricity could be operated with 15%–48% savings compared to conventional vehicles. Accordingly, transport cost for the driving public was projected to be significantly cheaper, net GHG emissions of the transport sector were also significantly lower and reserve currency savings could also be achieved.

Mutual learning by the workshop members highlighted many practical aspects of the imagined use of EVs. There was agreement that distances travelled were highly likely less than 30 km per day with speeds less than 60 km/hr. Status of car ownership, particularly larger 4x4 utility vehicles, was seen as an obstacle for ownership of small, compact EV sedans. Concerns were raised about battery recharge logistics and general lack of maintenance awareness by the general driving public. The current cost of EVs was also seen as limiting any roll-out to a small elite of the driving public. This creative and iterative discussion also spawned the possibility of retrofitting second hand cars and exploring the feasibility of electric buses on high use routes on Upolu.

The conceptualisation phase was complete with the production of the 'eCar Pilot Project Proposal' which was submitted to the National Energy Coordinating Committee (NECC). The workshop members felt that electro-mobility was a feasible option and there were sufficient grounds to commence testing some EVs in Samoa. Although no major outcomes have yet been seen – the conversion of Samoa's vehicle fleet from petrol to electricity is still some time away – the adoption of a policy by the Cabinet in Samoa demonstrated a forward-thinking and politically astute move. Adopting this policy on the eve of Samoa's hosting of the international SIDS conference in 2014 allowed it to be seen as being at the forefront of climate change-related policy making and it attracted the approval and interest of some donors. The case also illustrated the use of overseas advisors to provide specialist information, but the control of the initiative remained very much in the hands of the Government and people of Samoa.

A second key aspect of constructive engagement with the global aid environment is the way Pacific Island governments have sought to 'localise' the global aid agendas. In some cases these have been individual country responses but in many cases the Pacific region localises items on the global aid agenda by using regional institutions to good effect. The recent history of Pacific regionalism has been somewhat fraught with disagreement and friction. The lingering differences between the Pacific Islands Forum and the Pacific Community (SPC), the operation of the Melanesian Spearhead Group as a separate political organisation, the sometimes heavy hand of Australia and New Zealand in regional agreements, and the relative isolation of Fiji in the region (leading to Fiji's establishment of the Pacific Development Forum as an alternative to the Pacific Islands Forum) have all compromised the ability of the Pacific Islands Forum to lead efforts to develop and harmonise a Pacific regional approach to aid.

However, the Pacific Islands Forum has still managed to bring about a number of useful actions that have enabled national governments who comprise its membership to discuss and refine or localise the development agenda. The Pacific Plan of 2005 suggested a broad strategic approach to development in the region (PIFS 2007). There were strong neoliberal influences but there were also aspects that reflected the views of Pacific leaders for a greater liberalisation of labour markets (to allow access to New Zealand and Australia) and more regional cooperation. The Plan attempted to establish a vision of greater regional integration and cooperation. Although it established regular monitoring and evaluation and reporting procedures, it was thin on detailed activities and targets. More recently a review of the Pacific Plan, the result of the deliberations of a team led by Sir Makere Morauta, has reflected a more assertive and independent stance of Pacific countries vis-à-vis Australia and New Zealand. However it was also harshly critical of the limited impact of the Pacific Plan and pointed to poor governance (PIFS 2013).

The Koror Declaration – signed in Koror, Palau, in 2007 following a Forum leader's meeting – resulted from a close examination of, and reference to, the Paris Declaration of 2005 and discussion of how these principles would work in the Pacific context. The result was a set of 'Pacific Aid Effectiveness Principles', unmistakably derived from the Paris document yet with some modifications that had a Pacific imprint (PIFS 2010). The Koror document, with seven (rather than Paris' five) principles, strengthened the notion of recipient ownership (it emphasised the term 'country leadership'), it reinforced the call for 'multi-year commitments by development partners', it emphasised regional approaches, and it specified technical assistance as a key aspect. It showed a good understanding of global level agreements and concepts, but it chose and emphasised those which were of particular concern to the region.

The so-called Cairns Compact of 2009 was held against the backdrop of the Global Financial Crisis and it could be seen as a thinly-disguised effort by Australia to institute, through the Forum, a stronger neoliberal (or perhaps nascent retroliberal) development agenda in the region. The Compact made reference to the MDGs and the Paris Declaration but the first three of its six principles emphasised 'a recognition that a broad-based, private sector-led growth was essential to achieving faster development progress', 'improved governance and service delivery' and 'greater investment in infrastructure'. In addition the draft of the Compact was reputedly presented to

Pacific leaders by Australia with little or no prior consultation (see Box 5.3) and adopted rather weak greenhouse gas emission targets, reflecting the dominant view of Australia rather than the small island states (Fry 2015). Nonetheless, the Compact also gave impetus to the peer review process which attempted to share good practice with regard to development policy and aid coordination.

Out of the Koror Declaration and Cairns Compact has emerged a continuing regional approach to disseminating ideas and policies to promote aid effectiveness. One of the most crucial elements in this has been the Forum's oversight of the process of peer review. In this, teams of officials, usually chaired by a Forum official and including experienced staff from various Pacific Island governments, visit and review another country's aid policies and institutions. The aim is not to criticise or require countries to adhere to a rigid template but rather to share good practice. Discussions are held with politicians and officials in ways which uncover local policies and practices and then provide suggestions for change, if necessary. The role in such teams of experienced and able personnel such as Noumea Simi from Samoa (Box 6.6 below) is very important; such individuals may not only suggest more appropriate policies and procedures, but they can also encourage attitudinal shifts. This can shift policy away from a more passive stance to one which is simultaneously assertive and reflective. Peer reviews have been conducted across the region in many countries, including Nauru, Kiribati, Vanuatu, Tuvalu, Niue, Republic of the Marshall Islands, Tonga, PNG and FSM. In 2015, even New Zealand as a donor, was subject to a review by a team from the Pacific (PIFS 2015a). This represented an important symbolic shift – the first review of a donor by a team from recipient countries – and perhaps a real recognition of the Paris Declaration principle of mutual accountability.

Finally there has been considerable activity by individual countries and territories in order to adapt global aid agendas. Most prominently, Papua New Guinea drew up its own aid effectiveness principles, articulated in the Kavieng Declaration on Aid Effectiveness in 2008. This was an initiative led by the government of PNG to engage with its donors with the explicit aim to 'localise the Paris Declaration on Aid Effectiveness' (Government of Papua New Guinea 2008: 1). It directly mirrored the five principles of the Paris Declaration but in their elaboration there was more operational detail about the role of the government, its 'leadership', and its Mid-Term Development Strategy. There were interesting additions regarding consultation with NGOs and faith-based organisations, the need to build institutional capacity and the desirability of joint decision-making processes. As evidence that Pacific actors have their finger on the pulse of the times, it should be noted was these Kavieng principles of consultation with NGOs and faith-based organisations were endorsed in February 2008, it took the major donors until September of the same year to add similar principles to the Third High Level Forum that produced the Accra Agenda for Action (AAA).

A critical view of the Kavieng Declaration could be that it provided Australia, as PNG's main donor, with a rhetorical device to allow greater intervention and to impose stricter conditions. Perhaps it was example of how the global agenda was imposed on a country, pressuring it to adopt a set of principles that undermined, in practice, its real ability to manage its own development. Yet, at the same time, the Kavieng Declaration also remains an important example of how local officials and politicians engaged with the ideas of the Paris Declaration. There was evidence of a reworking so that some local concerns could be highlighted, and an alternative reading of the document could see some elements of potentially strong local control.

Elsewhere in the region, there has been similar activity with regard to localisation of the Paris Declaration and other agreements and understandings, though none were quite as explicit as the Kavieng Declaration. An example of such a declaration is the 'Joint Declaration on Aid Effectiveness' signed between the Government of Tonga and development partners Australia, New Zealand and the Asian Development Bank in October 2007. This agreement shaped the Paris Declaration Principles to the specific conditions in Tonga and aimed to foster more 'effective development coordination' in the country itself (Mountfort 2013). These national declarations on aid effectiveness mainly concerned the way local systems and institutions could be established and maintained to manage aid, and it is to these local systems that we now turn.

Paris in practice: recipient and donor relationships

We have seen that the Paris Declaration and other international agreements regarding development assistance, are well known within the Pacific region and Pacific officials have engaged actively with the global aid environment and in localising these understandings in ways which have fed into national aid management policies and systems. Donors have also been active in adopting these international agreements, especially in the first decade of the new millennium. As with their Pacific counterparts, aid officials from the both bilateral and multilateral aid agencies became imbued with ideas regarding aid effectiveness. The language of the Paris Declaration and the MDGs became part of the operating documents of aid agencies, at least until about 2010. In such an environment, it is possible to discern ways in which these global agreements have translated into some positive changes in the way aid relationships have been conducted.

The Paris Declaration principle of harmonisation addressed the issue of proliferation of donor institutions and modes of operation. The aim of the principle was to ensure that all major donors work together to reduce duplication and competition and harmonise, not only their operations but also their consultations so that they could speak directly – and by implication less often – with a single voice to recipient agencies. The role of the lead donor is critical in this regard: the nomination of one particular donor to collect, manage and speak for a range of donors. In terms of this lead donor mechanism, Australia and New Zealand along with the Asian Development Bank and the European Union, do seem to have worked at the principle. For some major aid activities, such as health SWAps in the Solomon Islands, some donor and recipient officials speak well of the process. The Cook Islands *Te Mata Vai* water project (Box 6.2 above) is a further example of the way that donor harmonisation and concerted action, coupled with strong local leadership and ownership, can lead to some innovative approaches to development assistance.

In Niue, for example, Talagi reports that local officials remarked on better and more effective relationships with New Zealand aid officials following the Paris principles (Box 6.4). Duplication has been reduced, there have been efficiency gains and programmes have been able to channel larger financial resources over a longer-term towards a more focused range of objectives.

Box 6.4 Paris in Alofi: changing aid relationships between Niue and New Zealand

Felicia Pihigia Talagi

Development assistance to Niue continues to be a critical component of Niue's development and New Zealand is by far the dominant donor. The assistance from New Zealand has not always come easily and there have been concerns expressed regarding New Zealand's heavy-handedness in the past. However, my research involving interviews with Niuean officials uncovered some positive recent changes (Talagi 2017).

One of these improvements was the working relationship between the officials of Niue and both the New Zealand High Commission staff in the country and MFAT staff in Wellington. Respondents felt it was no longer a relationship between the Government of Niue and New Zealand but a partnership. Officials said that paramount to the improvement was the management style of the aid programme and the aid officials who understood and were aware of the unique challenges in Niue. One respondent acknowledged the change and said: "*It's massive. It's because of the people that you see a drastic change in the relationship . . . it was extremely hard to do anything back then, everything was one way. Whenever we try to suggest something, we were always shut down*" (Respondent H).

As well as these changes in attitudes and engagement by officials on both sides, there have been important procedural and policy moves, many inspired by the Paris Declaration principles. Historically, the aid package from New Zealand was delivered on an annual basis. In 2011 this changed. The Governments of Niue and New Zealand signed the Joint Commitment for Development from 2011–2014 (JCfD – Government of Niue and Government of New Zealand 2011). The JCfD aligns to Niue's vision of '*Niue ke Monuina* – A Prosperous Niue' and it outlined the mutual commitments between the partners and the agreed priority sectors for assistance (population retention, sustainable economic development and financial stability/ good governance). One of the key features was that New Zealand's economic assistance to Niue was administered as part of the public revenues of Niue, in effect general budget support.

Another feature was the implementation of the Forward Aid Programme (FAP) introduced in 2011. The FAP outlines a three-year budget envelope with details of all activities under each of the JCfD priorities. The FAP is very useful because the Government of Niue uses it as a planning tool as opposed to earlier dealings where funding was provided to Niue only on an annual basis:

> *The FAP is great because we have a bit more certainty with it now. We can plan things ahead in three years. But also the budget is flexible. We can move monies from sector to sector and from year to year. We decide how much we can spend on the first, second and final year subject to discussions with New Zealand. But it's not hard like it used to be.* (Respondent H).
>
> *Flexibility has allowed us to achieve a lot more things, because some things didn't really become apparent initially, but only after we started down the path before we realised we needed other things, so that was very helpful.* (Respondent F)

These changes, and other measures, such as less rigid reporting systems, have reflected the Paris principles of ownership and alignment in particular but they also seem to have been underpinned by important attitudinal shifts and greater trust on the part of the donor. The Paris Declaration has thus provided a useful setting for country-level dialogue on how to promote effective aid. It is evident that the relationship between Niue and New Zealand has improved considerably. But there is still a lot to be done to enhance the relationship further.

Institutional sovereignty: developing national structures, policies and capabilities

State institutions, strategies and procedures represent the next key element in the way sovereignty can be exercised at the national level. We have seen how some countries have developed their own localised versions of global aid effectiveness principles, as in PNG's Kavieng Declaration, but more widespread have been efforts to produce cohesive aid coordination policies at the national level. Preceding such efforts, however, has been the push to produce national development plans. These have drawn from the model of PRSPs but since the late 2000s they appeared to take on a more distinctive and locally-grounded style.

The example of Tonga is illustrative as Mountford describes (2013):

> The first development plan by the Government of Tonga was published in June 1965. The development plan formed the basis for engagement with development partners however it was not expected that they would entirely fund the plan. "At this time, growth in foreign aid was neither expected nor envisaged, and the recognition of and aspiration for self-reliance was clearly evident" (Campbell 1992: 67). Although there was the potential for development partners to assist the government in achieving its development goals, the Government of Tonga had the intention of funding the majority of it."

The vision of the Government to fund Tonga's development plans, however, did not last. By the third development plan (DP3) covering the 1976–1980 period, nearly all was envisaged to be financed by donors' development aid. The government moved away from previous plans for development and adopted a strategic approach to planning by early 2009 under the National Strategic Planning Framework 2011–2014 The current second plan 'Tonga Strategic Development Framework (TSDF II), 2015–2025' includes seven national outcomes (Mountfort 2013).

Sector development plans have often followed, and been informed by, the wider national plans. These plans provide the strategic direction for sectors such as health, education, energy and transport, linking to global agreements such as the MDGs and defining local priorities and approaches. They have become the basis for SWAps and, whilst largely confined to particular core government ministries, they do map out a national plan of action that can be used to develop a list to table before donors for funding. In Tonga, the Tonga Energy Road Map (Box 6.5) is an example of one such sector plan and it shows an interesting interplay between the Tonga Government and donors. Despite some difficult negotiations along the way, the 'road map' was climate change and attracting funding for some quite substantial energy projects that should assist Tonga to become less dependent on imported fossil fuels. Further sectoral plans include the Police Programme, jointly between Australia and New Zealand; the Technical and Vocational Education and Training programme as well as the Tonga Education Support Programme (TESP). The latter sees the World Bank and New Zealand in a joint venture aligning with Tonga's education system development strategy over the medium -term of around fifteen years (Mountfort 2013).

Box 6.5 Tonga and the energy road map

Helen Mountfort

In 2008, the average cost of crude oil peaked, jumping from $US 25 per barrel in 2001 to $US 100 per barrel by 2008. The small Tongan economy that relied on imported petroleum for 100% of its grid electricity suffered disproportionately as a result of its extreme exposure to world fuel prices. Aid investments into the grid were primarily for diesel infrastructure or ad hoc and small-scale off-grid projects at the time. All lacked a sector wide approach and were not at a scale needed to alleviate the crippling prices of oil that Tonga faced.

While Tonga grappled with its rising electricity and fuel prices, development agencies such as the ADB, EU and World Bank began announcing funding for renewable energy in the Pacific, as countries around the world recognised the need to promote a low carbon development pathway to create new economic opportunities, increase energy access, and reduce carbon emissions. Over $US 2 billion dollars was promised to the developing countries by 2012. Lord Sevele, former Prime Minister of Tonga, questioned where this money was going and how Pacific Island countries could get access to it. As he recalled, "we would always receive an endless number of promises of help and access to the funds but at the end of the day the big three were only initially interested in large-scale projects that achieved international attention" (Sevele 2011). Lord Sevele gained global attention by announcing a 50% renewable energy target by 2012.

During the Pacific Energy Ministers' Meeting held in Tonga 2009, development partners approached the Government of Tonga about this 50% renewable energy goal, and began to offer their assistance. 'Akau'ola (Lord Sevele's energy advisor at the time) recalls that each development partner came to the table with different plans, offering various solutions to Tonga's energy problem. However, "all of the project proposals were ad hoc, none of them were coordinated and all of them were based upon the development partners' priorities" (personal communication, May 2012). The Prime Minister called a meeting requesting all donor partners in the one room. He walked into the room for five minutes and, as 'Akau'ola recalls the meeting:

> *He told everyone off. He said: 'I don't need your money. You come here and tell me [different things] . . . I have a big issue in Tonga, the price of electricity. What I want is for you tell me how I am going to reduce it, if you can't help me on that, get out of Tonga, I don't need you here, I don't need your money. Tell me how I can reduce my energy cost, and if you can help me with that, fine, we can talk money later. But I want you to come up with an idea of how I am going to reduce it'. And then he walked out, simple as that.* (Personal communication, May 2012).

The discussions that took place considered Tonga's energy problems, considered its 50% renewable energy target and came up with solutions and compromises that would allow Tonga to overcome its dependence on imported fuel. In April 2010, the Tonga Energy Road Map (TERM) was endorsed by Cabinet, setting out a ten-year plan to reduce the risks associated with oil price changes and increase sustainable energy access (Government of the Kingdom of Tonga 2010).

There were a number of factors that contributed to the establishment of the sector wide plan for energy. Lord Sevele's economic and strategy advisor Rob Solomon felt that, "the key to early success in getting the TERM document developed and signed off by donors was the Prime Minister sitting the TERM-IU within the Prime Minister's Office, ensuring the highest support of government when dealing with development partners" (Solomon, personal communication, May 2012). Lord Sevele also emphasised the importance of close relationships between government and respective development partner governments when promoting a plan such as TERM: "we drove that programme, and we drove that project and set the agenda. And they [development partners] saw the value of that and then came together."

As Sevele stated, whatever policy goal or direction a recipient country has it must, "sit as a priority at the highest level of government. You must tell development partners what you want, and then work with them to formulate credible framework, and finally, keep telling development partners what YOU as a country and people want" (Sevele 2011: 6).

Following such national and sector plans, have been efforts to establish national level aid coordination. In some cases these have taken the form of a national set of principles to be applied to all donors; in others it has been expressed in bilateral agreements with various donors. Samoa's *Development Policy Cooperation* document of 2010 was notable in the way it situated the Government of Samoa at the centre of development. Its first paragraph put the different parties in their place:

> The business of Government always involves partnership. Different agencies within Government need to work together; they also need to work with civil society and with the beneficiaries of public services. In Samoa, as in most developing countries, we also cooperate with the donors who provide a substantial share of public resources.
>
> (Government of Samoa 2010: 1)

This statement positioned the state as the lead actor in development policy and donors were added almost as a third element (after the state and civil society, even though their substantial contribution to development funding was acknowledged). It went on to emphasise the Paris Declaration principle of ownership quite explicitly and it pointedly steered a course towards general budget support (GBS):

> The ultimate objective of this process is to develop confidence in Government's capacity to plan and manage development programmes and to strengthen systems and procedures for utilisation of public funds as well as performance monitoring, which will encourage partners to provide budget support funding directly to Treasury.
>
> (Government of Samoa 2010: 1)

The document was concise, clear and assertive in the way it both established a framework within Samoa to incorporate all government agencies and community consultation, and asked donors to use grants (rather than loans) to support and align with the government's development efforts. It is perhaps the best and most explicit expression in the region of a national government asserting its sovereignty over its own development.

Samoa's aid coordination policy has been followed by others – such as the National Development Strategy in Tonga and in Cook Islands (Government of the Cook Islands 2011). Kiribati in 2015 produced its own *Development Cooperation Policy* document (Government of Kiribati 2015) and, again, the imprint of the Paris Declaration – together with Accra and Busan – was clear. Although it did not go as far as the Samoan policy in advocating GBS (SWAps however were promoted), the Kiribati policy was equally clear on the need for donors to align with national (state-defined) policies and procedures.

In order to support these national plans and aid coordination policies, institutional structure is of central relevance. Pacific Island states, in line with many other recipient countries, have established special aid coordination units to oversee relationships with donors and encourage a degree of conformity with the aid effectiveness principles. Rather than locate such units within a ministry of foreign affairs – where diplomatic interactions and protocols are so important – it is notable that they are instead mostly established within ministries of finance. This reflects the primacy of financial matters and systems within aid negotiations and disbursement. It also puts the ministry of finance at the centre of the local aid environment: it is responsible for financial management systems; it is usually the agency to which aid funds are transferred; and it is the agency that manages higher level aid modalities (SWAps and, if it eventuates, GBS). The arrangement also puts the ministry of finance on top of line ministries, such as health and education, which are responsible for the implementation of activities co-funded with development aid. So whilst the aid relationship is complicated by this role for the ministry of finance, it moves the institutional response of recipients towards structures which prioritise financial management.

In the Pacific, aid coordination units within the ministries of finance have come to play a critical role. They draw up the aid coordination policies, they are the principal point of contact between government and donors and they are concerned with ensuring that core ministries conform with their policies and procedures. In many ways they are the 'face' of the Paris Declaration principles: they embody and promote the ownership principle; they (arguably in practice more than donors) try to ensure that donors align with government policies and systems and harmonise their activities; and they have to demonstrate results-management and accountability.

Samoa has a clear structure regarding its aid coordination unit and its relationship to other government agencies. There three main elements to this. Oversight is provided by the Cabinet Development Committee (CDC) which is chaired by the Prime Minister and comprised of cabinet ministers, CEOs of government departments and there is an NGO representative. It is primarily concerned with overall development strategy of the country. More specifically concerned with ODA is the Aid Coordination Committee, a smaller body again chaired by the Prime Minister and charged with coordination across government and the identification of projects and activities that require external funding. There is also the Aid Coordination Debt Management Division of the Ministry of Finance, which is responsible for general aid coordination strategy and policies and it also acts as a secretariat for an Aid Coordination Committee. The active participation of the Prime Minister in the two committees means that aid policies have high level awareness and support. Aid coordination is not marginalised but rather given a central and strong position within the whole government structure. The result is that knowledge of aid, aid effectiveness principles and an adherence to a strongly assertive Samoan leadership of development is instilled throughout much of the government administration.

Other countries have a similar structure. In Cook Islands, there is a Development Coordination Division within the Ministry of Finance. Tonga provides another central example. The management of aid there was originally coordinated across multiple ministries. Before the Paris Declaration, the Ministry of Foreign Affairs coordinated the management of all bilateral assistance through the Central Planning Department and regional organisations (such as PFIS and SPC) directed their assistance directly through the sectoral ministries, for example the Ministry of Lands, Survey and Natural Resources etc. However, in July 2006 matters were reorganised and the Ministry of Finance and the Central Planning Department combined to become the Ministry of Finance and National Planning (MoFNP). MoFNP is currently responsible for the overall management of aid in Tonga and coordinates through the Aid Management Division (AMD). The Government of Tonga also has a Project and Aid Coordination Committee (PACC). Similar to Samoa, this committee has senior leadership: it is chaired by the Minister of Finance and includes chief executive officers of the public enterprises.

These national aid coordination units have been seeking to achieve control of the aid process. One of their concerns is the burden of consultation noted in Chapter Five. 'Aid coordination' in practice has involved not just coordination of agencies and departments within and outside government, but also of donor agencies. The development cooperation policy of Kiribati in 2015 contained clauses which made clear the government's frustrations with the proliferation of donor missions and sometimes fickle practices:

> 2.1.4. Development partners are required to harmonise and coordinate their support to facilitate interaction with Government and reduce transactions costs. This should include common reporting, undertakings or conditions, monitoring and evaluation systems wherever possible.
> 2.1.5. Development partners are encouraged to use Government systems and procedures to the largest extent possible. This includes use of Government banking and accounting systems, procurement, financial and progress reporting frameworks and external audit using the Kiribati National Audit Office.
> 2.1.6. Development partners should ensure predictability and timely flow of information on annual commitments and disbursements to allow alignment with budget and planning processes.
> 2.1.7. Development partners are requested to use simplified, clear and documented practices and procedures as closely aligned to Government systems as possible.
> 2.1.8. Development partners are requested to provide adequate advance warning of planned missions to Kiribati and to be prepared to make adjustments to the time frame to ensure effective Government participation. Partners should be aware of the budget calendar, which can affect availability of personnel, particularly in the Ministry of Finance and Economic Development, in the period October to December.
>
> (Government of Kiribati 2015: 8)

Samoa has been similarly explicit in its requirements on donors:

> In order to achieve . . . development partners are requested to provide details of their funding programmes to the Aid Coordination Division twice a year in February/March and September/October. Partners will be required to declare all assistance to Samoa including assistance that goes directly to NGOs and the Private Sector so that government is aware of all assistance provided to the country. This information should be provided in Government's standard format and should include actual disbursements for the previous financial year (1st July – 30th June) and forecasts for the coming three financial years including both committed and pipeline programmes and projects. It is important for the information to clearly differentiate between funds provided to the Government in cash and funds provided in kind through provision of goods or services paid directly by the donor.
>
> (Government of Samoa 2010: 10)

Again, these are clear signs of governments trying to manage the practice of aid delivery in ways which maintains a degree of control in their own hands. It is notable that requests to donors to conform to agreed principles and practices (such as alignment and harmonisation) are not relegated to the global agreements from which they were derived, nor to quiet words through interpersonal networks, but rather than are given explicit voice in national public policy documents throughout the Pacific.

These documents reveal some general strategies being adopted to manage aid at this scale. One of the most important is the drawing up of a calendar for visits from overseas aid officials. This annual picture identifies key pressure points for local agencies, such as times when budgets have to be finalised or over important breaks such as the Christmas holiday, and times when aid business could and should be conducted. Making this calendar clearly available to donors and insisting that they respect it, for example by having 'mission holidays' when external agencies are asked not to visit, is valuable in communicating and managing aid relationships. Giving adequate warning of visits is also necessary if local officials are to be available and prepared.

The above documents also reveal a plea to donors to improve their practices. Kiribati and Samoa – and others – ask for predictability in terms of the flow of funds and also of the requests from donors for information. Conversely they seek clear and regular information from donors concerning all their activities and plans, including funding of civil society. They also call for alignment in the form of using local systems and formats. And in another example of using the Paris Declaration principles assertively, the documents ask for greater donor harmonisation. These are instances of how Pacific Island recipients have adopted and used the global aid effectiveness principles to strengthen their position vis-à-vis donors: using the language of international agreements by recipients helps encourage donors to modify their practices and respect recipient ownership and leadership.

Behind these institutional arrangements, policies and practices lies the matter of capability. In many ways, this issue is framed as a deficit: Pacific Island institutions are often seen as lacking sufficient staff with appropriate skills and experience in order to manage aid effectively. This is usually seen in relation to financial and project management and as we saw in Chapter Five this can be a real constraint in small and stretched bureaucracies such as Tuvalu. Yet we will also see below that the reality is often more complex: expertise and experience are often very much in existence in the Pacific.

Where a lack of capability is identified as a problem, different strategies are adopted to deal with it. Samoa has identified the need for technical assistance but its overall tactic adopts a rather cautious approach: there is an overall principle of "preferential selection of national consultants unless there are clear skill gaps that cannot be met locally" and "technical assistance must be used for building institutional capacity through transfer of expertise and know-how wherever feasible" (GoS 2010: 9). As we saw in Chapter Five, technical assistance and training in the form of capacity building can be a way of ensuring that local officials are trained in the systems and even cultures of donors. Thus, governments such as Samoa are keen to ensure that technical assistance has the overall aim of building local expertise and that personnel trained remain grounded in Samoan culture, priorities and ways of working. It is revealing that Samoa's aid effectiveness document recommends specifically the use of local "government standards and rates for the payments of allowances and per diems" (GoS 2010: 7) in development cooperation. Presumably this is both to prevent large amounts of money being spent on expensive consultant fees at international rates, and to encourage the use of local consultants. It may also be a reflection of the way international agencies can offer pay rates well above local levels and thus attract able local officials away from their government jobs.

The use of 'outsiders' as consultants, volunteers and trainers presents both constraints and opportunities for Pacific institutions. On one hand, overseas personnel placed within government departments (when in the employ of aid donors) can be seen as owing their primary allegiance to donors and, at the most extreme, being merely external agents enforcing compliance of local systems with donor demands; working to bring those systems into line with donor practices and requirements. Such has certainly been the suspicion regarding placement of Australian staff in Papua New Guinea or Solomon Islands departments (*The Guardian* 2015). When such placements are on a short-term basis there is also limited opportunity for real transfer of skills or – perhaps just as important – learning of local systems and contexts by the overseas personnel.

On the other hand, external personnel can often prove to be important assets for Pacific Island agencies. Those on longer-term placements or people who have developed a lot of experience working in the region often have a strong empathy and affection for the region and its people – sometimes even ahead of their allegiance to their 'home' authorities. These are people who sometimes learn local languages and often become very well informed about local politics and society, even if they are not fully immersed in it. It is not uncommon for Australian, New Zealand, French, or American personnel to quietly criticise their own government's policies and work to strengthen Pacific systems and practices to counter such policies. In other cases, longer-term volunteers may be paid by their sending agencies, but they develop an affinity for, and responsibility to, their local host institutions and counterpart personnel. As an example, Nicola Wrighton had worked as a New Zealand aid official in Fiji and worked in several Pacific Island countries in that capacity, building a network of personal friendships and professional relationships. After leaving New Zealand government service she undertook work for the Government of Tuvalu, serving on the advisory committee of the Tuvalu Trust Fund and she was part of the Tuvalu official delegation to the Busan forum on aid effectiveness in 2011 as a trusted and valued resource person.

We also see throughout the region the use of relatively junior expatriate staff in local government agencies.[2] Such people can be useful, bringing a level of technical expertise, if not experience. They are not enmeshed in local social hierarchies and loyalties and may be a neutral voice. Yet they have no official affiliation back to their home governments and they may develop a general loyalty to the Pacific that they carry forward into their possible future work in donor agencies. This represents a useful form of long-term relationship building by Pacific agencies. It also may be useful in tapping human capital that can be used to advantage when it works but discarded (and even blamed!) when it does not.

The insider/outsider relationship can work in other ways. Donors often seek to do a similar thing, hiring local staff in order to tap into sources of local knowledge and personal networks to better inform and support their work. It can be a very effective means to provide continuity when turnover of expatriate staff on overseas postings is high and ensure that donors can link well to a range of local individuals and agencies. Yet the relationship between a locally employed donor official and local networks usually works both ways. Such officials are still immersed in their kin and social worlds and, whilst the donor agencies pay their wages, their primary affiliation is usually to their family, friends and country. In Pacific Islands where social networks amongst the local elites are often small and intimate, they, in turn, become a good source of information for local agencies, including government. Here the importance of the vernacular language is notable – local officials on both sides can communicate in languages and via channels that expatriate officials may rarely be able to access. The example outlined in Box 6.9 below shows how language and communication through these channels can be subtle, humorous and significant in complex ways.

Although the capacity and capability of Pacific governments may be constrained in some cases, often Pacific personnel can be seen as very able, experienced and astute in the way they manage aid. It is true that the demands on Pacific Island officials to attend international meetings are often heavy and it can compromise their ability to do their daily work. Yet, as we noted earlier, a positive consequence of these meetings, and the constant engagement with donors, aid agreements and aid strategy documents, is that many Pacific officials have developed considerable personal experience in, and knowledge of, the aid world. Furthermore, although some Pacific countries do experience a high degree of turnover of officials and see staff being snapped up with attractive offers from international agencies, there are also many instances of considerable continuity of service.

Perhaps a key person in Samoa's success in managing aid and asserting local leadership is Peseta Noumea Simi (see Box 6.6). She has been in the position of managing the country's aid relationships and systems for many years, she is very highly regarded and respected by both donors and local political leaders, and she has a deeper knowledge of the international aid world than most of the donor officials she meets with. Indeed, she has worked hard within Samoa to develop a cadre of skilled and experienced local officials and retain them with better salaries and conditions (Ulu 2013). Similar examples of skilled and experienced people can be found in Cook Islands (Finance Minister Mark Brown, head of the aid division Peter Tierney) and elsewhere. In addition, the knowledge and wisdom of such individuals is being disseminated throughout the region through personal networks (Pacific officials meet often at regional and international meetings) and mechanisms such as the PIFS peer review system.

Box 6.6 Peseta Noumea Simi: a profile

Avataeao Junior Ulu

I first met Peseta Noumea Simi in 2002 whilst working in Samoa for over a year. At the time she served as the Assistant Secretary for Aid (Economic) in the Samoa Department of Foreign Affairs where she served for over fifteen years. I held a locally engaged position at the New Zealand High Commission in Apia, so our offices regularly met. Noumea was known amongst the donor community as a no nonsense, direct, intelligent, formidable personality. I recall one donor saying, "no one ever crosses Noumea Simi, well I don't know anyone that is still living who has crossed her." It left me feeling proud that Noumea was not intimidated by donors, but equally fearful of the woman.

My position at the New Zealand High Commission was relatively junior level, so I personally had very little to do with Noumea. However, my manager wanted me to join their team to attend the high level meeting between Samoa and New Zealand. I jumped at the chance to meet this woman with my first duty to call Noumea to arrange the meeting. Wanting desperately to impress Noumea, I spent fifteen minutes preparing for the call, only for the phone call to last one minute flat, fortunately with a time and date secured. I attended that high level meeting and Noumea was impressive. She knew every detail about every project, budget lines, tables and percentages of contributions towards projects by donor, what other donors were doing in the same space as New Zealand and so on. It is no surprise that Noumea played central roles in revitalising the Aid Coordination Committee for the Samoan Government and developed its current national aid management framework. I walked out of that meeting as her number one fan!

I later learnt as a long serving public servant of the government Samoa, Noumea's earlier roles included head of the Rural Development Programme in the Office of the Prime Minister and Cabinet and in 1979 she established the first Office for Women's Affairs in Samoa.

When I returned to Samoa in 2012 to undertake my research in aid sovereignty, Noumea was the Assistant CEO of Finance (Aid

Coordination Debt Management) in Samoa's Ministry of Finance since 2003. She had also completed an MBA (Management) from the University of New England, a degree she undertook whilst working full-time. As my primary interviewee, Noumea really shaped the findings of my research. I got to see her personal side and her motivation for being the way she is. Noumea's message to small island states is for them to stand strong and proud of their cultural heritage so visitors do not take advantage of their hospitality and their land.

In 2014, Noumea was on the Review Team of New Zealand's OECD-DAC peer review. I was interviewed alongside other delegates of the NGO community about our interactions with the New Zealand aid programme. It was great to see that Noumea was not only being recognised regionally for her work, she was now on the international stage for aid and development. Noumea has been the Pacific Representative on the Global Partnership for Development Steering Committee and the Busan Global Partnership Group since 2012. She is a member of the SPC Independent Review team, and as a member of several development coordination (Cairns Compact) Peer Review of Federated States of Micronesia, Papua New Guinea and Niue. Also she represented the Pacific on the steering committee of the Global Partnership for Effective Development Cooperation from 2012–2014 and is a current member of the Advisory Group of the UN's Development Cooperation Forum. Since 2016, Noumea was appointed CEO for the Ministry of Foreign Affairs and Trade. I will continue to follow the career of this woman who proudly stands for not only Samoa, but her peers in the Pacific.

The interesting consequence of the existence of this growing body of Pacific expertise in aid is that knowledge of, and experience in, the Pacific aid world may well be tilted in favour of Pacific rather than donor officials. Donor officials may well have deep technical and specialist knowledge in some cases, and experience in international diplomacy in others, yet they will find that they are often sitting across a meeting table from a Pacific official with long and deep personal knowledge about the global aid agenda, who has attended many more regional meetings and who has seen aid projects and programmes evolve in their country over many years. Coupled with this is the issue of seniority. Aid officials in the Pacific, partly due to the relatively small size of their government agencies and partly due to the inter-connected nature of local societies, may be only a few steps short of, say, the Prime Minister in a decision-making sense and hold a relatively senior office in a ministry. The donor officials they deal with, however, are often relatively junior, having to report to the head of a unit within an aid division within a Ministry of Foreign Affairs that then has to report to a Minister who has aid as just one element of her or his portfolio.

The asymmetries of power within aid, then, are cast rather differently in the Pacific. In one way, donors are seen as having funds, specialist knowledge and global connections whilst on the other, recipients are regarded as poor, lacking capacity and overworked. A different reading of power, however, would see donors as suffering from a high turnover of staff without good local knowledge, being from relatively low bureaucratic ranks and with pressures from home to make agreements and ensure that strict conditions are met. They often face local officials who are of higher rank, experienced, with an excellent knowledge of aid and with the political nous to know when and how to apply pressure to donors.

These issues of capability have been recast following retroliberal changes in donor agencies. Whilst there has been relative continuity and growing experience within the Pacific, aid donors, especially in Australia and New Zealand, have experienced major changes in their staffing establishments. The DAC review of the New Zealand aid programme in 2015 noted turnover of personnel within the aid programme as one of the problems (OECD 2015: 52). Experienced staff have moved on in some cases, especially those who were committed to the former poverty alleviation agenda of the neostructural era. Many who had developed strong experience in the Pacific region are now working elsewhere. New staff have been recruited. They have been hired based on particular skills in tune with the new agenda, such as private sector development and economic development. Perhaps being in a wider foreign affairs ministry rather than a dedicated aid agency, many such officials see their future not in the Pacific working in poverty alleviation but as a trade envoy in Washington or a consulate official in London.

Furthermore, the new or updated policy documents which donor agencies have drafted following political shifts, have moved from an environment that had development aid and poverty alleviation as central tenets, to one which is much more explicitly supporting geopolitical foreign policy goals and commercial self-interest of their own countries. In these circumstances, aid relationships have been redrawn. Pacific officials may be more senior, have more knowledge and experience in the aid world and be confident in their country's abilities to lead and manage development; yet the ground has shifted under them and the advantages that they carefully developed and employed have been devalued by the new development discourse and associated aid regime.

State sovereignty and civil society

In Chapter Two we outlined how changing aid regimes resulted in the tilting of the balance between different players: donors and recipients; bilateral and multilateral agencies; and actors in state, market and civil society. In this section we examine how civil society has been involved in the Pacific with respect to aid sovereignty. Neoliberal restructuring in the 1980s and 1990s led generally to a significant increase in the participation of NGOs in the Pacific in aid-funded development activities – as they did in many parts of the Global South. States' administrations were reduced in size and scope and some of the slack, especially in terms of the delivery of basic services and welfare, was picked up by NGOs (Gordenker and Weiss 1995). The number of development NGOs increased throughout the world and important linkages were formed between funding organisations in the North with civil society in the South. The changes that occurred in the early years of the 2000s redrew this picture.

As we noted earlier, neostructural policies brought greater concern for poverty alleviation, inequalities and grew and rejuvenated the agency of the state administration in the delivery of basic services. In some respects, civil society now took a secondary role to the newly favoured state. However, the neostructural concern for public participation and consultation and the realisation that states were not yet able to deliver many key services as desired, meant that NGOs have still remained an important players in the aid landscape. They maintained many of their service delivery operations and they were supported to undertake community level participation as well as, in some cases, act as advocates for democracy and social justice (Wallace et al. 2007). The retroliberal turn since the late 2000s seems to have wrought some new changes, though the contours of these are not yet clear. It seems that there may be less interest by donors for the advocacy and activist role that NGOs have played but support of their role both as sub-contractors for service delivery and as relationship brokers between foreign businesses and local actors.

There are wider issues with regard to civil society and development sovereignty that lie outside the scope of this book. They relate to the relationships between NGOs in the Global South and those in the Global North: the flows of funds, the lines of accountability and the legitimacy of community organisations involved in development work. Some of these have been explored in a study of Pacific NGOs and the ownership principle by Pirnia (2016) (Box 6.7). In particular, Pirnia's study illustrated and analysed examples of ways some Pacific NGOs have developed practices which have aimed to build local ownership of development concepts as well as projects. Here we can briefly explore some the ways that engagement with CSOs[3] by donors and recipient governments has been linked to notions and practices of sovereignty.

Box 6.7 Civil society and definitions of ownership

Pedram Pirnia

The principle of ownership of development assistance and the idea that development processes should be owned by project beneficiaries is a relatively new phenomenon which has emerged largely within the last two decades, and is enshrined in the Paris Declaration. The issue of ownership however needs greater attention especially now in respect of the Sustainable Development Goals and as donors continue to dictate the development process.

The concept of ownership has been generally welcomed by the stakeholders and the terminology has been adopted by donor agencies but there is still no consensus amongst stakeholders as to what exactly the principle of 'ownership' of development aid actually means in policy and practice and how it can be measured. Ownership of development aid in practice is defined differently by different actors within the hierarchy of development aid. There are no agreed policies amongst donors and CSOs for cultivating ownership of development outcomes by the project beneficiaries at the grassroots, while there is a knowledge gap and lack of accord amongst aid and development actors as to how ownership of development outcomes can be incorporated into practice.

We can trace the evolution of the ownership principle and its definition globally. In 1998 the World Bank referred to 'borrower ownership' with regard to government officials who represent, agree, borrow and sign off development funds on behalf of the recipient population. This put emphasis on the state and its agencies as the operational mechanisms for ownership. In 2005 the Paris declaration deepened this notion with 'country ownership' seeing the key role for partner governments in designing and implementing development plans and priorities. The Accra Agenda for Action, however, began a widening of the definition with the idea that 'inclusive ownership' should involve the integration of all stakeholders into the formulation of the development agenda. Finally the Busan HLF outcome document in 2011 referred to 'democratic ownership', expanding the 'inclusive' principle to take into account the wider population (especially the private sector) and linking this to democratic processes.

The above definitions indicate the importance of ownership through continuous political statements by traditional donors. However the definitions are vague, confusing and do not precisely indicate how ownership is to be promoted, cultivated and measured in policy and practice. In short they have no teeth and are often circumvented by stakeholders. The question of ownership is thus side-tracked by many important and mainly political challenges such as partner priorities, resources, capacity and funding strings.

Although development aid is political, and donors are reluctant to specify and monitor ownership precisely, it does not mean that ownership cannot be defined. In fact, ownership has a range of ingredients that can be taken into consideration, incorporated into the development cycle and measured. Examining the work of effective CSOs in the Pacific that have managed to cultivate ownership of their projects by their project beneficiaries reveals that ownership is directly related to accountability and it occurs when project implementers commit from the start to ensure that the project outcomes are 'maintained' and 'protected' by the project beneficiaries. In short, development outcomes and the success of any development project are dependent on the quality of the partnerships, as well as the accountability of partners, and the authenticity and consistency of the dialogue.

Development results are more sustainable and effective when people identify the development change they desire, take ownership of the development process from the start, and especially when women and youth are properly engaged in decision-making. In short, development aid is effective when local intelligence and ideas are explored and included rather than superimposing of external ideas onto local problems.

Drawing on studies of CSOs in Fiji, Tonga and elsewhere in the Pacific, it is apparent that organisations which have been successful in promoting strong and effective forms of local community ownership are not only very deeply engaged with communities throughout the project cycle and cultivate their active roles in managing development, but also they take a long-term view, recognising the importance of lasting partnerships and maintaining relationships on a continuing basis. Therefore, we need to recognise that ownership can really only be considered effective when we appreciate the longer-term outcomes of development. The following statement thus encapsulates what might be considered the core indicator of ownership: *Ownership of development aid occurs when people make a commitment to maintain and protect development outcomes* (Pirnia 2016).

We have seen how donors in the Pacific have adjudged some states to be either illegitimate or lacking capacity to function effectively. In the case of Fiji between 2006 and 2014, and in earlier phases of military rule, donors did not regard the regime as legitimate, accusing it of having deposed democratically elected governments. As noted in Chapter Five, aid to government agencies was reduced or withdrawn and instead aid was strategically directed towards certain NGOs involved in development work or pushing for a return to democracy. Elsewhere, civil conflict and the fragility of the state, as in Bougainville and Solomon Islands has seen a willingness by donors to deal directly with civil society so that certain services, often health and education, could be delivered, particularly to remote or marginalised communities. Such aid has come not only from official sources – donor governments circumventing the institutions of the recipient state and making agreements directly with local NGOs – but also from international NGOs who have raised funding in the metropole (often with a high degree of official aid funding) and delivered it to their Pacific partner organisations.

Table 6.1 ODA Disbursements to civil society in Oceania 2004–2015 ($US million constant prices 2014, average per annum)

	2004–2006	*2007–2009*	*2010–2012*	*2013–2015*
Australia	6.35	45.35	71.49	53.46
New Zealand	10.52	20.28	25.35	29.73
Japan	5.19	6.66	14.39	10.70
EU institutions	1.44	2.43	12.03	11.06
Global Fund	0.00	0.91	14.86	20.35
Other donors	6.45	9.09	11.43	19.35

Source: OECD: Stat accessed 20 December 2016

Note: includes funding to NGOs working in Oceania, whether based in donor or recipient countries.

Two questions arise whether such aid builds or diminishes local sovereignty. First, circumventing the state, even if on the grounds that it lacks a democratic mandate, is a significant statement on the legitimacy of the government of the day and the refusal to acknowledge a government could be seen as undermining the sovereignty of the state. Whilst the intention of encouraging a return to a democratic rule, and thereby more broad-based national sovereignty, may lie at the heart of such strategies, the use of aid in this way means that it does become an instrument of political coercion. Civil society is caught in this broader political struggle, having to tread carefully between donors who provide funding and support their operations and their national authorities who control their legal existence. So whilst it enhances sovereignty at one level by promoting democracy or improving the provision of public services, aid used in this way can be seen as undermining a state's sovereignty.

Secondly, we can ask to what extent do CSOs in the Pacific represent a valid form of ownership of development? Development NGOs in the Pacific, as elsewhere, are often portrayed as community or 'grassroots' organisations, closely linked to local people, well in tune with their aspirations and socio-cultural conditions, able to articulate their resources and needs, and respond to their changing constraints and opportunities. There are certainly many examples of such NGOs in the Pacific (Pirnia 2016). Yet we also know from the international literature that there is a critical view of some NGOs in the Global South; seen as run by middle-class urbanites with significant business interests alongside other concerns (see for example Choudry and Kapoor 2013). Their leaders are adept at portraying themselves as representatives of 'the people' and at tapping into the concerns of donors in order to construct appropriate development projects that lead to a flow of funds, employment generation and the continued survival of their organisations. Again, it has been beyond the scope of this study to analyse such questions across the Pacific region as a whole.

Nonetheless, in contrast to this critical view, we can point to several examples of where local NGOs have been very effective in building effective and robust practices of local consultation and participation in development, often in locations, among groups and at a scale that state cannot reach. Pirnia has examined the examples of Ola Fou and MORDI in Tonga and both are NGOs that are committed to a high degree of participation and ownership of development activities by local communities (Pirnia 2016). These examples show how local NGOs and faith-based agencies, with some support from donor agencies, have contributed to greater development sovereignty in the Pacific. What is of interest here is not so much the actual development projects – whether that be water supplies, a local road or support for a school – but the processes that have been used firstly to garner local views on what is needed, possible and appropriate and then to put in place to ensure that the activity is sustainable (in Pirnia's terms to be 'maintained and protected' by the community).

On the other hand, there are examples internationally of how the work of NGOs may act to compromise a nation's sovereignty to some extent. This can happen when the heavy demands for compliance and reporting (as with ODA) can build aid relationships that see lines of accountability develop primarily upwards (between the local NGO and their overseas funders) rather than downwards (local NGOs respond and answer to their local constituencies) (Wallace et al. 2007). In more subtle ways, some civil society activities are predicated and funded on the basis of ideas, values or ideologies that have not originated locally and seek to build and support activities that change local Pacific attitudes and align them with global movements. This can take the form of thinly-disguised proselytising by faith-based NGOs undertaking community development projects; it could take the form of gender-based projects which challenge local patriarchal norms; or it could see environmental NGOs portray Pacific Island communities as vulnerable and helpless in the face of climate change (Barnett and Campbell 2010). However well-intentioned such activities might be, there is a strong element of Pacific 'development' being framed primarily by external world views and ideologies. They may bring benefits and improve well-being, but doubts must be raised about whether such development has been defined, managed and 'owned' by Pacific Island peoples and communities, let alone states.

The everyday exercise of sovereignty: cafés and kava

Thus far we have examined sovereignty in terms of concrete aspects: legal definitions, constitutional arrangements, institutional structures, policy documents and staff capabilities. Yet if sovereignty is to be given substance and to have tangible impact, it must also be reflected in the everyday ways people act to claim, assert and practice ownership of their own welfare and development. Sovereignty is not just a state of being; it is also a manner of thinking and a way of doing: It is 'performed'. In this context, power is exercised in complex and often subtle ways. Power is more than a matter of institutional relationships or relative wealth and status. Power is imbued and manifested in the way people relate to one another within institutional structures and social settings. In this section we examine examples of ways in which key actors within the Pacific Islands region – officials and politicians in particular – have developed innovative strategies to establish their authority and judiciously push out the boundaries of their power in relation to aid donors. *Mana* or *pule*, for example, are two Pacific words that convey this rather uniquely Pacific concept of 'authority'. In these ways the power that aid donors have, expressed through the funds they provide and the conditions they attach to those funds, is to various degrees balanced, resisted or redirected by a myriad of everyday practices by Pacific people at many levels in society.

The use of language

Language is a key marker of ownership. Language enables a political community to conduct the affairs of government; to communicate with the population; and to be able to express cultural meanings and values adequately all require the use of a vernacular language. Yet aid involves interactions with outsiders, who usually do not have a command of Pacific languages, let alone the complex epistemological and ontological foundations of these languages (Gegeo and Watson-Gegeo 2001). Aid negotiations and documents therefore usually resort to international languages, principally the language of donors. In the Pacific, English and French dominate. As a result, the outcomes of aid relationships have to be translated into Pacific languages if they are to be communicated more widely after agreements are reached. The need to communicate agreements widely is further complicated by the remarkable linguistic diversity of the Pacific region (Mugler and Lynch 1996). The issue of language is deeply embedded – and often contested – in the critical legal and constitutional debates within many Pacific Island countries (Box 6.8).

Box 6.8 Language, law and sovereignty

Alexander Mawyer

Some tensions and structural developments in economic and political neoliberalism and globalisation are readily analysable from the point of view of sovereignty. However, some factors are not so easily identified as entangled in sovereignty issues. The status of local and regional languages is among these easily overlooked dimensions.

On one hand, language change and loss joins some of the other most pressing issues facing Pacific states and their peoples. The endangerment and loss of languages within the Pacific Islands, comparable to patterns unfolding across the globe, is having a profound effect on local communities. Historically, Pacific Islands have been among the most linguistically diverse regions in the world. Today, 1,300 of the world's 7,100 languages still used by living peoples are found in Pacific Islands. However, fewer than 40% of the region's languages are thought to be reasonably safe for the foreseeable future, while 40% have been documented as vulnerable or endangered. Approximately 13% of these languages are critically endangered, a status defined by UNESCO as meaning that "the youngest speakers are grandparents and older, and they speak the language partially and infrequently." Finally, 7% of the region's languages have been lost over the last century; for these languages there are simply no speakers remaining. Issues of language shift and change, and community-based supports for language documentation, conservation and renewal bring many facets of Pacific Islands history and the contemporary Pacific into view, including local ecological knowledge (how local languages contain extraordinary understandings of regional environment, ecology and local plants and animals).

The fact of the ongoing erosion and loss of linguistic competence and the treasury of thought, art, history, mythology and everyday expression at stake given language loss is clearly a critical issues for cultural identity, community ties, and inter-generational transmission. However, language is not only an issue of 'culture'. In various ways, language is highly relevant to discussions of sovereignty and inverse sovereignty effects.

Across the region, the role and constitutional place of indigenous languages sits at the very foundation of many everyday and heightened moments of political and civic life. Rights to speak and be heard or to listen and be addressed in indigenous languages as opposed to global languages historically instituted during the colonial period cross-cut experience in children's classrooms, market phenomena, public signage, government bureaus, court proceedings, and debate on the floors of regional parliaments and assemblies. In French Polynesia, a semi-autonomous Pays d'Outre Mer now on the UN list of non-self-governing territories, The European Court of Human Rights' *Birk-Levy v. France* decision (2010) acknowledged indigenous language as "a fundamental element of cultural identity" but noted that the territory's "autonomous status" did not "guarantee 'linguistic freedom' as such, or the right of elected representatives to use the language of their choice when making statements and voting within an assembly" (ECHR, Registrar of the Court, "Application concerning the prohibition on addressing the Assembly of French Polynesia in Tahitian declared inadmissible," Press Release no. 727 06.10.2010). The implications for the execution of everyday and political agency by imperfectly bilingual (or multi-lingual speakers in French Polynesia to say nothing of island groups with a multitude of active languages) should be evident on reflection.

Moreover, linguistically encoded cultural-ontological distinctions inhering in the very articulation of persons, rights and things are also evidently sovereignty entangled. Beyond the ebb and flow of how individuals and groups articulate themselves in everyday civic and heightened political life, the role of language in providing a shared understanding of basic political categories. For instance, a great deal of political and social transformation can be seen to depend on what 'sovereignty' means as opposed to '*mana*' in the Waitangi documents which chartered political relationships between Maori iwi and the British State personified by the Queen. Alternately, the quotidean of everyday court affairs in the Cook Islands, the language-encoded meaning of particular *tikanga* established as *'ākono'anga ture 'enua* customary law bearing on language-encoded concepts or notions of land use and proprietorship rights, still have force under the Cook Islands Constitution and offer substantial opportunities for significant future upheavals of the social or political status quo. Combining elements of the cosmological and the everyday, a tremendous amount of recent scholarship is leading to an increasing sophistication of legal and technical arguments in American law in Hawai'i where the complex US annexation and subsequent development of the legal foundations of the State of Hawai'i has been shown to be layered over traditional Hawaiian Kingdom law, entirely encoded in Hawaiian, which can potentially be leveraged in state and federal courts to produce new individual rights and collective political possibility for Kanaka Ma'oli (Native Hawaiian) persons and groups.

Importantly, across contexts – everyday political and civic life, education and legal foundations for the rule of law and the nature of sovereignty itself – the tension between traditional indigenous languages and global languages particularly the languages of former or current colonial powers appear to be just as active and potentially transformative in independent sovereign states as well as enduringly dependent or semi-sovereign Pacific states.

The broadest possible range of factors bearing on human experience turn out to be entangled in the fact of sovereignty in the everyday. In this case, language is not only the medium of communication, a vehicle for intimate individual expression, but is thoroughly and perhaps surprisingly caught up in civic and political life and can make visible the cracks and fractures, the structural tensions in both dependent and independent states.

In many cases while one or two languages may have risen to dominance a Pacific territory or nation-state they often host dozens, or in the case of Melanesian countries, hundreds of languages, dialects and communalects. These dominant languages, including both Pacific languages as well as English or

French, may well be the official languages of a given territory or state, but this does not imply that the majority of the population will necessarily be able to access the documents concerning aid agreements written in these forms. In addition, the nature of diplomacy and agreements involves, ultimately, written rather than oral records of interactions (contracts, memoranda, policy documents, minutes of meetings). Emphasis is placed on precision of the written word – expressions and meanings that are circumscribed by legal and bureaucratic conventions – rather than giving prominence to the spoken word (as is the case in most Pacific cultures) and the subtleties of oral expression, gestures, humour, silences and conversation.

This observation regarding the dominance of non-local languages and written documents is not to imply that Pacific officials are not fluent in such languages as well as their own nor lacking in literary skills – many were well educated in English or French and spent time in education in metropolitan countries.[4] It does suggest though that language provides an arena where donors may be more comfortable and more able to employ familiar mechanisms of expression than local officials who have to straddle this formal written world of English (or French) and their own cultural context and language – and communicate across this divide.

Yet language issues may not be simply a matter of the imposition of external tongues and practices on passive local agents. Language can be used in interesting ways to express sovereignty. It is common in many aid interactions to insert local languages during the course of meetings. For example, the use of a customary welcome or prayer to open a meeting sets an important marker to acknowledge and 'ground' the place of a meeting in ways which establish the authority of local leaders or landowners. The insertion of certain local words or phrases into official documents may also be useful, not just as a symbol of locating it but also by embedding terms that can have complex layers of meaning locally (for example *mana, vanua, tapu, tangata*) but which can only be inadequately translated, conceptualised and understood in another language. There is also scope to use the way language is communicated, through metaphor and story-telling (concepts such as *talanoa* and *storian* are now much more common in research methodologies in the region), in the way aid practices, such as monitoring and evaluation, are applied in the region (Box 6.9).

Box 6.9 A local lens for monitoring and evaluation in Vanuatu

Mattie Geary Nichol

Increasing Ni-Vanuatu ownership of evaluations by applying a 'local lens' to the practice presents opportunities for increased effectiveness. Approaching evaluations from a local point-of-view, influenced by factors such as *kastom*,[5] requires all aspects of the practice to be re-examined. Fundamentally, this means changing the purpose of evaluations to inform local decision-making, rather than purely fulfil donor accountability needs. Design and delivery also need to be reconsidered as culturally appropriate approaches can improve effectiveness through more informative and reflective findings that can better inform evaluation recommendations.

For example, *kastom* and identity is heavily embedded and practiced through Vanuatu's languages and while Bislama is the national language, with English and French as official languages, Ni-Vanuatu have another 106 indigenous languages in which their identity and *kastom* is expressed. Recognising this means encouraging communities participating in evaluations to access the full extent of concepts and expressions available to them in their chosen language, allowing them to better articulate themselves and exert ownership over their responses.

Fluency and the ability to converse with ease, is key to the common Ni-Vanuatu practice of *storian*, which involves swapping stories, talking and yarning (Crowley 1995: 235). *Storian* stresses the importance of "building rapport with participants" (Warrick 2009: 83) and therefore requires those using it to take the time to build respectful relationships. Employing a *storian* approach in evaluations would challenge the traditional notion of evaluators' objectivity by recognising their individual background and personality and using this to build relationships with communities. *Storian* would require evaluators to offer themselves as people, not just as professionals.

Ni-Vanuatu have a strong relationship with land, a person's sense of being is related to her or his customary home and the social relationships there. A 'local lens' would recognise the need to respect a community's environment, history, *kastom* and power structures. Respecting this relationship means meeting people on their terms, including physically, which may involve travelling to isolated villages only accessible by dirt roads rather than basing interactions in air-conditioned offices.

Applying a local lens to evaluations aims to make evaluators more accessible to communities so that they are better able to exert their views and influence development activities. This can result in richer, more reflective information collected that can better inform evaluation recommendations. There are further advantages of employing local approaches and relationship building, such as strengthening networks with communities for collaboration on future activities.

These small examples illustrate the opportunities for evaluation to be more effective through tailoring the practice by the local context (Geary Nichol 2014). Local people need to drive the practice and determine the appropriate mix of methods and approaches required for evaluating different development activities with different communities. Such a shift will require trial and error. Unlike donor practices that have already had decades to develop, local approaches will require time for fine-tuning. There is space for moulding standard evaluation tools, such as the DAC Criteria for Evaluating Development Assistance, but what is important is that local epistemologies make the assessments, Ni-Vanuatu decide what success looks like and what steps need to be taken to achieve this.

Local officials may also be inclined to use their vernacular language at times when interacting with each other in settings where overseas officials are present. It may be because people are more comfortable talking to each other in their own language and be able to convey meanings more effectively, especially when not so fluent in the 'other' language. But it can also be useful to send messages that foreigners cannot easily understand. The example in Box 6.10 of the interactions in Samoan between a local official, a local working for an aid agency and the expatriate aid official is a striking example of how the use of language can convey important yet subtle messages about relative power. In such circumstances the use of local languages in front of overseas officials may be deliberate or not and it may lead to a degree of discomfort on the part of the visitor. Yet, carefully used, often with humour and tact, such performances of language convey important messages about the ability and authority of local agents and allow local agents to keep a degree of distance when needed.

Box 6.10 *Gagana* and donors in Samoa

Avataeao Junior Ulu

Officials in Samoa use *gagana* (language) Samoa to their advantage to convey messages amongst themselves but also sometimes to donors. One informant gave an example of negotiating with a donor after there had been severe damage following a cyclone. The official was coordinating relief and rebuilding efforts and was attempting to secure funding from donors to rebuild hotels severely damaged by the cyclone. The hotels were closed down and as a result local Samoans were without work.

The expatriate donor official, who had experience working in Samoa, was accompanied by a Samoan staff member employed by the local donor office. The donor official was unable to commit firm funding without gaining approval from head office and was reluctant to make a solid commitment of a certain sum. The senior Samoan official became increasingly frustrated with the donor, so the donor eventually tentatively committed to a figure.

At the end of the meeting, as the donor and local staff member were leaving, the Samoan official's office the latter said in Samoan to the locally engaged staff member: "don't come back with your boss if you haven't got any money." The donor laughed and responded: "I think I know what you're saying." The Samoan senior official replied: "good that was my intention" and closed the door.

All donors in Samoa speak English and follow international aid policy documents which are also in English. Thus, normally, donors are comfortable in the aid discourse and are in a position of relative power. However in the example above, the Samoan official cleverly controlled the process by introducing *gagana*, appreciating a degree of language competence by the donor official but being able to send a message about who was controlling the cyclone relief efforts.

The use of space

The location and setting of meetings is another aspect where we can see interesting practices taking place (Overton et al. 2013). Meetings between aid officials on both sides take place in a variety of places. Sometimes Pacific officials and politicians are invited to attend conferences and meetings overseas, often in the capital cities of major donors. In such events, diplomatic protocols are observed which recognise the status of the politician or official and this may confirm sovereignty in important ways.[6] Yet these are places where donors are 'at home' and naturally more comfortable and with access to significant levels of support from other staff. A common phenomenon is the existence of a permanent diplomatic presence in Pacific countries where aid is negotiated and monitored. Meetings occur most commonly in government offices and meeting rooms, familiar places where parties meet around a table and have access to means of communication. Both sides are close to their own offices and homes and meetings are constrained by time as well as space. Business can be conducted relatively efficiently.

But other spaces are also employed. Informal meetings are useful for establishing and maintaining personal relationships and they can be used to address issues outside of formal meetings. For expatriate officials such spaces may include the residences of the high commissioner or ambassador – used often for meals and get-togethers for donor and selected local officials and politicians. With comfortable surroundings and familiar food and drink, donors are in a position to act as host in another country. Similarly informal meetings involving expatriates and some members of the local elites can centre on certain cafes, bars and restaurants. The example of the Lime Lounge in Honiara (Overton et al. 2013) is illustrative: as one of the few places that combines air conditioning, metropolitan coffee and Wi-Fi access in Honiara, the Lime Lounge is renowned as a meeting place for expatriate officials, NGO workers and other visitors, occasionally with local counterparts. Again, a familiar and comfortable environment for an informal meeting allows business to be conducted by donors.

Conversely, space can also work well for local actors. Just as expatriate officials may like espresso coffee and air conditioning, some local officials and community representatives may feel more comfortable (or at least more in control) in community or church halls, a customary meeting house or in the open in a village setting. These are spaces where local languages, protocols and leaders hold greater sway and visitors are positioned – and welcomed – as outsiders. Whilst an overseas official may feel much less comfortable in such places, and feel less able to exert their influence over proceedings, local leaders and community members are more at ease and more likely to express their ideas, requests and suggestions. For local officials, the choice of venues for meetings is important. Offices and conference rooms in capital cities are the most efficient venues for meetings but there may be times where a more local or remote venue may be politically advantageous. Yet this, in turn, might have to be managed if it gives opportunity for local leaders and politicians to exert their influence and gravitas (*mana* or *pule* in Polynesian languages) ahead of a government's agenda.

On a more regional level, the choice of place for meetings has long been recognised as being of importance. Ratu Sir Kamisese Mara of Fiji was one of the founding figures of the Pacific Islands Forum and he was a leading proponent of rotating the leaders' meetings around each country in the Forum (Mara 1997). Although this added greatly to the expense of the meetings – it would have been much cheaper and more efficient for leaders and officials to meet in, say, Fiji all the time – rotation has meant that each country's *mana* is recognised as host and those who attend the meetings have the chance to see first-hand how, for example, Palau differs from Tonga or Vanuatu. This knowledge is as important across Pacific participants as it is with regard to outsiders and can foster a greater sense intra-regional shared experience. In a similar way, holding meetings between donors and local officials in locations outside the capital city – as was the case with the meetings behind the Kavieng Declaration of 2008 in PNG – ensures that visitors get to experience an element of development challenges in more remote areas.

The use of local protocols

The use of local protocols in meetings between donor and recipient officials has been noted above. As with the careful use of language and space, selective incorporation of protocols (such as *kava* ceremonies, meals, prayers and speeches) can help assert a sense of local sovereignty. However, this may not be a simple of question of whether to use such protocols or not, but rather a variety of approaches may be appropriate and used in different circumstances. In many instances, local protocols for meetings are not used. These include informal meetings in offices or cafés, or formal receptions at the home of the High Commissioner or Ambassador. Yet at other formal occasions (a ministerial reception, a conference or a leaders meeting) it is common to see some Pacific protocols observed.

This might be, for example, a *kava* ceremony in Fiji (or '*ava* in Samoa) where the rituals and speeches both define the position of hosts and guests and, through the order of speaking and presentation, recognise social order and the ranking of individuals involved. These are important events for they both respect and bestow status on visitors yet also assert the control of local authorities and ways of interacting. Although such ceremonies may soon move on to a meal and then a more conventional meeting and setting, the purpose has been served of 'locating' the meeting in place and marking and respecting the standing of the respective parties. Participants are recognised and seen as both representatives of their countries and as individuals – and the purpose of the meetings that follow are usually made explicit. A related issue in such ceremonies is the fact that they are often conducted with visitors sitting (often cross-legged) on mats with local participants and with speeches in the vernacular language. This is a less familiar and sometimes less comfortable position for *palagi* to be in. It all helps to challenge the otherwise dominant position visiting officials may occupy sitting around a meeting table, setting an agenda, and talking (in English) about funds to be dispersed.

The sharing of food is a critical element of Pacific social interaction and meetings frequently involve the provision of meals or generous morning and afternoon teas. Food can demonstrate the status and well-being of the hosts – generosity and abundance convey a message not of weakness and need but of capability and ownership. The sharing of meals, often soon after the start of a meeting, is also a good occasion for social interaction, for participants to forge or strengthen interpersonal relationships through everyday talk, whilst the formal business is postponed.

Religion is also an important element of many public interactions between aid officials. This can be an unfamiliar – even unwelcome – aspect of official interactions in the Pacific for visiting politicians and officials working with strongly secular institutions and often with personal views that are not strongly religious. The use of prayers and even opening sermons by clergy provides an explicit symbol of the power and presence of Christianity in most Pacific Island societies. It recognises the position of religious leaders and, symbolically at least, inserts them on to an official stage, if not into official dialogues.

In sum, local protocols and rituals are not just vestiges of past cultural norms in the Pacific; they are reminder to all parties in aid interactions that their deliberations and plans have to be worked through a present cultural milieu that is complex, ordered and contested (even if not immediately visible to the foreign eye). Local officials may be sometimes frustrated that local customary or religious leaders get to complicate the views that are placed before visitors; whilst visitors get to see that their world views and ways of operating have to operate through diverse cultures, social hierarchies and leadership structures. There is a need to find ways to incorporate these different 'paradigms' in development practices (Box 6.11).

Box 6.11 Hard and soft paradigms of project management

Siskula Sisifa

My doctoral research examined project actors' experiences in development projects in Tonga (Sisifa 2015). A recurring theme in the findings was the detachment between the philosophical assumptions that underpin project management practices and those of local project staff. Project management practices are based on the ideologies of Western advisors and consultants which are foreign to Tongan public servants and beneficiaries. Project management practices are based on positivist, quantitative and 'hard' paradigms that accentuate instrumentality, rationality and objectivity. In this light, project management effectiveness is determined by the tangible outputs and clearly defined project objectives. However, 'soft' project management elements, such as human resources management, require a shift in the epistemological assumptions and highlight the contextual complexities, cultural assumptions, content ambiguities and cyclical processes of development projects.

In revealing the true nature of development project management practices in Tonga, my research identified the pivotal role cultural values play in interpreting civil servants' behaviour. Tongan cultural values accentuate relationships and categorisation of knowledge and operate on vastly different philosophical assumptions than hard models. Relationships between project actors are critical and there is a need for interpersonal networks to be cultivated as part of managerial practices. This is especially pertinent for collective societies such as Tonga that value interpersonal connections.

Tongan cultural values emphasise collectivism through the principles of *tauhi'va* (reciprocity) and *mamahi'mea* (loyalty). Tongans also define themselves in relation to their family's connection to the monarch, which implicitly reveals an individual's social rank, and thus his or her relationship to others in the country. These principles ultimately guide an individual's actions and behaviour and heavily rely on the interaction and engagement in spaces that encourage relationships. Due to the heavy significance of collectivist, interdependent attributes, Tongans therefore define themselves based on the relationships they have cultivated.

There is a need to negotiate between the task and organisational perspectives and ensure that there is an adequate balance between soft and hard elements of project management. On the one hand, projects are conceptualised, scheduled and executed activities, which emphasise the hard, engineering components of the practice. However, projects are executed by humans and their behaviours and perceptions are shaped by their social and cultural context. I suggest a merging of hard and soft project models: a middle ground that appreciates technical tools and frameworks and is adapted to fit the philosophical assumptions of the individuals enacting the projects. With this in mind, international consultants should assume the role of facilitator rather than expert or superior. Advisors should empower and enable local public servants to perform project management effectively themselves. This highlights the need for donor agencies to make more of an effort to build rapport with civil servants and other project actors.

Furthermore, ultimately, a project will be deemed to be successful or effective (or not) not only on whether it delivers 'hard' predetermined outputs but also whether, in doing so, it builds good quality relationships that reflect and enhance reciprocity, trust and mutual respect.

(taken from and modified from thesis – Sisifa 2015: 235, 238, 249–250)

The use of rank

Most Pacific societies have a clear, albeit sometimes opaque and sometimes overly intricate to outsiders, sense of social order and rank. They can vary greatly from very fixed hierarchies based on genealogy and clan to seemingly fluid notions of 'achieved status' where one's place in society can rise and fall over time. There are also complex overlays of gender and age, but the recognition of relative rank is critical and to ignore or diminish the *mana* of any leader is generally experienced as great insult.

Within this context, the conduct of government business is intimately bound to questions of sovereignty. Pacific Island states may be very small and relatively poor compared to donor states such as Australia, the USA and Japan, yet they still have positions and roles (Prime Ministers, Permanent Secretaries and Deputy Directors) that are equivalent to their counterparts in larger states. For donors, the reality of having a portfolio of bilateral relationships means that aid negotiations have to be conducted by lower lever officials with relatively higher ranking recipient officers. Yet for a Pacific Island Minister of Finance, for example, it can be demeaning to have to sit at an official signing ceremony or participate in an important meeting sitting opposite a relative junior aid donor official. Their *mana* is compromised in the eyes of their constituents. Such inequalities of rank and title can underline the fundamental power imbalance in the aid relationship: wealthy donors dispense development aid to apparently poor and weak recipients. Sovereignty may be recognised but arguably not respected.

To counter this perception and assert a degree of local ownership over development, we have heard of attempts by Pacific leaders and officials to insist on a degree of equivalence of rank, at least at important public events. Thus Ulu (2013) recounts an occasion when a Samoan government official strongly admonished a junior donor staff member for going ahead with a signing ceremony involving a Samoan Cabinet Minister when the planned senior donor official was unavailable and was replaced by a junior one. And it seems as if some donors at least are well aware of these intricacies and their importance (Box 6.12).

Box 6.12 Respecting recipient status

Avataeao Junior Ulu

As Government of Samoa officials are well versed and keep abreast of high level aid effectiveness decisions they are able to hold donors accountable and therefore use donor resources to their benefit. Similarly most donors have built strong trusting relationships with the Government of Samoa over time which has given donors the freedom to release aid knowing the Government has an established track record.

This was proven when the 2009 tsunami hit Samoa. A senior Samoan official (GoS3) was away and their return a meeting was called between donors. When GoS3 arrived at the meeting there was a new donor representative that had only recently landed in Samoa for a posting. Whilst GoS3 was aware of the new donor representative they had never met face to face and a briefing was not scheduled until GoS3 had returned from overseas. With the tsunami recovery efforts under way the new donor representative decided to chair the meeting, not being aware of the personalities and procedures for the Government of Samoa when dealing with donors.

Australian officials joined the meeting on a teleconference call from Canberra and their representative asked who was chairing the meeting and also asked whether GoS3 was present in the meeting. They indicated they were not going to participate if GoS3 was not there. The recently arrived donor representative soon understood that the Government of Samoa was in charge of their aid programme and GoS3 took control of the meeting. "I advised donors what to do as we have been in a state of emergency before, so I was calm and I could direct people. The emergency advisory committee was activated, and we knew what to do" (GoS3).

This assertiveness is well known amongst donors. A representative from the private sector who held a senior position in the Government of Samoa (PS4) for many years said: "donors don't bully Samoa, they are in a position of maturity. Samoa also have officials that have worked for donors, so they know what their demands are and they can deliver on this."

(taken from and modified from Ulu 2013: 90–91)

Respect for rank is also demonstrated in the conduct of meetings. The setting of agendas is often a matter of joint discussion prior to meetings that allow for both sides to suggest, agree on and prepare for important items to be addressed. When this does not occur – as in the infamous Cairns meeting of the Forum Leaders in 2009 when Australia allegedly put forward the largely predetermined agenda and outcome document that became the Cairns Compact – it causes considerable disquiet and distrust. We suggest that many Pacific leaders and officials have come to recognise the importance of meeting agendas – at least in meetings with donors. Similarly, the role of meeting chairpersons has become recognised as critical, particularly as many aid meetings are chaired by Pacific hosts rather than by donors. Keeping meetings on track, sticking to the agenda, ensuring that all parties are able to have a say, and preventing powerful voices dominating is crucial for asserting local ownership of this important aspect of the conduct of aid business.

The power to say 'no'

Aid is often posited as a matter of benevolent donors giving to grateful recipients. There is a strong implication that, despite some benefits flowing back to donors, the net flow of resources is from donor to recipient. To refuse aid would seem to be foolish unless conditions were excessive and unbearable. In such a framing, donors seem to bear no negative consequences if recipients decline to receive aid and, in fact, gain from retaining unspent funds. Yet the political and managerial realities of aid are such that donors do indeed have incentives to ensure that aid allocations are disbursed to recipients and agreements are made.

Aid agencies in donor governments operate in a bureaucratic environment in which aid budgets have to be fought for, supported at ministerial level in competition with a myriad of other demands on a national budget. There is pressure to show not only that there are good arguments for giving aid – and no doubt including elements of self-interest for the donor – but also that the aid will be well and accountably spent and achieve verifiable results. Having received a budget allocation, aid agencies have to ensure that their budgets are spent. Underspending on a given aid budget may not only reflect poorly on the aid agency's professional ability to plan, but unspent budget lines may be lost in the following year, further reducing the bureaucratic status of the agency. Furthermore, such agencies must work within budget cycles, most typically annual, though longer-term triennial allocations may be in place. What this means is that there are distinct pressure points that donor officials have to contend with. They have to confront a budget planning process, justifying and forecasting aid deliveries, and they have to face the end of a budget cycle when it is optimal to ensure that their allocations are spent. Individual careers will be tied up with this process and the reputations and future plans of those wishing to climb the ranks may depend on efficient aid disbursement. There are disincentives for major overspending or underspending, both leading to cuts (and poor performance reports) in future. Pressures do not end with budget allocations being spent; donor officials are often under much pressure to demonstrate that clear results are being achieved and that risks are being minimised. These concerns were enshrined in the Paris Declaration principles of 'managing for results' (with a consequent demand for 'evidence' of results) and 'mutual accountability' (implying that donors share responsibility for any failures or inefficiencies). The neostructural emphasis on results and evidence appear to have been considerably strengthened within more recent retroliberal regimes: political leaders in Australia and New Zealand for example have been vocal in calls for demonstrable results and a 'return on investment'. As we noted in Chapter Two, this has led to a move away from longer-term poverty alleviation strategies in the education and health sectors, where more long-term and qualitative development outcomes are sought (literacy, aspirational shifts, achievement-orientation, trust in modern health care systems, a more responsive and efficient public sector etc). Instead there is greater emphasis on more quantifiable and shorter-term activities such as infrastructure provision, agricultural and tourism development and tertiary scholarships, where results can be measured in terms of kilometres of roads built, visitor hotel bed numbers or higher degrees completed. There is, as a result, a high demand on aid officials to gather data and present success stories.

These pressures constitute what amounts to a thinning of effective donor power. The 'power of the purse' in aid is not a simple donor-controlled phenomenon; instead it is a more two-way interplay. If recipients say 'no' to donors, or delay coming to agreements, donors may well be placed in a position of relative weakness, especially when the recipient has been identified as a key partner and awkward questions can be asked of the responsible minister at home why aid has not been delivered. Refusals or delays are also compounded closer to budget deadlines when aid agencies are under pressure to disburse. These pressures are theoretically less of an issue when higher level modalities have been agreed upon: SWAps or GBS agreements involve long-term commitments and, though there are important check points along the way, the opportunities for recipients to stall and apply further pressure is lessened. The apparent retroliberal shift back to more project-based modalities may have reversed the trend to more even and durable aid negotiations and agreements.

There is anecdotal evidence that some Pacific Island officials are aware of the power to say 'no' or to play a brinksmanship game in stalling agreements up to key pressure points in the budget cycle. An example told to Ulu of a new government building in Samoa may be partly apocryphal but it does illustrate an important point about recipient power. In this case the new government building was to be built by a donor using a construction company from their own country. The donor had been involved in many such projects elsewhere and decided to bring two possible plans so that Samoan officials could choose their preferred option. The officials looked closely at the plans and decided that neither was appropriate and explained why. Somewhat taken aback with a 'no' to both options, the donor delegation left and was not heard of for some time (the officials became somewhat worried that they had overstepped their authority and would be in trouble with their bosses who had signed off on the initial project agreement). Then, according to the officials, a donor delegation eventually returned some months later, requested a meeting and asked the Samoan officials what sort of building they wanted! This example is illustrative both of how some donors are poor at consultation and alignment and also how 'recipients' are not passive but can exercise assertiveness in declining to accept unsuitable projects

The consequence of saying 'no' for donors was a realisation that they needed to engage local officials fully in all aspects of the project cycle and respect their effective sovereignty. No doubt, though, there are also limits to refusing aid. There is a danger that donors will not return or that political masters within Pacific countries are not prepared to risk the loss of a new and visible project for their own political reasons. Outright refusal may instead be relatively rare – instead we may see more subtle manoeuvring involving delaying agreement or careful phrased disagreement over details.

Cultivation of personal relationships

Thus far in this chapter we have examined practices to assert sovereignty that have a strong institutional basis or framework, albeit ones which involve a high degree of individual astuteness, skill and even wit. Officials act as representatives of their government. Sovereignty is expressed when those government agencies achieve degrees of control over their own affairs and development resources. However, we argue that sovereignty is also performed and expressed by individuals in their own right. Personal relationships, affiliations, affections and world views on both sides of the aid world are important in the way aid is negotiated, dispensed and evaluated and this is, we think, especially the case in the Pacific Islands where small population size and in general demonstrative cultures that rely on and promote a high degree of interpersonal interaction.

People are known widely in a society; kinship networks are tightly interwoven and the personal and professional faces of an official are often impossible to separate in practice. Furthermore, even donors become part of this web of relationships, especially those who are posted in-country or who are frequent visitors. They are recognised, liked or disliked, trusted or treated with suspicion and often their personal lives (their families, their sporting interests, religion, out-of-hours behaviour etc.) are well known locally. Secrets may be hard to hide.

Most parties in aid relationships in the Pacific recognise the value of, and seek to develop, personal relationships. Friendships are made, social time is shared and fun is had. These relationships help to build trust and understanding but they can also blur the lines separating donor and recipient. Friendly and empathetic donor officials with a good knowledge of local people, cultures, political intrigues and even local language are much more likely to accord respect where it is due, listen to and appreciate different local perspectives and then act as subtle advocates for Pacific views. They may also develop an appreciation of local complexities and realities, for example by being able to discriminate between justifiable and widely-held requests for assistance from the self-serving machinations of a leading politician. On the other hand Pacific officials who have a solid grounding in the donor world (for example by attending university in Australia or having relatives in New Zealand) can appreciate that donors do not speak with a single voice, but they are subject to changes in political ideology, they have their own elements of cultural diversity, come from equally contested political landscapes which they must as diplomats and bureaucrats paper-over, and have personal lives they can understand.

This overlap of individual personalities and relationships creates a complex and interesting stage for aid to be played out. Friendships and socialising outside of work become vehicles for quiet and informal discussions that may not directly focus on current business, but they build understanding and create avenues for communication when formal procedures and channels fail. Both donors and Pacific officials seem to be happy to develop such relationships – and see the benefits. In some cases, they may become more intimate, long-term and public: the New Zealand High Commissioner to Niue from 2010 to 2014, Mark Blumsky, married Pauleen Rex from Niue and stayed on in Niue to live after his term finished. Indeed, it could be suggested that relationships, such as marriage, is a very long-standing strategy in Pacific diplomacy that cements a political connection with a personal one and symbolises long-term commitment of both parties, as well as opening a new means of communication with an opposite party.

Personal relationships that bridge the donor-recipient divide, though, are not simply matters of diplomatic and personal crossover. We contend that they constitute an interesting element of the performance of sovereignty. On one hand the cultivation of empathetic donor officials helps Pacific countries understand and soften what might otherwise be hard and inexplicable positions on the other side. It helps access and if need be redirect the power of the donor. And this can happen, in various ways, through the expatriate communities, from a high ministerial and diplomatic level where relationships are usually more formal and guarded, through to lower official ranks and volunteers (where some expatriates may feel greater affiliation to Pacific rather than their own home institutions and politicians). In summary, then, we have seen how the everyday exercise – performance – of sovereignty is manifested in a wide variety of ways: in language, space, relationships and cultural protocols. These are neither random nor accidental, these practices are often deliberate and well known. In the hands of some Pacific politicians and officials, a well-timed assertive chairing of a meeting, the moving of a meeting to a village hall, the use of another language to convey a message, the symbolic use of one's social status, or an informal chat to a friend on the other side can not only reap benefits in terms of the outcome of negotiations but also indicate that he or she is in control, and that sovereignty is being exercised.

Conclusions

In this chapter we have moved from the position in Chapter Five which hypothesised that an 'inverse sovereignty' effect existed in the Pacific Islands: that the aid regimes of the past twenty years have adopted a rhetoric of recipient ownership yet, in reality, imposed a range of conditions and practices that have undermined the ability of Pacific people and institutions to control their own development. In this chapter we do not reject that hypothesis – the effect is indeed apparent in many cases. Rather we suggest that there is a countervailing set of practices operating, largely – though not exclusively – driven by Pacific actors. These make up what might be termed a 'subversive' sovereignty effect: they are important local strategies and practices which reassert local control and power.

These practices are in part explicit and institutional – the flying of flags, attending and engaging with global meetings, the formulation of local policy documents – but they are also subtle, obscure and personal – the use of language, being assertive at meetings, developing friendships. In many cases these practices involve external agents as sometimes complicit in the process of asserting local sovereignty. The practices blur the distinction between locals and foreigners, donors and recipients. Officials, both the expatriate ones and the Pacific Island ones, are constrained by their structures (governments, cultures) yet there is considerable room for these officials to reconfigure and operationalise relationships and activities within the domain of aid. In this way, we challenge views of donors as being simply interested in imposing their own policies and priorities and always being in a position of power; or views of recipients as being passive, in need, powerless and incapable. Instead, in the complex arenas in which aid is negotiated and implemented, there is considerable room for quiet resistance, astute negotiations, as well as compliance; for power to be exercised in several different directions, and for individuals and personal relationships to shape outcomes that differ from institutional blueprints and stereotypes.

This, we argue, amounts to a much more complex view of sovereignty than defined in Westphalian frameworks. The interaction of local agency and global structures in the Pacific aid world is fluid; sometimes they move in contradiction to one another and sometimes in relative harmony. Sovereignty is not imposed, static and absolute; instead it is constantly played out, asserted, challenged, renegotiated, subverted and performed. In many cases, the daily practices of sovereignty draw on local cultures, behaviours and rituals; in other ways they draw on global understandings and ways of operating. Neither does the assertion of local sovereignty automatically lead to good development (as the Paris Declaration first principle of 'ownership' suggests). It certainly can significantly help local communities and individuals engage with and have a sense of responsibility for their own development. We believe it should remain a key principle of good development practice, for it is politically much more preferable – as well as operationally more effective and durable – to imposed and top-down strategies. Sovereignty can only ever be partial and contested. Ownership of development can be captured by elites or self-serving leaders or groups with their own ideologies. Yet, as is often the case, able and astute local politicians and officials in the Pacific can do a great deal to put in place aid-supported development strategies which are appropriate, durable and just.

Notes

1 The authors acknowledge the particular input from Helen Mountfort in the drafting of this chapter. Her work with the Government of Tonga and subsequently for ODI in London has provided some important insights into our thinking.
2 Contributors to this book who have worked in this capacity include Helen Mountford and Klaus Thoma.
3 We use the term NGOs, civil society and CSOs more or less interchangeably. Within the Pacific Island region this covers a wide range of organisations. Some are secular, others faith-based; many are engaged in service delivery and local development projects, others have a more overt political and advocacy role; and some are solely local and small-scale, others are linked to very large global organisations.
4 In the case of Japanese or Chinese aid also, it may be that both sides use a second (or third) language such as English in order to communicate.
5 a concept closely tied to Ni-Vanuatu identity, often used to represent custom, culture and tradition (Bolton 2003)
6 The visit of, say, the Foreign Minister of a Pacific country of 20,000 people or so to Wellington or Canberra requires protocols that would also be accorded to the foreign minister of France or Mexico.

Government Buildings, Funafuti, Tuvalu

7 Conclusions – aid and Oceanic sovereignty

Introduction

In this book we sought to analyse the impacts of the global aid environment, and the shifting aid regimes which characterise it, on 'development sovereignty' in the Pacific Islands. It is important here to recap the broad concept of sovereignty we have adopted in this study. We see sovereignty not as a simple one-dimensional legalistic definition circumscribed by a Westphalian concept of an independent state. Rather, we suggest, there are 'layers' of sovereignty from that broad constitutional formulation of a state, through sets of government policies and institutions, to daily practices. It is these performances of sovereignty that give meaning and substance to the way Pacific agents in the aid sector exercise control. In this sense, we are particularly interested in 'development sovereignty' – the way Pacific people and agencies exercise effective control over the development and well-being of their citizens and communities.

As was discussed extensively in Chapter 2, the mid-2000s saw the rise of the neostructural aid regime, embodied most explicitly in the Millennium Development Goals of 2000 and the Paris Declaration of 2005. At the core of this global approach to development aid and poverty alleviation were the principles of promoting recipient ownership, reinforced by donor alignment and harmonisation. Whilst laudable as goals we noted that in practice the high burden of compliance and consultation associated with implementing and reporting had the potential to reduce independent development policy making – what we termed the 'inverse sovereignty' effect.

In order to explore the 'inverse sovereignty effect' we posed three questions in order to explore the hypothesis which are restated here:

1. How do size and scale affect the ability of Pacific states to engage effectively with aid donors and secure benefits from aid?
2. To what extent does political status relate to the ability to establish policy sovereignty?
3. What 'cultures' of aid – in terms of official and unofficial policies and practices – have developed and how can these help secure policy sovereignty in the region?

We do not intend to revisit these questions at great length here, but it is useful to summarise our discussion of them in brief. In doing so we identify a number of central themes that run through our analysis and that should be borne in mind in future work in this area. The main contribution of this chapter is to suggest what the future might hold in terms of the relationship between the most recent aid regime – retroliberalism – and other evolving scenarios. In this regard we discuss the possible rise of inward-looking nationalism that is currently taking shape, principally in Europe and the United States. We discuss its potential rise among donors in the Pacific and how this has the potential to undermine the pursuit and practice of sovereignty among aid recipients. Finally we reflect on the possible alternatives to retroliberalism and autarkic populism, embodied in the form of the SDGs, and discuss how we might facilitate this, thereby stimulating Oceanic sovereignty and associated concepts of connectedness, inclusion and permanence.

Size and scale and Pacific development sovereignty

In this book, we began with the macro-scale practices and negotiations that define sovereignty in terms of the relationships between Pacific Island states and their metropolitan partners. At this level, we then considered the ways in which Pacific officials engage with the global aid environment and adapt these concepts and principles to regional and national contexts. The discussion then moved onto what might be termed meso-scale sovereignty: the way state policies and institutions are developed to enhance local control and the way civil society has been involved, or not, in 'owning' development. Finally at the micro-scale, we explored the attitudes, practices and daily interactions that illustrated the ways in which Pacific Island agency is exercised and performed with regard to development sovereignty.

Whilst it is true at a general level that larger states are better equipped to face the compliance and implantation of donor policy associated with the neostructural regime, smaller states are able to use their size in ways that unlock greater levels of aid and policy sovereignty. In Chapter Six we explored this countervailing set of processes and practices and suggested that sovereignty is, in many ways, being asserted through practices and performances in the aid sector. Small states, often cast as marginal, of doubtful economic and political viability and with limited resources, are seemingly some of the most successful in the Pacific region in securing relatively high material standards of living for their population. To be able to do this, their sovereign status, and the ability to assert it, are crucial. This, we contend, is because they have been able to pursue astute and complex strategies in relation to development sovereignty across the macro-, meso- and micro-scales.

Political status – sweet spots of sovereignty and de-marginalisation

The inverse sovereignty argument predicts that there is a negative relationship between aid flows and political dependence. Those territories that remain associated with metropolitan patrons exhibit higher living standards and greater relative levels of aid (even if these may be highly unevenly distributed). In this sense we have argued that it is possible to identify a 'sweet spot' of sovereignty, the location of which will vary according to the political history and socio-economic context of each relationship. This is a space where some degrees of close association are maintained with a metropolitan patron – yielding benefits in terms of aid and free movement of people – yet the rhetoric and reality of independence are retained and local ownership is practiced strongly through capable and experienced public sectors and assertive political leadership. Through the witting assertion of sovereignty supposedly diminutive territories can gain access to wider economic and social networks that for larger and more independent territories may not be possible. Paradoxically, then, the pursuit of full constitutional independence may diminish the options for exercising some forms of development sovereignty. In this sense the sovereignty conferred by independence in larger territories is something of a myth. They are apparently just as susceptible to donor pressures and conditionalities (witness the influence of Australia on Papua New Guinea) and the benefits of full political sovereignty (such as a seat at the United Nations and the exercise of independent foreign policy) may be outweighed by the loss of potential resources and options (such as migration opportunities) that come from closer association. Furthermore, we conclude that degrees of close association in states such as Cook Islands and Wallis and Futuna do not lessen the ability of local agents to exercise high degrees of effective local control and even influence over donor policy in some cases. In terms of aid relationships, then, smallness and political association of some nature may be desirable – although there is an optimum point (or sweet spot) on the continuum in each case. In this sense global structures can be subverted by the seemingly small, dependent and weak. Developing close relationships 'over the horizon', then, is not a matter of compromising effective sovereignty; it can be a way of strengthening it. As Jean-Marie Tjibaou, the Kanak independence leader, once eloquently explained, independence provided a means for Kanak people to negotiate their interdependencies.

Performing sovereignty and the role of aid culture

The evolution of aid 'cultures' is very important in determining development sovereignty. There are certain components of Pacific culture in terms of common linguistic, historic and sociological traits that can and have been used to reassert agency to face the top-down power of aid structures. A strong sense of pride, independence and confidence has been evident in Samoa, for example, in the way it has been able to assert considerable control over and occasional resistance to perceived donor interference. But it has been the exercise of daily sovereignty and how this is performed that has provided the greatest countervailing force to the deterministic rolling out of global aid policy and metropolitan control. It is this set of evolving cultural practices that leaves scope for the assertion of agency. The relationship between former colonies and the historic powers remains deeply intertwined, variable, dynamic and complex. Therefore the postcolonial landscape is one where a variety of institutions and individuals interact in multifaceted ways. Relationships occur at many levels, not just between the nation-states. This, and perhaps every, postcolonial moment is not monolithic. Nor have all states and territories in the Pacific region developed the same culture or degree of development sovereignty: some have succeeded in establishing strong systems and performances of sovereignty; others appear to be more limited in their ability to lead, respond or resist external control. Yet, across the region in different ways, global structures, signals and flows are interpreted, ignored, adjusted and reformed by agency on the ground. This involves much more holistic and flexible performances of sovereignty that go beyond formal constitutional arrangements and that ultimately, in the day to day affairs of the region, are of at least as much influence. Pacific Island officials are not passive or overwhelmed (even if they are often seriously pressured and overworked) but they are frequently assertive, smart, knowledgeable and strategic. Over time they have gained experience, formed personal and professional networks within and outside the region, and developed sets of policies and practices that embed high degrees of local ownership. Sovereignty, then, is as much a set of attitudes and practices as it is a constitutional fiat.

Pacific currents, global tides – new aid regimes

This book began with a suspicion regarding the Paris Declaration and the neostructural aid regime. Ostensibly promoting participation and ownership whilst at the same time creating dependence and imposing prescribed forms of development seemed at best contradictory, and at worst duplicitous. However, through the course of our research we grew to appreciate the neostructural aid regime in a number of ways. This was because we saw how Pacific agency could be exercised and strengthened through their strategic engagement with the regime. Despite the new demands and pressures that it brought, Pacific officials often embraced the new way of operating in the first decade of the 2000s. They learnt and gained much experience – often more than their donor counterparts – in the agreements, targets and modes of operation. Neostructuralism came with a considerable set of conditions and new ways of working but it also provided an explicit set of principles, norms and rules that had global recognition and commitment. These could – and have been – effectively used by recipients to hold donors to account. Pacific officials could quote the Paris and Koror Declarations, which donors had endorsed, and use these to insist that local government leadership and ownership be respected and that donors align with this and harmonise their diverse operations and funding. In addition, whilst the specific goals may be questioned with regard to the relevance to conditions in the Pacific region, the MDGs and the neostructural poverty agenda did at least help define a clear set of objectives that demanded concerted and substantial global action and long-term commitment. They constructed an aid regime predicated on the conditions and (albeit narrowly defined) needs within recipient countries. Donor self-interest and tied aid were at least submerged in the rhetoric of the poverty agenda. And new respect and support for the role of states in development – through forms of budget support – marked a major shift from the austerity and state roll backs of the earlier blunt neoliberal regime. The promise of the neostructural regime and the rhetoric of ownership were undermined by the various inversions of sovereignty that we have identified. However, the practice and performance of sovereignty has in some cases acted as a countervailing force to this top-down process – a kind of 'structuration' of aid.

We therefore concluded that neostructuralism had limitations, it continued to promote neoliberal-styled globalisation and did create significant new pressures and impositions on Pacific Island economic and political systems. But, in retrospect, it was relatively benign. It provided a framework that could be made to work to attract external resources and promote some limited forms of development sovereignty in the region.

If we felt that neostructuralism offered limited prospects for development sovereignty then its successor – retroliberalism – offers even less. There are uncertainties with regard to what might replace retroliberalism and the implications of this in terms of development ownership. When we consider the current rise of populist inward-looking strategies the outlook becomes even less positive. However, there are progressive possibilities and the rise of the SDGs as a focal point brings positive prospects. Below we first consider the current regime and the two potential successors – autarkic populism and progressive internationalism – and the implications for development sovereignty.

Retroliberalism and the private buccaneers of capitalism

In a sense aid regimes are artificial constructs, and in some ways hide more than they reveal. Principles and modalities are as dynamic and polymorphous as the geopolitical and economic trends that give rise to them. Furthermore, in any given locality they manifest the unique combination of local and global trends in the context of specific configurations of historic, environmental and social conditions. In this sense we do not subscribe to a deterministic view of the impacts of regimes as determined solely by a set of processes from 'above'. There is always room for agency to resist and reform the system and thereby play a role in the evolution and reconfiguration of the broader structures of regimes. In brief, the concept of the regime is a structurationist intellectual construct that allows us to make sense of specific aid moments and trends. Notwithstanding this, there can be no doubt that the history of aid in the Pacific shows that the discourses and agendas of the donors are often inordinately influential, and that, in the short-term at least, aid policy has been largely determined from above and outside.

The concept of retroliberalism is a way of conceptualising the post-GFC aid regime that has evolved across the world (see Chapter Three and Murray and Overton 2016a). The previous neostructural period was characterised by an implicit international consensus with regard to aid policy and its implementation, arising in large part from the agreement surrounding the pursuit of the Millennium Development Goals. Although it unfolded in different ways across the globe, and indeed the Pacific, it was nevertheless characterised by a number of common principles and targets and therefore there was a certain level of consistency and certainty that underpinned aid relationships and negotiations in a way that is not exhibited currently. The exact theoretical and policy contours of retroliberalism are perhaps less explicitly apparent than in the case of previous aid regimes. There is a degree of consistency amongst most donors in the new aid policies they are adopting: chief amongst these are the promotion of the private sector, the pursuit of economic growth, and a much more explicit statement of donor self-interest. Yet we now lack a specific global agreement and a specified set of principles and modalities that accompany this new approach. There have been some attempts (such as the international conference in Addis Ababa in 2015 – United Nations 2015) to reach consensus on new models for financing development assistance, combining state and private resources but there is no broad agreement on a concerted global approach. In addition, whilst the poverty agenda, support for recipient state-building and budget support have not ended, they have fallen from prominence in the articulation of aid strategies by donors. We argue that, as a result of these changes, the prospects for development sovereignty are greatly diminished under the current retroliberal regime. Recipient ownership is not emphasised nor, we suggest, is it as respected. In the adoption of the rhetoric of 'shared prosperity', development sovereignty has been constrained and shifted more to the realm of opaque business deals. Retroliberalism hands control explicitly back to the donors, and the corporations that back them, in ways that we have not seen since the colonial era. In this sense we like to think of the current era as one dominated by the private 'buccaneers of capitalism' in a way that echoes Firth with respect to the impacts of globalisation in the region (Firth 2000). We believe that this is not conducive to holistic, appropriate and sustainable development in the Pacific. Analysing the asymmetry in aid power relations that characterises this new regime is sorely required if meaningful economic, cultural and environmental goals are to be attained.

Autarkic populism – 'lifeboat ethics' in the Pacific?

The retroliberal regime is undoubtedly leading to a retreat in the space available for the assertion of development sovereignty in the Pacific given the state-supported capitalist model that privileges the economic and the commercial interests of donors. The negative implications of this, however, pale in comparison to the potential impact of the rise of nationalist geopolitical trends that appear to be unfolding across the Western world at the time of writing. The rise of austerity, the bailing out of the banks and restoration of the privileges of the economic elite following the GFC of 2008 onwards has polarised political opinion. It appears that 'centrist' political viewpoints have receded and that more populist and, in some cases, increasingly radical viewpoints from the left and right of the spectrum have gained ground. This split was exemplified in the UK in the mid-2010s, for example, with the rise of the UKIP (United Kingdom Independence Party) in local elections in 2014 on the right and the ascent of union-backed Jeremy Corbyn to the leadership of the Labour Party in 2015 at the other end of the spectrum. A similar process led to the rise of two opposing candidates in the 2016 US election: Democrat Bernie Sanders and Republican Donald Trump. The UK referendum vote of 2016, leading to the proposed withdrawal of the UK from the EU, and the US Presidential victory for Donald Trump suggest that, for the moment at least, right-wing nationalistic, inward-looking and isolationist policies are winning elections in the West. At the time of writing, we can see this in the evolution of politics across the European Union where right-wing parties in France, the Netherlands and Germany are gaining popularity.

This new populist and conservative nationalism taps into considerable discontent among voters who experienced the worst material effects of the GFC and perhaps feel that those who precipitated it – private banks, multinationals and complicit neoliberal governments – were not punished effectively. Although the new populists themselves often gained through neoliberalism, they blame inequality on 'globalisation' (free-markets and economic integration in this discourse) as well as immigrants, social welfare, and big government. This internally incoherent account is combined with instability in the Middle East which, whilst not created by the GFC, was certainly fanned by it. This led in part to the Arab Spring followed by the crushing opposition to democracy across the region and the partly consequent consolidation of ISIL and other radical groups. This has created a situation where arguments concerning reducing the terrorist threat, stemming immigration and protecting jobs for domestic workers have been conflated and isolationism posited as a blanket solution.

It is uncertain what the effects of the trends discussed above will be on international assistance. There are some indications, at least from the USA and the UK, that it will lead to a reduction in aid flows and an even more explicit focus on donor interests than even retroliberalism provides. Within the retroliberal approach there is an undertaking to engage with the world – it promotes global investment and trade. However, under the new autarkic populism international connectivity and multilateralism are under serious threat. In this sense it mirrors 'lifeboat ethics' – the idea that the survival of the better off is threatened by the demands of the poor and they should be cast adrift to fend for themselves (Hardin 1974). This approach is clearly in the interests of those that already hold power. In the Pacific, it is difficult to see what might happen in this regard. The rise of autarkic populism in

Australia is a worrying trend for the region for example and we have already seen this, the largest donor in the region, cut its aid budget substantially. It may mean that non-traditional donors such as China gain more influence in the region. It may mean that development assistance, like climate change, slips far down – or off – the agenda of metropolitan donors who see nationalist isolationism as preferable to concerted global agreements and action to combat global problems. The Pacific region is in danger of being ignored in such a scenario. Under this scenario it will not be in the interests of Pacific territories to be inward-looking in economic, social or political senses. Oceanic sovereignty, as we have argued, depends on Pacific people maintaining and extending economic, social and political networks across the globe. Local sovereignty is enhanced by global engagement.

Progressive internationalism – a new waka fleet?

Given the rise of retroliberalism and the evolving spectre of autarkic populism it may be difficult to be optimistic about the future of development sovereignty in the Pacific. Yet perhaps one inherent characteristic of the majority of the Pacific territories offers some hope for the future. One of the greatest resources the Pacific territories have is relative smallness in demographic and economic senses – but, critically, this has to be coupled with forms of effective sovereignty. In a multilateral world, where small territories sometimes gain equal representation to their larger counterparts (in some institutions and in theory at least), it remains very much in the interests of the Pacific to promote multilateralism. We have seen how Pacific Island territories have creatively pursued opportunities to make the most of the chinks in the armour of the global aid regime; is it possible that the Pacific can set the agenda in order to create a multilateral, connected world constructed to face key global issues? Climate change offers such an opportunity – it is a problem that can command consensus and requires a multilateral, international and globalised response – and Pacific leaders have been successful in raising the issue in, and contributing significantly to, global meetings and agreements.

More significantly in terms of development assistance and development sovereignty, we see the Sustainable Development Goals (SDGs) as an opportunity to construct a renewed global project – a new 'fleet' that combines the energies and resources of many different countries, each navigating in the same direction (even if by different celestial means!) (Box 7.1). Although open to criticism for being too diffuse and thus difficult to both monitor and achieve, it might be that the breadth of the SDGs offers just the antidote required to growing nationalism. They encompass a range of serious global issues, all inter-related. And, they have emerged, unlike the MDGs, following a long period of international consultation and consensus. They have the potential to offer the development world the sort of framework and vision that can be a catalyst for concerted action. It may be that the SDGs could provide the basis for a reinvigoration of a reformed neostructural approach. The incorporation of climate change goals in the SDGs is significant for the Pacific territories. Although there may be an unfortunate tendency in the climate change and sea level rise debates to portray the Pacific region as helpless and passive victims of global environmental change, it would be wise for Pacific leaders to continue to be active in the international arena to build consensus with regard to the importance of the SDGs and action to combat climate change. Similarly, it is pleasing to see the way some Pacific countries, such as Cook Islands, have already developed their own sustainable development goals linked to their own values, resources and aspirations. In doing so, they are well placed to join – and help lead – the new global sustainable development fleet.

Box 7.1 The new project: the sustainable development goals and the Pacific

Adele Broadbent

A new global project to replace the MDGs, the Sustainable Development Goals (SDGs), came into effect on January 2016. The Pacific region was a strong supporter of the SDGs in the UN deliberations. The SDGs are seen to continue the work of the MDGs while placing focus on important issues for the Pacific that were not covered by the older goals, such as access to energy and management of ocean resources. In fact, the inclusion of a separate oceans goals (SDG 14) was as a direct result of advocacy by the Pacific and other small island states. However, the question now is will these goals prove to be what they promised in the Pacific? Or will they create the same storm of complexity and reporting that dogged the MDGs, especially for smaller nation states?

The promise of the SDGs is much bigger and more inclusive than the MDGs. The seventeen goals of the SDSGs:

- provide a coherent, holistic framework for addressing multiple interlinked environmental, social and economic challenges;
- require all member states to address the social, economic and environmental dimensions of sustainable development in a balanced manner;
- declare that implementation must embody the principles of inclusiveness, integration and 'leave no one behind';
- should be supported by an effective means of implementation;
- and must promote social justice, effective institutions, and durable peace. (UNA-UK 2017)

These points are not lost on those on the frontline of reporting back to donors from Pacific nations. While there has been a warm welcome to the universality of the SDGs, there is also a focused concern from developing nations that they need to outline clearly their context if they are not to be dragged down a relentless track of coming up short on data and capacity.

On the UN's Sustainable Development Knowledge Platform, Samoa (which has been at the head of the response for the Pacific) has very clearly laid out the challenges of the SDGs from lessons learnt from the MDGs (UN 2017). In that country's Voluntary Review Process it says:

- There is a need to localise the indicators for relevance and greater accountability.
- The government may set its own national targets and indicators to take into account national peculiarities.
- Given the breadth and depth of the seventeen goals and 169 targets, the demand for quality statistics will increase significantly and will be a challenge for many of the small island countries.
- All countries need to tailor the SDG indicators to their respective contexts and prioritise the goals for implementation that are most relevant and suited to their capacities.
- In selecting the relevant global SDGs indicators to the Pacific context, it is important to ensure an open and inclusive consultation process that ensures country ownership of the SDGs.
- Tracking progress on an annual basis will likely prove difficult given the increased number of goals, targets, and indicators. To mitigate this, a thematic area could be considered each year, with complete reporting done every five years.

It is still far from certain who will pay for the multi-trillion dollar SDGs let alone what indicators for each of the 169 targets will, in the end, be reported on. However, on a planet facing a raft of multidimensional challenges, the SDGs are the best universal bet at covering all the bases for better world.

Both the political left and the right are rejecting 'globalisation' for different reasons, but this is not the answer in the context of the Pacific. It is in the interests of the region to promote a more inclusive, open, outward looking form of engagement with the wider world. In many ways this mirrors what has always been practiced in Pacific diaspora – through the creation of mobile transnational corporations of kin. The multilateral goals encapsulated the SDGs offer promise in this regard. They may be flawed in terms of their complexity and universality and it would be naive to expect them to quickly disrupt and replace both the current retroliberal and emerging autarkic aid regimes. However progressive internationalism, as we saw with (imperfect) neostructuralism, offers the sort of environment where Pacific development sovereignty can be established and asserted.

Oceanic sovereignty

In this book we have learnt that sovereignty is complex, multi-layered, dynamic, performed, and divisible. The performing of sovereignty in Pacific Island territories is a vibrant, sometimes opportunistic, and always long-term endeavour. In this sense the narrow Westphalian definition of sovereignty is neither sufficient as an explanation nor a singular goal in the Pacific context. Rather we have witnessed an Oceanic sovereignty that is built upon connections, interactions, mutual optimisation, and reciprocity. It looks outwards and forwards. It is critical that Oceanic sovereignty is protected and that space for it is permanently prized-open in the aid arena. This protection will take witting and committed action based on genuine partnership where the binaries of donor and recipient are broken down and where the notions of marginality and smallness are both inverted and subverted.

It is also important to look backwards and inwards. We have seen that Pacific governments and people have succeeded in the global development arena by having a strong sense of identity and values. These derive from history, language, culture and religion. And whilst we might tend to romanticise these – and be naive with regard to some of the inherent power structures (and forms of inequality) within them – they are critical as a catalysing force for articulating local development strategies and for binding diasporic communities together. Many Pacific people are very successful 'global citizens', active and comfortable in many occupations and parts of the world. But for many if not most, culture and identity endure: the ties of Pacific family, kinship, language and land cross oceans, continents – and generations. Oceanic sovereignty thus is at once something which extends beyond the horizon and maintains and seeks new relationships – and something which is firmly rooted in place and culture.

We are confident, notwithstanding the current global malaise and the rise of retroliberalism and populism, that we can look forward to a brighter future. That future will be shaped by both global forces and Pacific attributes. Pacific human and cultural resources continue to equip Pacific societies with the ability to adapt, engage and thrive in the wider world. Globally, a change will be likely based on the global frameworks such as the SDGs and it will require public rejection of the regressive elements of the new right-wing politics. It should also be wary of the anti-globalist arguments of the left-wing, and seek to alter globalisation, not reject it. On the other hand, if the Western world does indeed slide into inward-looking nationalism then the Pacific region will have to diversify its engagement in a geopolitical sense (as it is already doing and extent – see Chapter 4), and this may well open up new and promising avenues for interaction under the right conditions.

Development sovereignty is made by local agents of change in a global political-economic structure that is largely not of their choosing. But there are chinks in the structure that allow the re-constitution of the system from below as we have seen extensively in the case studies here. To optimise this restructuring from below requires the pursuit of an intricate set of practices, attitudes, protocols and relationships in and of the region. Such traits and practices are far from new in the Pacific. They have existed across millennia. Oceanic development sovereignty requires that they are identified, appreciated, understood, and perpetuated. Sovereignty is asserted and celebrated when the frames, conditionalities and demands of the aid world are rejected and replaced by a confident and proud expression of Oceanic identity and ownership.

Hau'ofa's (1993: 16) poetic conclusion is particularly apposite:

> *Oceania is vast, Oceania is expanding, Oceania is hospitable and generous, Oceania is humanity rising from the depths of brine and regions of fire deeper still, Oceania is us. We are the sea, we are the ocean, we must wake up to this ancient truth and together use it to overturn all hegemonic views that aim ultimately to confine us again, physically and psychologically, in the tiny spaces which we have resisted accepting as our sole appointed place, and from which we have recently liberated ourselves. We must not allow anyone to belittle us again, and take away our freedom.*

References

ABC (2016) 'PNG's plan to hire Cuban doctors criticised amid budget cuts', www.abc.net.au/news/2016-11-25/png-looks-to-cuban-doctors-to-fix-rural-health/8058338, accessed 20 December 2016.

ADB (2014) 'Key indicators for Asia and the Pacific 2014', www.adb.org/publications/key-indicators-asia-and-pacific-2014, accessed 10 June 2016.

ADB (2016) 'ADB Sovereign projects 2016', www.adb.org/projects/, accessed 11 May 2016.

Agg, C. (2006) *Trends in Government Support for Non-Governmental Organizations: Is the "Golden Age" of the NGO Behind Us?*, Geneva, UNRISD.

Agnew, J. (2009) *Globalization and Sovereignty*, Lanham, Rowman & Littlefield Publishers.

Aiafi, M. A. P. R. (2016) 'Public policy processes in the Pacific Islands: A study of policy initiation, formulation and implementation in Vanuatu, the Solomon Islands, Samoa and regional inter-governmental organisations', unpublished PhD thesis, Wellington, Victoria University of Wellington.

Alfini, N. and Chambers, R. (2007) 'Words count: Taking a count of the changing language of British aid', *Development in Practice*, 17(4–5), 492–504.

Allen, T. and Stremlau, N. (2005) 'Media policy, peace and state reconstruction', Discussion Paper No. 8, London, Crisis States Research Centre, London School of Economics.

American Samoa Government (2014) *American Samoa Statistical Yearbook 2014*, Pago Pago, Department of Commerce, Statistical Division.

Anders, M. (2016) 'ODA redefined: What you need to know', *Devex*, www.devex.com/news/oda-redefined-what-you-need-to-know-87776, accessed 29 July 2016.

Anderson, T. (2008) *The Limits of RAMSI*, Sydney, AID/WATCH.

Anderson, T. and Lee, G. (eds.) (2010) *In Defence of Melanesian Customary Land*, Sydney, AID/WATCH.

Angelo, A. (1997) 'Tokelau: The last colony?', *New Zealand Studies*, 7(3), 8–12.

Angelo, A. (2002) 'To be or not to be . . .: Integrated, that is the problem of islands', *Revue Juridique Polynesienne*, 2, 87–108.

Armitage, D. (2005) 'The contagion of sovereignty: Declarations of independence since 1776', *South African Historical Journal*, 52(1), 1–18.

Armon, J. (2007) 'Aid politics and development: A donor perspective', *Development Policy Review*, 25(5), 653–656.

Asante, A. D., Negin, J., Hall, J., Dewdney, J. and Zwi, A. B. (2012) 'Analysis of policy implications and challenges of the Cuban health assistance program related to human resources for health in the Pacific', *Human Resources for Health*, 10, 10.

AusAID (2001) *Reducing Poverty: The Central Integrating Factor of Australia's Aid Program*, Canberra, Australian Agency for International Development.

Australian Overseas Aid Program (1997) *One Clear Objective: Poverty Reduction through Sustainable Development*, Canberra, Australian Agency for International Development.

Australian Overseas Aid Program (1998) *Australia and Pacific Island Countries: Partners in Development*, Canberra, Australian Agency for International Development.

Bahn, P. G. and Flenley, J. (1992) *Easter Island Earth Island*, London, Thames and Hudson.

Baldacchino, G. (2006a) 'Managing the hinterland beyond: Two ideal-type strategies of economic development for small island territories', *Asia Pacific Viewpoint*, 47(1), 45–60.

Baldacchino, G. (2006b) 'Innovative development strategies from non-sovereign island jurisdictions? A global review of economic policy and governance practices', *World Development*, 34(5), 852–867.

Baldacchino, G. (2010) *Island Enclaves: Offshoring Strategies, Creative Governance, and Subnational Island Jurisdictions*, Montreal, McGill-Queen's University Press.

Baldacchino, G. and Hepburn, E. (2012) 'A different appetite for sovereignty? Independence movements in subnational island jurisdictions', *Commonwealth & Comparative Politics*, 50(4), 555–568.

Baldacchino, G. and Milne, D. (2006) 'Exploring sub-national island jurisdictions: An editorial introduction', *The Round Table*, 95(386), 487–502.

Ball, M. M. (1973) 'Regionalism and the Pacific Commonwealth', *Pacific Affairs*, 46(2), 232–253.

Banks, G., Murray, W. E., Overton, J. and Scheyvens, R. (2012) 'Paddling on one side of the canoe? The changing nature of New Zealand's development assistance programme', *Development Policy Review*, 30(2), 169–186.

Barcham, M., Scheyvens, R. and Overton, J. (2009) 'New Polynesian triangle: Rethinking Polynesian migration and development in the Pacific', *Asia Pacific Viewpoint*, 50(3), 322–337.

Barker, J. (1995) 'For whom sovereignty matters', in Barker, J. (ed.) *Sovereignty Matters: Locations of Contestation and Possibility in Indigenous Struggles for Self-Determination*, Lincoln, University of Nebraska Press, 1–31.

Barnett, J. and Campbell, J. (2010) *Climate Change and Small Island States: Power, Knowledge, and the South Pacific*, London, Earthscan.

Batibasaqa, K., Overton, J. and Horsley, P. (1999) 'Vanua: Land, people and culture in Fiji', in Overton, J. and Scheyvens, R. (eds.) *Strategies for Sustainable Development: Experiences from the Pacific*, London and New York, Zed Books, 100–108.

Bedford, C., Bedford, R. and Ho, E. (2010) 'Engaging with New Zealand's recognized seasonal employer work policy: The case of Tuvalu', *Asian and Pacific Migration Journal*, 19(3), 421–445.

Bedford, R., Bedford, C., Wall, J. and Young, M. (2017) 'Managed temporary labour migration of Pacific Islanders to Australia and New Zealand in the early twenty-first century', *Australian Geographer*, 48(1), 37–57.

Beeson, M. (2003) 'Sovereignty under siege: Globalisation and the state in Southeast Asia', *Third World Quarterly*, 24(2), 357–374.

Bertram, I. G. (1986) '"Sustainable development" in Pacific micro-economies', *World Development*, 14(7), 809–822.

Bertram, I. G. (1993) 'Sustainability, aid, and material welfare in small South Pacific Island economies, 1900–1990', *World Development*, 21(2), 247–258.

Bertram, I. G. (2004) 'On the convergence of small island economies with their metropolitan patrons', *World Development*, 32(2), 343–364.
Bertram, I. G. (2006) 'Introduction: The MIRAB model in the twenty-first century', *Asia Pacific Viewpoint*, 47(1), 1–13.
Bertram, I. G. and Watters, R. F. (1985) 'The MIRAB economy in South Pacific microstates', *Pacific Viewpoint*, 26(3), 497–519.
Bertram, I. G. and Watters, R. F. (1986) 'The MIRAB process: Earlier analyses in context', *Pacific Viewpoint*, 27(1), 47–59.
Blok, P. and Molhuysen, P. (eds.) (1924) *Nieuw Nederlandsch Biografisch Woordenboek* (Vol. 6), Leiden, Sijthoff NV.
Bolton, L. (2003) *Unfolding the Moon: Enacting Women's Kastom in Vanuatu*, Honolulu, University of Hawai'i Press.
Braithwaite, J., Dinnen, S., Allen, M., Braithwaite, V. and Charlesworth, H. (2010) *Pillars and Shadows: Statebuilding as Peacebuilding in Solomon Islands*, Canberra, Australian National University.
Brant, P. (2016) 'Mapping Chinese aid in the Pacific', in Powles, M. (ed.) *China and the Pacific: The View from Oceania*, Wellington, Victoria University Press, 173–175.
Brickenstein, C. (2015) 'Impact assessment of seasonal labor migration in Australia and New Zealand: A win-win situation?', *Asian and Pacific Migration Journal*, 24(1), 107–129.
Briguglio, L. (1995) 'Small island developing states and their economic vulnerabilities', *World Development*, 23(9), 1615–1632.
Broadbent, A. (2012) 'Power, media and development: A study of the Solomon Islands', unpublished Master of Development Studies thesis, Wellington, Victoria University of Wellington.
Brown, M. (2015) 'Te Mato Vai: Cook Islands-China and NZ Delivering Water to Rarotonga', presentation to 'China and the Pacific Conference: The View from Oceania', 25–27 February, Apia, www.victoria.ac.nz/chinaresearchcentre/programmes-and-projects/china-symposiums/china-and-the-pacific-the-view-from-oceania/26-Hon-Mark-Brown-The-Tripartite-China,-NZ,-Cook-Islands-A-Cook-Island-Perspective.pdf, accessed 20 February 2016.
Brown, M. (2016) 'The tripartite China/NZ/Cook Islands project in the Cook Islands', in Powles, M. (ed.) *China and the Pacific: The View from Oceania*, Wellington, Victoria University Press, 201–207.
Brown, R. and Connell, J. (1993) 'The global flea market: Migration, remittances and the informal economy in Tonga', *Development and Change*, 24(4), 611–647.
Brown, W. (2010) *Walled States, Waning Sovereignty*, New York, Zone Books.
Bryant-Tokalau, J. (2010) *The Fijian Qoliqoli and Urban Squatting in Fiji: Righting an Historical Wrong?*, Geelong, Alfred Deakin Research Institute, Deakin University.
Buiter, W. H. (2007) '"Country ownership": A term whose time has gone', *Development in Practice*, 17(4–5), 647–652.
Campbell, I. C. (1989) *A History of the Pacific Islands*, Christchurch, University of Canterbury Press.
Campbell, I. C. (1992) 'A historical perspective on aid and dependency: The example of Tonga', *Pacific Studies*, 15(3), 59–75.
Chile, L. (2006) 'The historical context of community development in Aotearoa New Zealand', *Community Development Journal*, 41(4), 407–425.
Choudry, A. and Kapoor, D. (eds.) (2013) *NGOization: Complicity, Contradictions and Prospects*, London, Zed Books.

Christ, K. G. (2012) '*Hakarongo mai*! Rapanui women and decolonisation for development', unpublished Master of Development Studies thesis, Wellington, Victoria University of Wellington.
Connell, J. (1988) *Sovereignty and Survival: Island Microstates in the Third World* (No. 3), Sydney, Department of Geography, University of Sydney.
Connell, J. (1991) 'Island microstates: The mirage of development', *The Contemporary Pacific*, 3(2), 251–287.
Connell, J. (1993) 'Island microstates: Development, autonomy and the ties that bind', in Lockhart, D. G., Drakakis-Smith, D. and Schembri, J. (eds.) *The Development Process in Small Island States*, London, Routledge, 117–147.
Connell, J. (2003) 'Losing ground? Tuvalu, the greenhouse effect and the garbage can', *Asia Pacific Viewpoint*, 44(2), 89–107.
Connell, J. (2008) 'Niue: Embracing a culture of migration', *Journal of Ethnic and Migration Studies*, 34(6), 1021–1040.
Connell, J. (2014) 'The two cultures of health worker migration: A Pacific perspective', *Social Science & Medicine*, 116, 73–81.
Connell, J. (2016) 'Last days in the Carteret Islands? Climate change, livelihoods and migration on coral atolls', *Asia Pacific Viewpoint*, 57(1), 3–15.
Connell, J. and Brown, R. P. C. (2005) *Remittances in the Pacific: An Overview*, Manila, ADB.
Cooke, B. and Kothari, U. (2001) *Participation: The New Tyranny*, London, Zed Books.
Coopmans, J. P. A. (1983) 'Het plakkaat van verlatinge (1581) en de Declaration of Independence (1776)', *BMGN-Low Countries Historical Review*, 98(4), 540–567.
Craig, D. and Porter, D. (2006) *Development Beyond Neoliberalism? Governance, Poverty Reduction and Political Economy*, Abingdon and New York, Routledge.
Crocombe, R. G. (1972) 'Land tenure in the South Pacific', in Ward, R. G. (ed.) *Man in the Pacific Islands*, Oxford, Clarendon, 219–251.
Crocombe, R. G. (ed.) (1987) *Land Tenure in the Pacific* (3rd ed.), Suva, University of the South Pacific.
Crocombe, R. G. (1995) *The Pacific Islands and the USA*, Suva, Rarotonga and Honolulu, East West Center, Institute of Pacific Studies, University of the South Pacific and Pacific Islands Development Program.
Crocombe, R. G. (2007) *Asia in the Pacific: Replacing the West*, Suva, University of the South Pacific and Institute of Pacific Studies.
Crocombe, R. G. and Meleisea, M. (eds.) (1994) *Land Issues in the Pacific*, Christchurch, MacMillan Brown Centre for Pacific Studies, University of Canterbury.
Crowley, T. (1995) *A New Bislama Dictionary*, Suva, University of the South Pacific.
Curtain, R., Dornan, M., Doyle, J. and Howes, S. (2016) *Pacific Possible: Labour Mobility: The Ten Billion Dollar Prize*, Canberra, The World Bank and Australian National University.
DAC (n.d.) 'Definition of ODA', www.oecd.org/dac/stats/officialdevelopmentassistancedefinitionandcoverage.htm#Definition, accessed 9 September 2015.
DAC Secretariat (2016a) *The Scope and Nature of 2016 HLM Decisions Regarding the ODA-Eligibility of Peace and Security-Related Expenditures*, Paris, OECD-DAC, www.oecd.org/dac/HLM_ODAeligibilityPS.pdf, accessed 15 May 2017.
DAC Secretariat (2016b) *ODA Reporting of In-Donor Country Refugee Costs: Members' Methodologies for Calculating Costs*, Paris, OECD-DAC, www.oecd.org/dac/stats/RefugeeCostsMethodologicalNote.pdf, accessed 15 May 2017.

Daes, E.-I. A. (2008) 'An overview of the history of indigenous peoples: Self-determination and the United Nations', *Cambridge Review of International Affairs*, 21(1), 7–26.
Daley, L. (2010) 'Hijacking development futures: "Land development" and reform in Vanuatu', in Anderson, T. and Lee, G. (eds.) *In Defence of Melanesian Customary Land*, Sydney, AID/WATCH, 34–39.
De Carvalho, B., Leira, H. and Hobson, J. M. (2011) 'The big bangs of IR: The myths that your teachers still tell you about 1648 and 1919', *Millennium: Journal of International Studies*, 39(3), 735–758.
DFAT (2014) *Aid Program Performance Report 2013–14: Solomon Islands*, https://dfat.gov.au/about-us/ publications/Documents/solomon-islands-appr-2013-14.pdf, accessed April 2016.
Dinnen, S. (2004) 'Lending a fist? Australia's new interventionism in the Southwest Pacific', *State, Society and Governance in Melanesia*, Discussion Paper 2004/5, Canberra, Research School of Pacific and Asian Studies, Australian National University.
Dinnen, S. and Firth, S. (eds.) (2008) *Politics and State Building in Solomon Islands*, Canberra, Asia Pacific Press and Australian National University.
Dixon, M. (2006) 'Crime and justice', in Mulholland, M. and contributors, *State of the Maori Nation: Twenty-First-Century Issues in Aotearoa*, Auckland, Reed Publishing (NZ) Ltd, 189–203.
Dollar, D. and Pritchett, L. (1998) *Assessing Aid: What Works, What Doesn't, and Why*, Oxford, Oxford University Press.
Dommen, E. (1980) 'Some distinguishing characteristics of island states', *World Development*, 8(12), 931–943.
Downer, A. (1998) 'Australia and the Pacific Islands: Strategies for development: Speech by the Minister for Foreign Affairs The Hon: Alexander Downer MP' (transcript), Suva, Fiji, 17 December.
Dugay, C. (2015) 'What does Japan's new charter mean for development?', *Devex*, www.devex.com/news/what-does-japan-s-new-charter-mean-for-development-85595, accessed 16 December 2016.
Duncan, R. and Nakagawa, H. (2006) *Obstacles to Economic Growth in Six Pacific Island Countries*, Washington, DC, World Bank.
Dunn, L. (2011) 'The impact of political dependence on small island jurisdictions', *World Development*, 39(12), 2132–2146.
Durie, M. (1998) *Te Mana, Te Kawanatanga: The Politics of Māori Self-Determination*, Oxford, Oxford University Press.
Eaton, C. (1988) 'Vakavanua land tenure', in Overton, J. (ed.) *Rural Fiji*, Suva, Institute of Pacific Studies, University of the South Pacific, 19–30.
Ellis, F. (1985) 'Employment and incomes in the Fiji sugar economy', in Brookfield, H. C., Ellis, F. and Ward, R. G. (eds.) *Land, Cane and Coconuts*, Canberra, Department of Human Geography, Research School of Pacific Studies, Australian National University, 65–110.
Escobar, A. (1995) *Encountering Development: The Making and Unmaking of the Third World*, Princeton, Princeton University Press.
Eyben, R. (2007) 'Harmonisation: How is the orchestra conducted?', *Development in Practice*, 17(4–5), 640–646.
Faleomavaega, E. (1994) 'American Samoa: A unique relationship in the South Pacific', in vom Busch, W., Crocombe, M. T., Crocombe, R., Crowl, L., Deklin, T.,

Larmour, P. and Williams, E. W. (eds.) *New Politics in the South Pacific*, Rarotonga and Suva, Institute of Pacific Studies, University of the South Pacific, 113–122.
Farrelly, T. and Nabobo-Baba, U. (2014) '*Talanoa* as empathic apprenticeship', *Asia Pacific Viewpoint*, 55(3), 319–330.
Ferguson, J. (2009) 'The uses of neoliberalism', *Antipode*, 41, 166–184.
Field, M. (2010) 'Samoan PM savages Air NZ', www.stuff.co.nz/travel/destinations/pacific-islands/4338180/Samoan-PM-savages-Air-NZ, accessed 7 April 2017.
Firth, S. (2000) 'The Pacific Islands and the globalization agenda', *The Contemporary Pacific*, 12(1), 178–192.
Fisk, E. K. (1980) 'The island of Niue: Development or dependence for a very small nation', in Shand, R. T. (ed.) *The Island States of the Pacific and Indian Oceans*, Canberra, Development Studies Centre, Australian National University, 441–457.
Fleras, A. and Spoonley, P. (1999) *Recalling Aotearoa: Indigenous Politics and Ethnic Relations in New Zealand*, Auckland, Oxford University Press.
Fry, G. (1991) 'The politics of South Pacific regional cooperation', in Thakur, R. (ed.) *The South Pacific: Problems, Issues and Prospects*, New York, St. Martin's Press, 169–181.
Fry, G. (1997) 'Framing the islands: Knowledge and power in changing Australian images of the South Pacific', *The Contemporary Pacific*, 9(2), 305–344.
Fry, G. (2015) 'Recapturing the spirit of 1971: Towards a new regional political settlement in the Pacific,' SSGM Discussion Paper 2015/3, Canberra, Australian National University.
Fukuyama, F. (1992) *The End of History and the Last Man*, New York, Avon Books.
Fullilove, M. (2006) 'RAMSI and state building in Solomon Islands', *The Defender*, Autumn, 31–35.
Furlong, K. and Hamano, A. (2014) *Territorial Economic Accounts for American Samoa, the Commonwealth of the Northern Mariana Islands, Guam and the U.S. Virgin Islands*, BEA Briefing Paper, Bureau of Economic Analysis, U.S., Department of Commerce, www.bea.gov/scb/pdf/2014/01%20January/0114_territorial_economic_accounts.pdf, accessed 20 August 2016.
The Guardian (2015) 'PNG will cancel contracts of Australians advising government, says Peter O'Neill', www.theguardian.com/world/2015/jul/31/png-will-cancel-contracts-of-australians-advising-government-says-peter-oneill, accessed 17 March 2017.
Geary Nichol, M. (2014) 'Principles in practice? Ownership in monitoring and evaluation in Vanuatu', unpublished Master of Development Studies thesis, Wellington, Victoria University of Wellington.
Gegeo, D. W. (1998) 'Indigenous knowledge and empowerment: Rural development examined from within', *The Contemporary Pacific*, 10(2), 289–315.
Gegeo, D. W. and Watson-Gegeo, K. A. (2001) '"How we know": Kwara'ae rural villagers doing indigenous epistemology', *The Contemporary Pacific*, 13(1), 55–88.
Gibson, J., Boe-Gibson, G., Rohorua, H. T. S. and McKenzie, D. (2007) 'Efficient remittance services for development in the Pacific', *Asia-Pacific Development Journal*, 14(2), 55–74.
Gibson, J. and Mckenzie, D. (2012) 'The economic consequences of "brain drain" of the best and brightest: Microeconomic evidence from five countries', *The Economic Journal*, 122(560), 339–375.
Global Indigenous Caucus (2007) *Report of the Global Indigenous Peoples' Caucus Steering Committee*, www.humanrights.gov.au/sites/default/files/content/social_justice/declaration/screport_070831.pdf, accessed 15 June 2015.

Goldsmith, A. and Dinnen, S. (2007) 'Transnational police building: Critical lessons from Timor-Leste and Solomon Islands', *Third World Quarterly*, 28(6), 1091–1109.

Gordenker, L. and Weiss, T. G. (1995) 'Pluralising global governance: Analytical approaches and dimensions', *Third World Quarterly*, 16(3), 357–388.

Gould, J. (2005) *The New Conditionality: The Politics of Poverty Reduction Strategies*, London, Zed Books.

Government of the Cook Islands (2011) *Cook Islands Official Development Assistance Policy*, Avarua, Government of the Cook Islands.

Government of the Cook Islands (2013) *Government Financial Statistics*, www.mfem.gov.ck/economic-statistics/gov-fin-stats, accessed 20 December 2014.

Government of the Kingdom of Tonga (2010) *Tonga Energy Road Map 2010–2020*, https://sustainabledevelopment.un.org/content/documents/1330tongaEnergy%20Strategy.pdf, accessed 9 June 2017.

Government of Kiribati (2015) 'Development cooperation policy', www.mfed.gov.ki/sites/default/files/Kiribati%20Development%20Cooperation%20Policy.pdf, accessed 20 May 2017.

Government of Niue and Government of New Zealand (2011) 'Joint commitment between New Zealand and Niue 2011–2014', www.mfat.govt.nz/assets/_securedfiles/Aid-Prog-docs/Commitment-for-Development/NZ-Niue-Joint-Commitment-for-Development-Niue.pdf, accessed 24 May 2017.

Government of Papua New Guinea (2008) 'Kavieng declaration on aid effectiveness: A joint commitment of principles and actions between the Government of Papua New Guinea and development partners', Port Moresby, Government of Papua New Guinea.

Government of Samoa (2010) *Development Cooperation Policy: Partners in Development: Promoting Aid Effectiveness*, Apia, Ministry of Finance.

Government of Samoa and MFAT (2011) *New Zealand-Samoa: Joint Commitment for Development*, Apia, Government of Samoa.

Grant, T. D. (1998) 'Defining statehood: The Montevideo Convention and its discontents', *Columbia Journal of Transnational Law*, 37, 403–458.

Grynberg, R. (1993) 'Trade liberalisation in the post-Cold War era and its implications for the Fiji sugar industry', *Journal of Pacific Studies*, 17, 132–160.

Grynberg, R. (1995) *The Impact of the Sugar Protocol of the Lomé Convention on the Fiji Economy*, Economics Division Working Paper 95/8, Canberra, Research School of Pacific Studies, Australian National University.

Hadley, S. and Miller, M. (2016) *PEFA: What Is It Good for? The Role of PEFA Assessments in Public Financial Management Reform*, London, Overseas Development Institute, www.odi.org/publications/10403-pfm-public-finance-management-reform-pefa, accessed 15 May 2017.

Hagan, S. (1987) 'Race, politics, and the coup in Fiji', *Bulletin of Concerned Asian Scholars*, 19(4), 2–18.

Hameiri, S. (2007) 'The trouble with RAMSI: Re-examining the roots of conflict in Solomon Islands', *The Contemporary Pacific*, 19(2), 409–441.

Hameiri, S. (2012) 'Mitigating the risk to primitive accumulation: State-building and the logging boom in Solomon Islands', *Journal of Contemporary Asia*, 42(3), 405–426.

Hanks, M. F. (2011) 'Aid, sanctions and civil society: An analysis of the impacts of targeted sanctions on Fiji's non-governmental organisations', unpublished Master of Development Studies thesis, Wellington, Victoria University of Wellington.

Hardin, G. (1974) 'Lifeboat ethics: The case against helping the poor', *Psychology Today*, 8, 38–43.
Harvey, D. (2005) *A Brief History of Neoliberalism*, New York, Oxford University Press.
Hau'ofa, E. (1983) *Tales of the Tikongs*, Honolulu, University of Hawai'i Press.
Hau'ofa, E. (1993) 'Our sea of islands', in Waddell, E., Naidu, V. and Hau'ofa, E. (eds.) *A New Oceania: Rediscovering Our Sea of Islands*, Suva: University of the South Pacific, 1–16.
Hau'ofa, E. (2000) 'The ocean is us', in Hooper, A. (ed.) *Culture and Sustainable Development in the Pacific*, Canberra, Asia Pacific Press and Australian National University, 32–43.
Hau'ofa, E. (2008) *We Are the Ocean: Selected Works*, Honolulu, University of Hawai'i Press.
Hayes, G. (1991) 'Migration, metascience, and development policy in island Polynesia', *The Contemporary Pacific,* 3(1), 1–58.
Hayward-Jones, J. (2014a) 'Australia's costly investment in Solomon Islands: The lessons of RAMSI', *Lowy Institute Analysis*, www.lowyinstitute.org/sites/default/files/hayward-jones_australias_costly_investment_in_solomon_islands_0.pdf, accessed 17 March 2017.
Hayward-Jones, J. (2014b) 'Australia-Papua New Guinea emerging leaders dialogue: Outcomes report', *Lowy Institute Analysis*, www.lowyinstitute.org/sites/default/files/australia_png_emerging_leaders_dialogue_web_0.pdf, accessed 17 March 2017.
Helu, I. F. (1993) 'Identity and change in Tongan society since European contact', *Journal de la Société des Océanistes*, 97(2), 187–194.
Helu, I. F. (1994) 'Thoughts on political systems for the Pacific Islands', in vom Busch, W., Crocombe, M. T., Crocombe, R., Crowl, L., Deklin, T., Larmour, P. and Williams, E. W. (eds.) *New Politics in the South Pacific*, Rarotonga and Suva, Institute of Pacific Studies, University of the South Pacific, 319–332.
Hendrikse, R. P. and Sidaway, J. D. (2010) 'Commentary', *Environment and Planning A*, 42, 2037–2042.
Hepburn, E. (2012) 'Recrafting sovereignty: Lessons from small island autonomies', in Gagnon, A. and Keating, M. (eds.) *Political Autonomy and Divided Societies: Imagining Democratic Alternatives in Complex Settings*, Basingstoke: Palgrave Macmillan, 118–133.
Herr, R. A. (1975) 'A minor ornament: The diplomatic decisions of Western Samoa at independence', *Australian Journal of International Affairs*, 29(3), 300–314.
Herr, R. A. (1986) 'Regionalism, strategic denial and South Pacific security', *The Journal of Pacific History*, 21(4), 170–182.
Hintjens, H. and Hodge, D. (2012) 'The UK Caribbean overseas territories: Governing unruliness amidst the extra-territorial EU', *Commonwealth & Comparative Politics*, 50(2), 190–225.
Hooper, A. (1993) 'The MIRAB transition in Fakaofo, Tokelau', *Pacific Viewpoint*, 34(2), 241–264.
Hooper, A. (2008) 'Tokelau: A sort of "self-governing" sort of "colony"', *Journal of Pacific History*, 43(3), 331–339.
Howe, K. R. (1984) *Where the Waves Fall: A New South Sea Islands History from First Settlement to Colonial Rule* (Vol. 2), Honolulu, University of Hawai'i Press.

Howell, J. and Lind, J. (2008) 'Changing donor policy and practice in civil society in the post-9/11 aid context', *Third World Quarterly*, 30(7), 1279–1296.

Hughes, H. (2003) *Aid has Failed the Pacific*, Issue Analysis No. 33, Sydney, The Centre for Independent Studies.

Iati, I. (2010) *Reconsidering Land Reform in the Pacific*, Research Paper, Wellington, Council for International Development.

Iati, I. (2016) 'China in the Pacific: Alternative perspectives', in Powles, M. (ed.) *China and the Pacific: The View from Oceania*, Wellington, Victoria University Press, 128–138.

IEOM (2014) 'Nouvelle-Calédonie', Rapport Annuel 2013, Nouméa, Institut d'Emission de l'Outre-Mer (IEOM).

Ingersoll, K. A. (2016) *Waves of Knowing: A Seascape Epistemology*, Durham and London, Duke University Press.

ISEE (2013) *Tableau de l'Economie Calédonienne*, Nouméa, Institut de la Statistique et des Etudes Économiques Nouvelle Calédonie (ISEE), www.isee.nc/population/population.html, accessed 16 May 2017.

ISEE (2015) *Tableau de l'Economie Calédonienne (version abrégée, 2015)*, Nouméa, Institut de la Statistique et des Etudes Économiques Nouvelle Calédonie (ISEE), www.isee.nc/publications/tableau-de-l-economie-caledonienne-tec, accessed 16 May 2017.

Jacobs, A. C. (2016) 'Exploring the role of education in a MIRAB economy: Brain drain or brain gain? The case of Wallis and Futuna', unpublished Master of Development Studies thesis, Wellington, Victoria University of Wellington.

Jacobsen, T., Sampford, C. and Thakur, R. (2008) *Re-Envisioning Sovereignty: The End of Westphalia?*, Aldershot, Ashgate Publishing Ltd.

JICA (2009) *We Are Islanders! For the Future of the Pacific*, www.jica.go.jp/english/publications/jica_archive/brochures/pdf/islanders.pdf, accessed 11 November 2016.

JICA (2010) 'Launch of the new JICA and Japan's ODA structure', www.iist.or.jp/wf/magazine/0734/pdf/100222-jica-oda-e.pdf, accessed 11 November 2016.

Jotia, A. L. (2011) 'Globalization and the nation-state: Sovereignty and state welfare in jeopardy', *US-China Education Review*, 2, 243–250.

Kaufmann, D., Kraay, A. and Zoido-Lobaton, P. (1999) *Governance Matters*, World Bank Policy Research Working Paper No. 2196, Washington, DC, World Bank.

Kelly, A. (2015) 'Restoring democracy: Australian responses to military coups in Fiji', *Journal of International Studies*, 11, 1–13.

Kiddle, F. B. (2014) 'A prologue to development: Establishing the legitimacy of the rule of law in post-conflict Solomon Islands', unpublished Master of Development Studies thesis, Wellington, Victoria University of Wellington.

Kidu, C. (2009) 'Maternal health in Papua New Guinea: Reality, challenges, and possible solutions', in Boston, J. (ed.) *Eliminating World Poverty: Global Goals and Regional Progress*, Wellington, Institute of Policy Studies, 179–186.

Knapman, B. (1987) *Fiji's Economic History, 1874–1939: Studies of Capitalist Colonial Development*, Pacific Research Monograph No. 15, Canberra, National Centre for Development Studies, Australian National University.

Kochenov, D. (2012) 'Dutch Caribbean territories facing EU law', *West Indian Law Journal* (Jamaica 50th jubilee special edn), 147–153.

Koeberle, S. and Stavreski, Z. (2006) 'Budget support: Concepts and issues', in Koeberle, S., Stavreski, Z. and Walliser, J. (eds.) *Budget Support as More Effective Aid*, Washington, DC, World Bank, 3–27.

Kothari, U. (ed.) (2005) *A Radical History of Development Studies*, London, Zed Books.
Krasner, S. D. (2005) 'The case for shared sovereignty', *Journal of Democracy*, 16(1), 69–83.
Ku, J. and Yoo, J. (2013) 'Globalization and sovereignty', *Berkeley Journal of International Law*, 31(1), 210–235.
Lal, B. V. (1986) 'Politics since independence: Continuity and change, 1970–1982', in Lal, B. V. (ed.) *Politics in Fiji: Studies in Contemporary History*, Sydney, Allen and Unwin, 74–106.
Lal, B. V. and Fortune, K. (2000) *The Pacific Islands: An Encyclopedia* (Vol. 1), Honolulu, University of Hawai'i Press.
Lal, B. V. and Pretes, M. (eds.) (2001) *Coup: Reflections on the Political Crisis in Fiji*, Canberra, Pandanus Books.
Larmour, P. (1997) 'Corruption and Governance in the South Pacific', State, Society and Governance in Melanesia, Discussion Paper 97/5, Canberra, Research School of Pacific and Asian Studies, Australian National University.
Larmour, P. (2002) 'Conditionality, coercion and other forms of "power": International financial institutions in the Pacific', *Public Administration and Development*, 22(3), 249–260.
Larmour, P. (2003) 'The foreignness of the state in the South Pacific', *The New Pacific Review*, 2(1), 24–33.
Larmour, P. and Barcham, M. (2005) 'National Integrity Systems in Small Pacific Island States', Policy and Governance Program, Discussion Paper 05–9, Canberra, Asia Pacific School of Economics and Government and Australian National University.
Legifrance (1789) *Déclaration des Droits de l'Homme et du Citoyen de 1789*, www.legifrance.gouv.fr/Droit-francais/Constitution/Declaration-des-Droits-de-l-Homme-et-du-Citoyen-de-1789, accessed 30 July 2015.
Legifrance (1905) *Loi du 9 Décembre 1905 Concernant la Séparation des Eglises et de l'Etat, Version consolidée au 19 mai 2011*, www.legifrance.gouv.fr/affichTexte.do?cidTexte=LEGITEXT000006070169&dateTexte=20080306, accessed 28 August 2015.
Leiva, F. (2008) 'Toward a critique of Latin American neostructuralism', *Latin American Politics and Society*, 50, 1–25.
Lewis, O. (2017) 'Advocates claim clearer guidelines needed for Recognised Seasonal Employer scheme worker deductions', *Marlborough Express*, 20 March 2017, www.stuff.co.nz/business/farming/agribusiness/90611107/Advocates-claim-clearer-guidelines-needed-for-RSE-worker-deductions, accessed 7 April 2017.
Liu, S. (2016) 'China's engagement with the South Pacific: Past, present and future', in Powles, M. (ed.) *China and the Pacific: The View from Oceania*, Wellington, Victoria University Press, 53–61.
Llewellyn-Fowler, M. and Overton J. (2010) 'Bread and butter human rights: NGOs in Fiji', *Development in Practice,* 20(7), 827–839.
Locke, K. (2015) 'NZ to assist Cuban medical programme in the Pacific', *The Daily Blog*, http://thedailyblog.co.nz/2015/02/23/nz-to-assist-cuban-medical-programme-in-the-pacific/, accessed 20 December 2016.
Lockhart, D. G., Drakakis-Smith, D. and Schembri, J. (eds.) (1993) *The Development Process in Small Island States*, London, Routledge.

Lotti, A. (2011) *Le Statut de 1961 à Wallis et Futuna: Genèse de Trois Monarchies Républicaines (1961–1991)*, Paris, Editions L'Harmattan.

Maaka, R. and Fleras, A. (2000) 'Engaging with indigeneity: Tino rangatiratanga in Aotearoa', in Ivison, D., Patton, P. and Sanders, W. (eds.) *Political Theory and the Rights of Indigenous Peoples*, Oakleigh, Cambridge University Press, 89–109.

Maaka, R. and Fleras, A. (2005) *The Politics of Indigeneity: Challenging the State in Canada and Aotearoa New Zealand*, Dunedin, Otago University Press.

Mackintosh, H. (2011) *Perspectives of Sovereignty in Aotearoa and the Pacific*, Briefing Paper May 2011, School of Geography, Environment and Earth Sciences, Wellington, Victoria University of Wellington.

Mara, Ratu Sir K. (1997) *The Pacific Way: A Memoir*, Honolulu, University of Hawai'i Press.

Marcus, G. E. (1993) 'Tonga's contemporary globalizing strategies: Trading on sovereignty amidst international migration', in Harding, T. and Wallace, B. (eds.) *Contemporary Pacific Societies*, Englewood Cliffs, NJ, Prentice Hall, 21–33.

Marsters, E., Lewis, N. and Friesen, W. (2006) 'Pacific flows: The fluidity of remittances in the Cook Islands', *Asia Pacific Viewpoint*, 47(1), 31–44.

Mawdsley, E. (2010) 'The non-DAC donors and the changing landscape of foreign aid: The (in)significance of India's development cooperation with Kenya', *Journal of Eastern African Studies*, 4(2), 361–379.

Mawdsley, E., Murray, W. E., Overton, J., Scheyvens, R. and Banks, G. A. (2015) *Sharing Prosperity? A Comparative Analysis of Aid Policy in New Zealand the United Kingdom in the 2010s*, NZADDS Working Paper 2015(1), Wellington, NZADDS.

Mawdsley, E., Murray, W. E., Overton, J., Scheyvens, R. and Banks, G. A. (2018) 'Exporting stimulus and "hared prosperity": Re-inventing aid for a retroliberal era', *Development Policy Review*,36 O25-O43.

Mawdsley, E., Savage, L. and Kim, S.-M. (2014) 'A "post-aid" world? Paradigm shift in foreign aid and development cooperation at the 2011 Busan High Level Forum', *Geographical Journal*, 180(1), 27–38.

May, R. J. (2003) 'Weak states, collapsed states, broken-backed states and kleptocracies: General concepts and Pacific realties', *The New Pacific Review*, 2(1), 35–58.

McCawley, P. (2010) 'Aid objectives: The hole in the aid review', *Devpolicy Blog*, http://devpolicy.org/aid-review-effectiveness20101125/, accessed 11 November 2016.

McCully, M. (2009). 'New Priorities for New Zealand Aid', Speech to the New Zealand Institute of International Affairs, 1 May http://www.national.org.nz /Article.aspx?ArticleID=29843, accessed 5 May 2009.

McElroy, J. L. (2006) 'Small island tourist economies across the life cycle', *Asia Pacific Viewpoint*, 47(1), 61–77.

McElroy, J. L. and Hamma, P. (2010) 'SITEs revisited: Socioeconomic and demographic contours of small island tourist economies', *Asia Pacific Viewpoint*, 51(1), 36–46.

McElroy, J. L. and Parry, C. (2012) 'The long-term propensity for political affiliation in island microstates', *Commonwealth & Comparative Politics*, 50(4), 403–421.

McElroy, J. L. and Pearce, K. (2006) 'The advantages of political affiliation: Dependent and independent small-island profiles', *Round Table*, (386), 529–539.

McGregor, A., Challies, E., Overton, J. and Sentes, L. (2013) 'Developmentalities and donor-NGO relations: Contesting foreign aid policies in New Zealand/Aotearoa', *Antipode*, 45(5), 1232–1253.
McMichael, P. (2017) *Development and Social Change* (6th ed.), Thousand Oaks, Pine Forge.
Meerts, P. and Beeuwkes, P. (2008) 'The Utrecht negotiations in perspective: The hope of happiness for the world', *International Negotiation*, 13(2), 157–177.
Melbourne, H. (1995) *Maori Sovereignty: The Maori Perspective*, Auckland, Hodder Moa Beckett Publishers Ltd.
MFAT (2014a) *Cook Islands*, www.mfat.govt.nz/Countries/Pacific/Cook-Islands.php, accessed 28 August 2015.
MFAT (2014b) *Recent Official Visits: Papua New Guinea to New Zealand*, Wellington, Ministry of Foreign Affairs and Trade, www.mfat.govt.nz/en/countries-and-regions/pacific/papua-new-guinea/, accessed 10 September 2016.
Moon, P. (2000) 'Maori sovereignty and concepts of state', *He Tuhinga Aronui*, 4, 46–59.
Morgan, M. G. and McLeod, A. (2006) 'Have we failed our neighbour?', *Australian Journal of International Affairs*, 60(3), 412–428.
Mountfort, H. (2013) 'An analysis of the aid effectiveness agenda in Tonga', unpublished Master of Development Studies thesis, Wellington, Victoria University of Wellington.
Mrgudovic, N. (2012a) 'Evolving approaches to sovereignty in the French Pacific', *Commonwealth & Comparative Politics*, 50(4), 456–473.
Mrgudovic, N. (2012b) 'The French overseas territories in transition', in Clegg, P. and Killingray, D. (eds.) *The NonIndependent Territories of the Caribbean and Pacific*, London, Institute of Commonwealth Studies, University of London, 85–103.
Mugler, F. and Lynch, J. (eds.) (1996) *Pacific Languages in Education*, Suva, Institute of Pacific Studies, University of the South Pacific.
Munro, D. (1990) 'Transnational corporations of kin and the MIRAB system: The case of Tuvalu', *Pacific Viewpoint*, 31(1), 63–66.
Murray, W. E. (2001) 'The second wave of globalisation and agrarian change in the Pacific Islands', *Journal of Rural Studies*, 17(2), 135–148.
Murray, W. E. and Overton, J. (2011a) 'Neoliberalism is dead, long live neoliberalism: Neostructuralism and the new international aid regime of the 2000s', *Progress in Development Studies*, 11(4), 307–319.
Murray, W. E. and Overton, J. (2011b) 'The inverse sovereignty effect: Aid, scale and neostructuralism in Oceania', *Asia Pacific Viewpoint*, 52(3), 272–284.
Murray, W. E. and Overton, J. (2015) *Geographies of Globalization* (2nd ed.), London, Routledge and Taylor and Francis.
Murray, W. E. and Overton, J. (2016a) 'Retroliberalism and the new aid regime of the 2010s', *Progress in Development Studies*, 16(3), 1–17.
Murray, W. E. and Overton, J. (2016b) 'Peripheries of neoliberalism: Impacts, resistance and retroliberalism as reincarnation', in Springer, S., Birch, K. and MacLeavy, J. (eds.) *Handbook of Neoliberalism*, London, Routledge, 422–432.
Murray, W. E. and Storey, D. (2003) 'Political conflict in postcolonial Oceania', *Asia Pacific Viewpoint*, 44(3), 213–224.
Murray, W. E. and Terry, J. (2004) 'Niue's place in the Pacific', in Terry, J. and Murray, W. E. (eds.) *Niue Island: Geographical Perspectives on the Rock of Polynesia*, Paris, INSULA, UNESCO, 9–30.

Naidu, V. (2009) 'Changing gears on the Millennium Development Goals in Oceania', in Boston, J. (ed.) *Eliminating World Poverty*, Wellington, Institute of Policy Studies, 103–126.

Naidu, V. and Wood, T. (2008) *A Slice of Paradise? The Millennium Development Goals in the Pacific: Progress, Pitfalls and Potential Solutions*, Working Paper No. 1, Apia, Oceania Development Network, National University of Samoa.

Narayan, D., Chambers, R., Shah, M. K. and Petesch, P. (2000) *Voices of the Poor: Crying Out for Change*, Washington, DC and New York, World Bank Publications and Oxford University Press.

The National (2016) 'Cuban doctors for PNG', www.thenational.com.pg/cuban-doctors-png/, accessed 20 December 2016.

Negin, J. (2012) 'Cuba in the Pacific: More than rum and Coke', *DevPolicy Blog*, http://devpolicy.org/cuba-in-the-pacific-more-than-rum-and-coke-2-20120224/, accessed 20 December 2016.

New Zealand Foreign Affairs and Trade Aid Programme (2015a) *New Zealand Aid Programme: Strategic Plan 2015–19*, Wellington, Ministry of Foreign Affairs and Trade.

New Zealand Foreign Affairs and Trade Aid Programme (2015b) *New Zealand Aid Programme: Investment Priorities 2015–19*, Wellington, Ministry of Foreign Affairs and Trade.

New Zealand Foreign Affairs and Trade Aid Programme (2016) *Evaluation of New Zealand's Development Cooperation in Tonga: Final Report*, Wellington, Adam Smith International and MFAT.

New Zealand Parliament (2010) *Ministerial Statements: UN Declaration on the Rights of Indigenous Peoples-Government Support*, Wellington, New Zealand Parliament, www.parliament.nz/en-nz/pb/debates/debates/speeches/49HansS_20100420_00000077/power-simon-ministerial-statements-%E2%80%94-un-declaration-on, accessed 13 June 2016.

New Zealand Parliament, Foreign Affairs, Defence and Trade Committee (2010) *Inquiry into New Zealand's Relationships with South Pacific Countries: Report of the Foreign Affairs, Defence and Trade Committee*, Wellington, New Zealand House of Representatives.

Oberst, A. and McElroy, J. L. (2007) 'Contrasting socio-economic and demographic profiles of two, small island, economic species: MIRAB versus PROFIT/SITE', *Island Studies Journal*, 2(2), 163–176.

OECD (n.d.) 'In-donor refugee costs in ODA', www.oecd.org/dac/financing-sustainable-development/In-donor-refugee-costs-in-ODA.pdf, accessed 16 December 2016.

OECD (2008) *The Paris Declaration on Aid Effectiveness and the Accra Agenda for Action*, Paris, OECD, www.oecd.org/dac/effectiveness/34428351.pdf, accessed 20 February 2014.

OECD (2010) *Paris Declaration Survey: Draft Summary of Progress*, Nuku'alofa, OECD.

OECD (2012) *Better Aid: Aid Effectivness 2011: Progress in Implementing the Paris Declaration*, Paris, OECD.

OECD (2015) *OECD Development Co-operation Peer Reviews: New Zealand*. Paris, OECD.

Office of Insular Affairs (2016) *Budget Justifications and Performance Information, Fiscal Year 2016*, United States Department of the Interior, www.doi.gov/sites/

doi.gov/files/migrated/budget/appropriations/2016/upload/FY2016_OIA_Greenbook.pdf, accessed 20 August 2016.

O'Meara, T. (1987) 'Samoa: Customary individualism', in Crocombe, R. (ed.) *Land Tenure in the Pacific* (3rd ed.), Suva, University of the South Pacific, 74–113.

Oostindie, G. (2006) 'Dependence and autonomy in sub-national island jurisdictions: The case of the Kingdom of the Netherlands', *The Round Table*, 95(386), 609–626.

Orange, C. (2004) *An Illustrated History of the Treaty of Waitangi*, Wellington, Bridget Williams Books Ltd.

Osborne, J. B. (2014) 'Democratic transition in the development context: The case study of Tonga', unpublished Master of Development Studies thesis, Wellington, Victoria University of Wellington.

Osiander, A. (2001) 'Sovereignty, international relations, and the Westphalian myth', *International Organization*, 55(2), 251–287.

O'Sullivan, D. (2006) 'Needs, rights, nationhood, and the politics of indigeneity', *MAI Review, Target Article*, 1(1).

Otero, G. (2011) 'Neoliberal globalization, NAFTA, and migration: Mexico's loss of food and labor sovereignty', *Journal of Poverty*, 15(4), 384–402.

Ovendale, R. (1995) 'Macmillan and the wind of change in Africa, 1957–1960', *The Historical Journal*, 38(2), 455–477.

Overton, J. (1987) 'Fijian land: Pressing problems, possible tenure solutions', *Singapore Journal of Tropical Geography*, 8(2), 139–151.

Overton, J. (1989) *Land and Differentiation in Rural Fiji*, Pacific Research Monograph No. 19, Canberra, National Centre for Development Studies, Australian National University.

Overton, J. (1999) '*Vakavanua, vakamatanitu*: Discourses of development in Fiji', *Asia Pacific Viewpoint*, 40(2), 173–186.

Overton, J. (2009) 'Reshaping development aid: Implications for political and economic relationships', *Policy Quarterly*, 5(3), 3–9.

Overton, J. (2011) 'Owning the Millennium Development Goals: Aid and development policy in the Pacific', in Lynch, B. and Hassall, G. (eds.) *Resilience in the Pacific: Addressing Critical Issues*, Wellington, New Zealand Institute of International Affairs, 93–107.

Overton, J. (2016) 'The context of overall aid in the Pacific: And its effectiveness', in Powles, M. (ed.) *China and the Pacific: The View from Oceania*, Wellington, Victoria University Press, 161–172.

Overton, J. and Murray, W. E. (2014) 'Sovereignty for sale? Coping with marginality in the South Pacific: The example of Niue', *Hrvatski Geograkski Glasnik (Croatian Geographical Bulletin)*, 76(1), 5–25.

Overton, J. and Murray, W. E. (2016) 'Aid and the "circle of security"', in Grugel, J. and Hammett, D. (eds.) *The Palgrave Handbook of International Development*, Oxford, Palgrave Macmillan, 433–450.

Overton, J., Murray, W. E. and McGregor, A. (2013) 'Geographies of aid: A critical research agenda', *Geography Compass*, 7(2), 116–127.

Overton, J., Prinsen, G., Murray, W. E. and Wrighton, N. (2012) 'Reversing the tide of aid: Investigating development policy sovereignty in the Pacific', *Journal de la Société des Océanistes*, 135, 229–242.

Oxfam (n.d.) 'Killer facts: Poverty in the Pacific', www.oxfam.org.nz/what-we-do/issues/millennium-development-goals/killer-facts-pacific-poverty, accessed 20 November 2015.

Patman, R. G. and Rudd, C. (2005) *Sovereignty Under Siege? Globalization and New Zealand*, Aldershot, Ashgate Publishing Ltd.
Peck, J. (2010) 'Zombie neoliberalism and the ambidextrous state', *Theoretical Criminology*, 14(1), 104–110.
Peck, J., Theodore, N. and Brenner, N. (2010) 'Postneoliberalism and its malcontents', *Antipode*, 41, 94–116.
Perez, J., Gistelinck, M. and Karbala, D. (2011) 'Sleeping lions: International investment treaties, state-investor disputes and access to food, land and water', *Oxfam Policy and Practice: Agriculture, Food and Land*, 11(1), 119–156.
Pérez, M. (2003) 'Australia's "Pacific solution": Regional imapct, global questions', *The New Pacific Review*, 2(1), 89–104.
Pérez Silva, V. (2010) 'Los derechos del hombre, sociedades secretas y la conspiración de los pasquines', *Credencial Historia*, (241), www.banrepcultural.org/blaa virtual/revistas/credencial/enero2010/derechos.htm, accessed 7 November 2015.
PIFS (2001) 'Forum economic ministers meeting', Record of Meeting Discussions 18–20 June 2001, Rarotonga, Cook Islands, PIFS.
PIFS (2004) *Social Impact Assessment of Peace Restoration Initiatives in Solomon Islands*, www.forumsec.org.fj/resources/uploads/attachments/documents/Social%20 Impact%20of%20Peace%20Restoration%20Initiatives%20in%20Solomon%20 Islands%202004.pdf, accessed 12 September 2015, accessed 14 May 2015.
PIFS (2007) *The Pacific Plan for Strengthening Regional Cooperation and Integration*, Revised Version 2007, Suva, PIFS.
PIFS (2010) 'Pacific aid effectiveness principles', www.forumsec.org.fj/resources/ uploads/attachments/documents/Pacific_Aid_Effectiveness_Principles_ Final_2007.pdf, accessed 30 November 2010.
PIFS (2013) *Pacific Plan Review 2013: Report to Pacific Leaders*, Suva, PIFS.
PIFS (2015a) *New Zealand's Development Cooperation in the Pacific: Report of the Forum Compact Peer Review 2015*, Suva, PIFS.
PIFS (2015b) *Pacific Regional MDGs Tracking Report 2015*, Suva, PIFS, www. forumsec.org/resources/uploads/attachments/documents/2015%20Pacific%20 Regional%20MDGs%20Tracking%20Report1.pdf, accessed 16 December 2016.
Pirnia, P. (2016) 'Cultivating ownership of development aid: The role of civil society in the Pacific', unpublished PhD thesis, Wellington, Victoria University of Wellington.
Pitty, R. and Smith, S. (2011) 'The indigenous challenge to Westphalian sovereignty', *Australian Journal of Political Science*, 46(1), 121–139.
Poirine, B. (1994) 'Rent, emigration and unemployment in small islands: The MIRAB model and the French overseas departments and territories', *World Development*, 22(12), 1997–2009.
Poirine, B. (1998) 'Should we hate or love MIRAB?', *The Contemporary Pacific*, 10(1), 65.
Pourmokhtari, N. (2013) 'A postcolonial critique of state sovereignty in IR: The contradictory legacy of a "West-centric" discipline', *Third World Quarterly*, 34(10), 1767–1793.
Powles, M. (ed.) (2016) *China and the Pacific: The View from Oceania*, Wellington, Victoria University Press.
Prasad, B. (2016) 'Soft loans and aid: China's economic influence in the Pacific', in Powles, M. (ed.) *China and the Pacific: The View from Oceania*, Wellington, Victoria University Press, 176–179.
Prescott, S. M. (2008) 'Using *talanoa* in Pacific business research in New Zealand: Experiences with Tongan entrepreneurs', *AlterNative: An International Journal of Indigenous Peoples*, 4(1), 127–148.

Prime Minister's Office of Papua New Guinea (2014) *Report 4 June 2014*, Port Moresby, Office of the Prime Minister, www.officeofprimeminister.com/#!june-2014/c1j7, accessed 12 September 2016, accessed 14 May 2016.
Prinsen, G. and Blaise, S. (2017) 'An emerging "islandian" sovereignty of non-self-governing islands', *International Journal*, 72(1), 56–78.
Prinsen, G., Lafoy, Y. and Migozzi, J. (2017) 'Showcasing the sovereignty of non-self-governing islands: New Caledonia', *Asia Pacific Viewpoint*, 58(3), 331–346.
Radio New Zealand (2017) 'Taiwan and Tuvalu remain solid', www.radionz.co.nz/international/pacific-news/325786/taiwan-and-tuvalu-remain-solid, accessed 25 May 2017.
Rasmijn, A. (2013) 'Arubaanse regering wil geen homohuwelijk (Aruba Government does not want gay marriage)', *Caribisch Netwerk*, 13 June, http://caribischnetwerk.ntr.nl/2013/06/13/arubaanse-regering-wil-geen-homohuwelijk/, accessed 20 April 2015.
Ratuva, S. (2011) 'The Chinese lake', *The Listener*, 2 May.
Ravuvu, A. (1983) *Vaka i Taukei: The Fijian Way of Life*, Suva, Institute of Pacific Studies, University of the South Pacific.
Ravuvu, A. (1988) *Development or Dependence: The Pattern of Change in a Fijian Village*, Suva, Institute of Pacific Studies, University of the South Pacific.
Reus-Smit, C. (2011) 'Struggles for individual rights and the expansion of the international system', *International Organization*, 65(2), 207–242.
Richmond, O. P. (2011) 'De-romanticising the local, de-mystifying the international: Hybridity in Timor-Leste and the Solomon Islands', *The Pacific Review*, 24(1), 115–136.
Rist, G. (1997) *The History of Development*, London, Zed Books.
Robertson, R. T. (1988) 'Vanuatu: Fragile foreign policy initiatives', *Development and Change*, 19(4), 617–647.
Robertson, R. T. and Sutherland, W. (2001) *Government by the Gun: The Unfinished Business of Fiji's 2000 Coup*, Annandale, NSW, Pluto Press.
Robertson, R. T. and Tamanisau, A. (1988) *Fiji: Shattered Coups*, Sydney, Pluto Press.
Robie, D. (1989) *Blood on Their Banner: Nationalist Struggles in the South Pacific*, London, Zed Books.
Rocha Menocal, A., Denney, L. and Geddes, M. (2011) *Informing the Future of Japan's ODA, Part One: Locating Japan's ODA within a Crowded and Shifting Marketplace*, London, Overseas Development Institute.
Ronnås, P. (1993) 'The Samoan farmer: A reluctant object of change?', *Development and Change*, 24(2), 339–362.
Routledge, D. (1985) *Matanitū: The Struggle for Power in Early Fiji*, Suva, Institute of Pacific Studies, University of the South Pacific.
Scheyvens, R. and Overton, J. (1995) '"Doing well out of our doing good": A geography of New Zealand aid', *Pacific Viewpoint*, 36(2), 192–207.
Scheyvens, R. (2002) *Tourism for Development: Empowering Communities*, Pearson Education.
Shubuya, E. (2004) 'The problems and potential of the Pacific Islands Forum', in Shibuya, E. and Rolfe, J. (eds.) *Honolulu, Asia-Pacific Center for Security Studies*, 102–115.
Scott, D. (1993) *Would a Good Man Die? Niue Island, New Zealand and the Late Mr Larsen*, Auckland, Hodder and Stoughton.
Selwyn, P. (1980) 'Smallness and islandness', *World Development*, 8(12), 945–951.
Sevele, F. (2011) 'Speech for the launch of the Vanuatu Energy Road Map', Manuscript, Port Vila.

Shaw, B. (1982) 'Smallness, islandness, remoteness and resources: An analytical framework', in Higgins, B. (ed.) *Regional Development Dialogue Special Issue*, Nagoya, UNCRD, 95–109.
Simpson, G. (2008) 'The guises of sovereignty', in Jacobsen, T., Sampford, C. and Thakur, R. (eds.) *Re-Envisioning Sovereignty: The End of Westphalia?*, Aldershot, Ashgate Publishing Ltd, 51–69.
Sisifa, S. P. (2015) 'The project management practices used in development projects: Reflections from Tonga', unpublished PhD thesis, Auckland, University of Auckland.
Smith, S. E. (2010) 'Uncharted waters: Has the Cook Islands become eligible for membership in the United Nations?', *NZJPIL*, 8, 169.
Spate, O. H. K. (1979) *The Pacific since Magellan*, Canberra, Australian National University.
Stephen, S. (2001) 'Howard's "Pacific solution" is neo-colonialism', *Green Left Weekly*, www.greenleft.org.au/content/howards-pacific-solution-neo-colonialism, accessed 17 May 2017.
Storey, D. (2004) *The Fiji Garment Industry*, Auckland, Oxfam New Zealand.
Storey, D., Bulloch, H. and Overton, J. (2005) 'The poverty consensus: Some limitations of the "popular agenda"', *Progress in Development Studies*, 5(1), 30–44.
Suaalii-Sauni, T. and Aiolupotea, S. M. (2014) 'Decolonising Pacific research, building Pacific research communities and developing Pacific research tools: The case of the talanoa and the faafaletui in Samoa', *Asia Pacific Viewpoint*, 55(3), 331–344.
Tabutaulaka, T. T. (2005) 'Australian foreign policy and the RAMSI intervention in the Solomon Islands', *The Contemporary Pacific*, 17(2), 283–308.
Talagi, F. Pihigia (2017) 'Paris in Niue: An analysis of the aid effectiveness agenda in Niue', unpublished Master of Development Studies thesis, Wellington, Victoria University of Wellington.
Taylor, M. (1987) 'Issues in Fiji's development: Economic rationality or aid with dignity', in Taylor, M. J. (ed.) *Fiji: Future Imperfect?*, Sydney, Allen and Unwin, 1–13.
Te Mato Vai (n.d.) 'Questions and answers', www.tematovai.com/images/files/TMV_QA_September.pdf, accessed 28 May 2017.
Thaman, K. H. (1993) 'A conversation about development', in Walsh, A. C. (ed.) *Development That Works! Lessons from Asia-Pacific*, Palmerston North, Amokura Publications, A4.1–4.4.
Thaman, K. H. (2003) 'Decolonizing Pacific studies: Indigenous perspectives, knowledge, and wisdom in higher education', *The Contemporary Pacific*, 15(1), 1–17.
Thoma, K. (2014) 'Electric vehicles in the Pacific Islands? An investigation of the possibilities of electro-mobility in Samoa', Master's thesis, Wellington, Victoria University of Wellington.
Thompson, R. C. (1994) 'Britain, Germany, Australia and New Zealand in Polynesia', in Howe, K. R., et al. (eds.) *Tides of History: The Pacific Islands in the Twentieth Century*, St, Leonards, NSW, Allen and Unwin, 71–92.
Thorne, A. and Raymond, R. (1989) *Man on the Rim: The Peopling of the Pacific*, Sydney, Angus and Robertson.
Ulu, A. J. (2013) '*Pule*: Development policy sovereignty in Samoa', unpublished Master of Development Studies thesis, Wellington, Victoria University of Wellington.
UN (1945) *Charter of the United Nations*, www.un.org/en/documents/charter/index.shtml, accessed 18 May 2014.
UN (2013) *General Assembly Adds French Polynesia to UN Decolonization List*, 18 May, www.un.org/en/decolonization/site-news.shtml, accessed 20 May 2014.

UN (2015) *Addis Ababa Action Agenda of the Third International Conference on Financing for Development (Addis Ababa Action Agenda)*, New York, United Nations.

UN (2017) *Sustainable Development Knowledge Platform: Samoa*, https://sustainabledevelopment.un.org/hlpf/2016/samoa, accessed 3 June 2017.

UNA-UK (2017) *Sustainable Development Goals: From Promise to Practice*, London, United Nations Association, UK, www.unglobalcompact.org/docs/publications/UNA-UK%20SDGS%202017.pdf, accessed 3 June 2017.

UNFCC (2010) *Report of the Conference of the Parties on Its Sixteenth Session, Held in Cancun from 29 November to 10 December 2010: Addendum: Part Two: Action Taken by the Conference of the Parties at Its Sixteenth Session*, https://unfccc.int/resource/docs/2010/cop16/eng/07a01.pdf, accessed 15 May 2017.

United Nations General Assembly (1960) *Declaration on the Granting of Independence to Colonial Countries and Peoples*, New York, United Nations.

United Nations General Assembly (2000) *We the Peoples: The Role of the United Nations in the Twenty-First Century: Report of the Secretary-General*, New York, United Nations, http://unpan1.un.org/intradoc/groups/public/documents/un/unpan000923.pdf.

United Nations General Assembly (2007) *Resolution 61/295: United Nations Declaration on the Rights of Indigenous Peoples*, www.un.org/esa/socdev/unpfii/documents/DRIPS_en.pdf, accessed 13 September 2009.

US-GAO (2013) *Compacts of Free Association: Micronesia and the Marshall Islands Continue to Face Challenges Measuring Progress and Ensuring Accountability*, 20 September, www.gao.gov/assets/660/658031.pdf, accessed 20 August 2015.

Vaioleti, T. M. (2006) '*Talanoa* research methodology: A developing position on Pacific research', *Waikato Journal of Education*, 12, 21–34.

Van Beverhoudt, A. E. (2003) *America's Tropical Isles*, www.sandcastlevi.com/travel/isles/intro.html, accessed 8 November 2015.

Voi, M. (2000) 'Vaka moana: The ocean roads', in Hooper, A. (ed.) *Culture and Sustainable Development in the Pacific*, Canberra, Asia Pacific Press and Australian National University, 207–220.

Voigt-Graf, C. (2008) 'Migration and transnational families in Fiji: Comparing two ethnic groups', *International Migration*, 46(4), 15–40.

Wallace, T., Bornstein, L. and Chapman, J. (2007) *The Aid Chain: Coercion and Commitment in Development NGOs*, Rugby, Intermediate Technology Publications and Practical Action Publications.

Ward, R. G. (1965) *Land Use and Population in Fiji: A Geographical Study*, London, H.M.S.O.

Ward, R. G. (1967) 'The consequences of smallness in Polynesia', in Benedict, B. (ed.) *Problems of Smaller Territories*, London, Althone Press, 81–96.

Ward, R. G. (1985) 'Land, land use and land availability', in Brookfield, H. C., Ellis, F. and Ward, R. G. (eds.) *Land, Cane and Coconuts*, Canberra, Department of Human Geography, Research School of Pacific Studies, Australian National University, 15–64.

Ward, R. G. (1987) 'Native Fijian villages: A questionable future?', in. Taylor, M. J. (ed.) *Fiji: Future Imperfect?*, Sydney, Allen and Unwin, 33–45.

Ward, R. G. and Kingdon, E. (eds.) (1995) *Land, Custom and Practice in the South Pacific*, Cambridge, Cambridge University Press.

Warner, N. (2003) *"100 Days of RAMSI": Message to the People of Solomon Islands*, www.ramsi.org/Media/docs/031031-Special-Coordinator-Nick-Warner-100-days-of-RAMSI-9cd1906a-907f-40da-aaba-0b96082a2ea0-0.pdf, accessed 16 May 2016.

Warrick, O. (2009) 'Ethics and methods in research for community based adaptation: Reflections from rural Vanuatu', in Reid, H. (ed.) *Community-Based Adaptation to Climate Change*, London, International Institute for Environment and Development, 76–87.

Watters, R. (1987) 'MIRAB societies and bureaucratic elites', in Hooper, A., Britton, S., Crocombe, R., Huntsman, J. and MacPherson, C. (eds.) *Class and Culture in the South Pacific*, Auckland and Suva, Centre for Pacific Studies, University of Auckland, New Zealand and Institute of Pacific Studies, University of the South Pacific.

Wesley-Smith, T. (2007) 'Self-determination in Oceania', *Race and Class*, 48(3), 29–46.

Wesley-Smith, T. (2016) 'Reordering Oceania: China's rise, geopolitics, and security in the Pacific Islands', in Powles, M. (ed.) *China and the Pacific: The View from Oceania*, Wellington, Victoria University Press, 98–110.

Wood, T. (2015) 'The ups and downs of New Zealand aid: Budget 2015', *Blog*, www.cid.org.nz/news-old/the-ups-and-downs-of-new-zealand-aid-budget-2015/, accessed 20 May 2017.

World Bank. (2000). *Partners in Transforming Development: New Approaches to Developing Country-Owned Poverty Reduction Strategies*. Washington DC: World Bank. Retrieved from http://www.imf.org/external/np/prsp/pdf/prspbroc.pdf, accessed 7 January 2017.Wrighton, N. (2010a) 'So what's the problem? International development arrivals in Tuvalu', *Just Change*, 18, 9.

Wrighton, N. (2010b) 'Participation, power and practice in development: A case study of theoretical doctrines and international agency practice in Tuvalu', unpublished Master of Development Studies thesis, Wellington, Victoria University of Wellington.

Wrighton, N. and Overton, J. (2012) 'Coping with participation in small island states: The case of Tuvalu', *Development in Practice*, 22(2), 244–255.

Young, I. M. (2004) 'Two concepts of self-determination', in May, S., Modood, T. and Squires, J. (eds.) *Ethnicity, Nationalism, and Minority Rights*, Cambridge, Cambridge University Press, 176–196.

Zwart, P. (2016) 'The tripartite China/New Zealand/Cook Islands project: A New Zealand perspective', in Powles, M. (ed.) *China and the Pacific: The View from Oceania*, Wellington, Victoria University Press, 208–214.

Index

Page numbers in *italics* indicate figures and in **bold** indicate tables on the corresponding pages.